AF720164

Biofilm Science and Technology

Biofilm Science and Technology

Editor: Claudia Wayne

FOSTER ACADEMICS
www.fosteracademics.com

www.fosteracademics.com

FOSTER ACADEMICS

Cataloging-in-Publication Data

Biofilm science and technology / edited by Claudia Wayne.
p. cm.
Includes bibliographical references and index.
ISBN 978-1-64646-560-6
1. Biofilms. 2. Bacteria--Physiology. 3. Bacterial diversity.
4. Emerging infectious diseases. I. Wayne, Claudia.
QR100.8.B55 E44 2023
579.17--dc23

Foster Academics,
118-35 Queens Blvd., Suite 400,
Forest Hills, NY 11375, USA

ISBN 978-1-64646-560-6 (Hardback)

Contents

Preface

In my initial years as a student, I used to run to the library at every possible instance to grab a book and learn something new. Books were my primary source of knowledge and I would not have come such a long way without all that I learnt from them. Thus, when I was approached to edit this book; I became understandably nostalgic. It was an absolute honor to be considered worthy of guiding the current generation as well as those to come. I put all my knowledge and hard work into making this book most beneficial for its readers.

A biofilm includes any syntrophic consortium of microorganisms in which cells stick to each other as well as to a surface. The microorganisms in a biofilm can demonstrate notable organization and can collaborate, communicate, and coordinate with one another. Biofilms can form on various surfaces in natural, industrial and hospital environments. The biofilms challenge the theory that bacteria are single-celled organisms, which act in isolation instead of working together as a group of cooperating cells to improve their own fitness and the biofilm's efficiency. The biofilm bacteria may share nutrients and are protected from the damaging effects of environmental factors such as antibiotics, desiccation, and the immune system of the host body. A biofilm is generally formed as soon as a free-swimming bacterium attaches to a surface. This book is compiled in such a manner, that it will provide in-depth knowledge about biofilm science and technology. It will serve as a reference to a broad spectrum of readers.

I wish to thank my publisher for supporting me at every step. I would also like to thank all the authors who have contributed their researches in this book. I hope this book will be a valuable contribution to the progress of the field.

Editor

1

Controlling Persister and Biofilm Cells of Gram-Negative Bacteria with a New 1,3,5-Triazine Derivative

Ali Adem Bahar [1,2,†], Zhigang Liu [3,†], Meagan Garafalo [1,2], Neville Kallenbach [3,*] and Dacheng Ren [1,2,4,5,*]

[1] Department of Biomedical and Chemical Engineering, Syracuse University, Syracuse, NY 13244, USA; E-Mails: abahar@syr.edu (A.A.B.); meagan8390@yahoo.com (M.G.)
[2] Syracuse Biomaterials Institute, Syracuse University, Syracuse, NY 13244, USA
[3] Department of Chemistry, New York University, New York, NY 10003, USA; E-Mail: zl264@nyu.edu
[4] Department of Civil and Environmental Engineering, Syracuse University, Syracuse, NY 13244, USA
[5] Department of Biology, Syracuse University, Syracuse, NY 13244, USA

† These authors contributed equally to this work.

* Authors to whom correspondence should be addressed; E-Mails: nrk1@nyu.edu (N.K.), dren@syr.edu (D.R.).

Academic Editor: Jean Jacques Vanden Eynde

Abstract: Infections caused by multidrug-resistant bacteria have been on the rise. This important issue presents a great challenge to the healthcare system and creates an urgent need for alternative therapeutic agents. As a potential solution to this problem, antimicrobial peptides (AMPs) have attracted increasing attention due to their broad spectrum of targeted microbes. However, most AMPs are expensive to synthesize, have relatively high cytotoxicity to mammalian cells, and are susceptible to proteolytic degradation. In order to overcome these limitations, novel synthetic AMPs are desired. Using 1,3,5-triazine (TN) as a template, several combinatorial libraries with varying cationic charge and lipophilicity were designed and screened by the Kallenbach lab. From this screening, TN-5 was identified as a potent lead. In the present study, this compound was tested for its antimicrobial activities on *Escherichia coli* and *Pseudomonas aeruginosa*. In addition to regular planktonic cells, the effects on biofilms and persister cells (metabolically inactive and antibiotic tolerant subpopulation) were also investigated. TN-5 was found to have a minimum inhibitory concentration (MIC) of 12.8 μM for both species and kill regular planktonic cells of both species dose dependently. TN-5 is also effective against persister cells of both *E. coli* and *P. aeruginosa*. The killing of biofilm cells of the mucoid *P. aeruginosa* PDO300 was enhanced by alginate lyase.

Keywords: antimicrobial peptide; biofilm; persister cells; TN-5; *Escherichia coli*; *Pseudomonas aeruginosa*

---

1. Introduction

Since the discovery of penicillin in 1928 [1] and the achievement of economical production of this antibiotic in the 1940s [2], the use of antibiotics has been a crucial step in controlling infectious diseases with numerous lives saved [3]. However, the emergence and spread of antibiotic resistant microorganisms have rendered many antibiotics ineffective [4]. Such rapid development of multidrug resistant bacteria coupled with the insufficient investment in antimicrobial research has led to a concerning decline in effective therapies against bacterial infections, which presents a serious public health problem [5].

In addition to antibiotic resistance based on drug resistance genes [6], bacteria also exhibit high level antibiotic tolerance by forming persister cells (dormant subpopulation of phenotypic variants [7] and biofilms (surface attached structures with bacterial cells embedded in an extracellular matrix secreted by attached cells [8]). Persister cells and biofilms are not based on drug resistance genes; however, they allow bacteria to survive the treatment with potent antibiotics and facilitate the development of drug resistant strains through mutation and horizontal gene transfer [9].

According to the U.S. Centers for Diseases Control and Prevention, 65% of all infections in developed countries are caused by biofilms [10]. The biofilm matrix serves as a protective barrier, making bacterial cells more tolerant to antibiotics as well as host defense [11]. Biofilms are also enriched in persister cells. Thus, even if an antibiotic can penetrate the biofilm matrix, it might only kill normal cells within a biofilm population. After the course of antibiotic treatment, persister cells revive and repopulate the biofilm, which in turn causes an infection to relapse [10].

An additional concern with biofilms of *Pseudomonas aeruginosa* is their ability to convert to mucoid variants [12]. A mucoid strain is characterized by overproduction of the exopolysaccharide alginate, with increased tolerance to some antibiotics [13] and phagocytosis by human macrophages [14], as well as enhanced protection from dehydration [15]. In cystic fibrosis patients, the leading cause of mortality is a respiratory failure due to chronic lung infection with *P. aeruginosa* strains that undergo mucoid conversion [12]. Mucoid isolates typically coincide with persistent chronic infection in cystic fibrosis patients.

The challenges of drug tolerant infections have created an urgent need for new antimicrobials and treatment strategies. Recent research has shown the great potential of antimicrobial peptides (AMPs) as a class of powerful agents against both Gram-positive and Gram-negative bacteria [16,17]. A wide range of AMPs are naturally produced by the innate immune system of multicellular organisms in response to infections [18]. These AMPs are a unique group of molecules with a varying number of amino acids (generally from 12 to 50), including positively charged residues (such as arginine, lysine, or histidine) and a large proportion of hydrophobic residues. In humans, AMPs are found mainly in the tissues and organs that are exposed to airborne pathogens [19].

The structure and charge of an AMP play a major role in the mechanism of its actions. AMPs are generally cationic molecules with both hydrophobic and hydrophilic faces [20]. This amphipathic characteristic helps these molecules to integrate into the lipid bilayer membranes [21]. Membrane integrity disruption (via interaction with negatively charged cell membrane), inhibition of macromolecule (protein, DNA and RNA) synthesis, or interaction with certain intracellular targets are thought to be the primary mechanisms in AMP lethality [22]. Positively charged side chains in AMPs enable an initial nonspecific electrostatic interaction with the negatively charged cell membrane. This process is followed by the insertion of AMP molecule into cell membrane with the help of hydrophobic residues [23]. For example, in the AMPs LL-37 and β-defensin, the cationic face is positioned on the opposite side of the hydrophobic face, which helps the penetration into the membrane [20].

AMPs afford promising candidates for novel therapeutic agents and complement traditional antibiotic therapies because of some unique advantages, including broad-spectrum activity (antibacterial, antiviral, and antifungal), less resistance by microbes, and related broad anti-inflammatory activities [24]. A number of AMPs and derivatives have been developed as therapies for infectious diseases such as oral mucositis [25], pulmonary infections associated with cystic fibrosis [26], and some sexually transmitted diseases [27]. AMPs can be used alone, in co-treatment with antibiotics, or as stimulators for immune system and toxin inhibiting agents in septic shocks [24].

Despite the aforementioned advantages, wide applications of AMPs are still limited by several factors. They are generally expensive to synthesize [28], vulnerable to proteolytic degradation upon intravenous administration [29] and sensitive to environmental factors such as salt concentration, pH, and the presence of plasma and serum proteins [30]. Another challenge is that some AMPs are cytotoxic to host cells [31].

In order to overcome these challenges, a number of synthetic AMPs and AMP-mimetics have been developed. Based on the concept that cationic charge, size, and lipophilicity are recognized major factors determining the antibacterial activity of AMPs, recently, the Kallenbach lab designed and screened several combinatorial libraries based on 1,3,5-triazine as a scaffold. Several lead compounds with good antimicrobial activity and low hemolytic activity were identified from the screening of hundreds of triazine compounds [32]. With further structure activity relationship analysis, the compound TN-5 was identified as a potent antimicrobial compound (the screening results will be published elsewhere). In this study, TN-5 (Figure 1) was tested for its antimicrobial activity on *E. coli* and *P. aeruginosa*, including regular planktonic cells, persister cells, and biofilms.

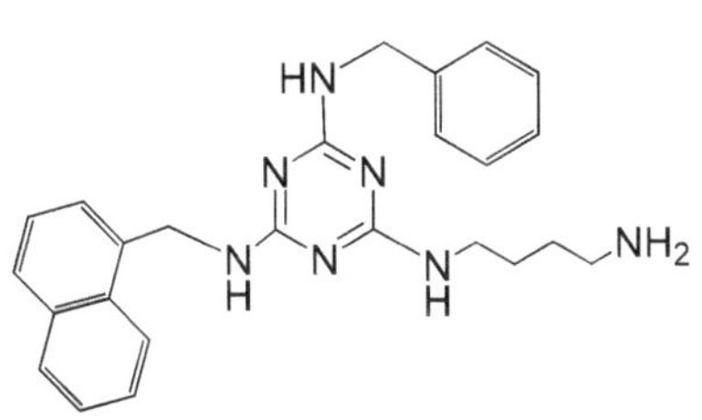

TN-5

Figure 1. Chemical structure of TN-5.

2. Results

2.1. Minimum Inhibitory Concentration (MIC) and Minimal Bactericidal Concentration (MBC) of TN-5

TN-5 was found to completely inhibit the growth of *E. coli* RP437, *P. aeruginosa* PAO1, and *P. aeruginosa* PDO300 at the concentration of 12.8 μM (MIC, Table 1). The MBC value was found to be higher than 96 μM for all three strains.

Table 1. Minimum Inhibitory Concentration (MIC) and Minimal Bactericidal Concentration (MBC) values of TN-5 on bacterial strains used in this study (based on three biological replicates).

Antimicrobial Test	Bacterial Strains		
	***E. coli* RP437**	***P. aeruginosa* PAO1**	***P. aeruginosa* PDO300**
MIC (μM)	12.8	12.8	12.8
MBC (μM)	>96	>96	>96

2.2. Antimicrobial Effects of TN-5 on Planktonic Cells

To further study the killing activity of TN-5, exponential cultures were used to test the effects of TN-5 on the viability of planktonic cells. TN-5 was found effective in killing all bacterial strains tested in this study (*E. coli* RP437, *P. aeruginosa* PAO1 and PDO300) dose dependently in 3 h (Figure 2). Longer incubation times (up to 24 h) did not cause additional killing (Figure S1). For *E. coli* RP437, TN-5 showed 20.3% ± 2.5% (p = 0.07), 55.6% ± 8% (p < 0.001), 84.6% ± 8.5% (p < 0.001), and 99.9% ± 0.1% (3.4 log; p < 0.001) killing of the total population (>99% as normal cells) at concentrations of 5, 20, 50, and 100 μM, respectively (Figure 2A). These results show that TN-5 is highly effective against normal planktonic cells of *E. coli* RP437.

Significant killing effects were also observed on planktonic cells of *P. aeruginosa* PAO1 and PDO300 strains. For example, killing of 22.7% ± 6.7% (p < 0.001), 61.8% ± 4.5% (p < 0.001), 74.7% ± 11.8% (p < 0.001), and 97.8% ± 10% (1.6 log; p < 0.001) of *P. aeruginosa* PAO1 was observed when TN-5 was added at 5, 20, 50, and 100 μM respectively (Figure 2B). For *P. aeruginosa* PDO300, significant killing was observed at 50 and 100 μM with 44.1% ± 3.1% (p < 0.001) and 94.1 ± 6.8% (1.23 log; p < 0.001) of the total population (>99% as normal cells) killed, respectively (Figure 2C). These results show that TN-5 is also effective against normal planktonic cells of *P. aeruginosa*.

2.3. Antimicrobial Effects of TN-5 on Persister Cells

TN-5 was found to kill persister cells of *E. coli* RP437, *P. aeruginosa* PAO1 and *P. aeruginosa* PDO300 (Figure 3). The killing of *E. coli* RP437 persister cells was 33.8% ± 0.8% (p <0.001), 43.8% ± 2.8% (p < 0.001), and 96.3% ± 3.0% (1.35 log; p < 0.001) when TN-5 was added at 50, 100, and 200 μM, respectively (Figure 3A).

Similarly, TN-5 was also able to reduce the viability of *P. aeruginosa* PAO1 persister cells where 79.6% ± 2.9% (p < 0.001) and 89.9% ± 4.5% (p < 0.001) of cells were killed by 100 μM and 200 μM of TN-5, respectively (Figure 3B). In comparison, TN-5 was less effective on persister cells of the mucoid strain *P. aeruginosa* PDO300, with only 33.2% ± 5.6% (p < 0.001) and 36.4% ± 5.2% (p < 0.001) killed by 100 μM and 200 μM of TN-5, respectively (Figure 3C). This is likely due to the presence of alginate on the surface of the mucoid *P. aeruginosa* PDO300 cells [33], which may reduce the penetration by TN-5.

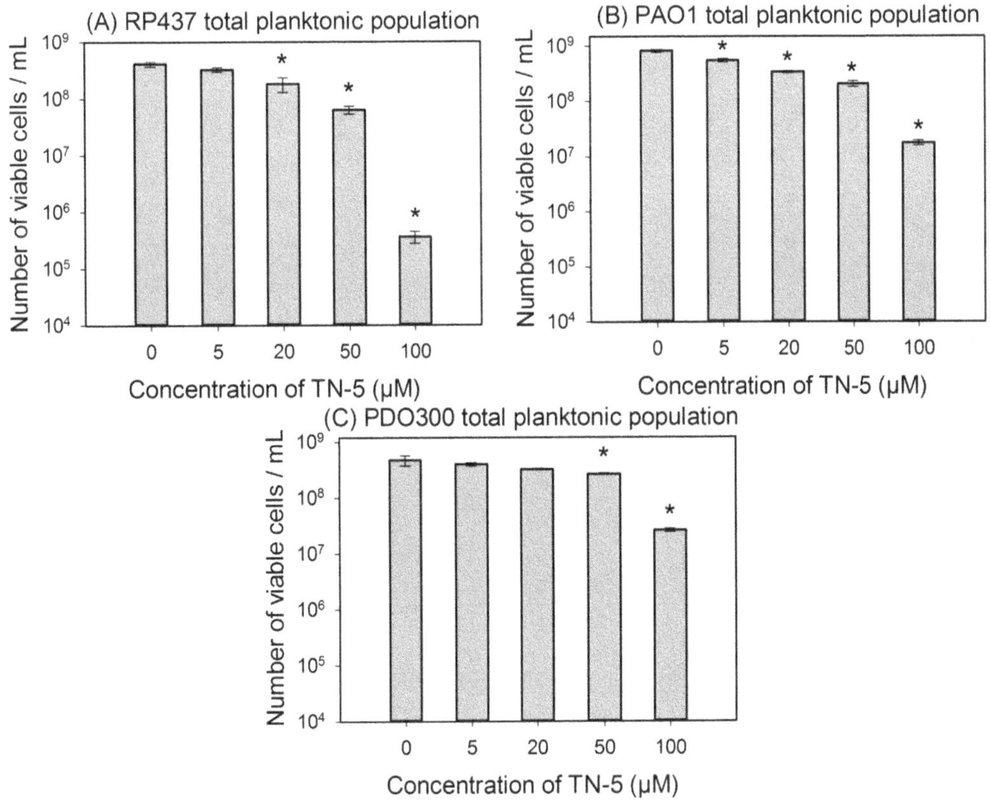

Figure 2. Effects of TN-5 on planktonic cells of *E. coli* RP437 (**A**), *P. aeruginosa* PAO1 (**B**), and *P. aeruginosa* PDO300 (**C**). TN-5 was added in exponential phase cultures at different concentrations and the viability after treatment was determined by counting colony forming unit (CFU). All significant differences (compared to the TN-5 free control) with $p < 0.01$ are marked with an asterisk.

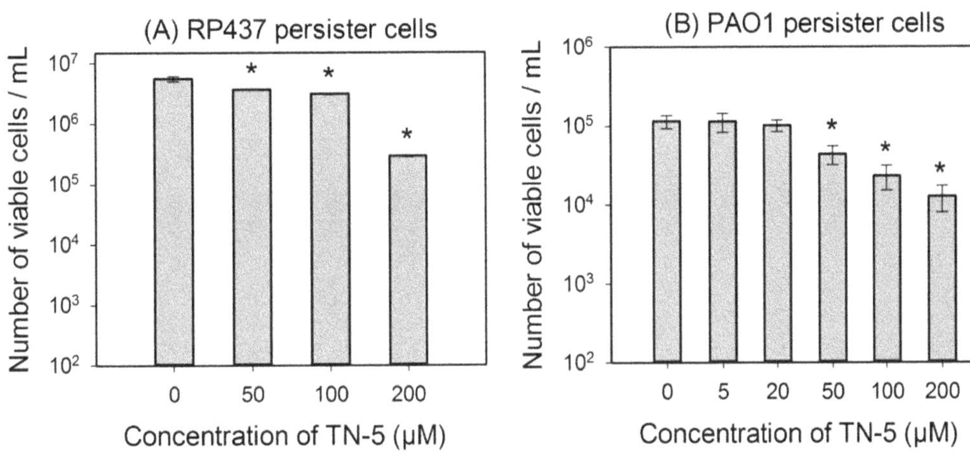

Figure 3. *Cont.*

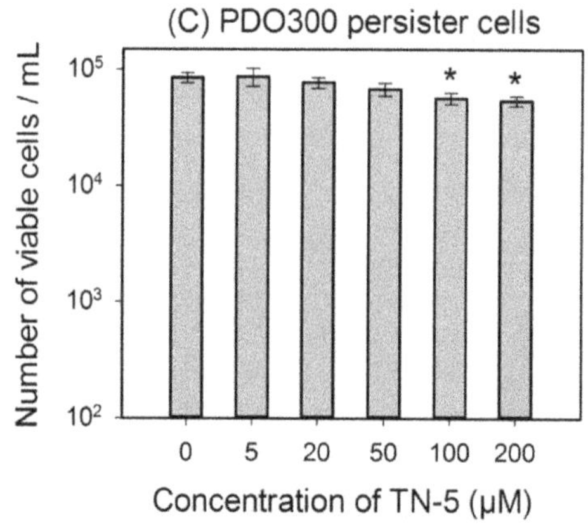

Figure 3. Effects of TN-5 on persister cells of *E. coli* RP437 (**A**), *P. aeruginosa* PAO1 (**B**), and PDO300 (**C**). The persister cells were isolated by treating exponential cultures of *E. coli* RP437 with 100 μg/mL ampicillin and overnight *P. aeruginosa* cultures with 200 μg/mL ciprofloxacin (both for 3 h). All significant differences (compared to the TN-5 free control) with $p < 0.01$ are marked with an asterisk.

2.4. Antimicrobial Effects of TN-5 on Biofilm Cells

TN-5 exhibited effective and dose dependent killing of *E. coli* RP437 biofilms. Killing of 78.5% ± 30.2% ($p < 0.001$) and 98.9% ± 9.6% (1.24 log; $p < 0.001$) of *E. coli* RP437 biofilm cells was achieved when TN-5 was added at 100 and 200 μM, respectively (insignificant at 50 μM) (Figure 4). In addition to the effects on established biofilms of *E. coli* RP437, TN-5 effectively prevented biofilm formation of all the strains used in this study. Complete biofilm inhibition of *E. coli* RP437 and *P. aeruginosa* (both PAO1 and PDO300) was achieved with 10 and 20 μM TN-5, respectively (Figure S2). However, 100 μM TN-5 alone did not show significant killing of cells in established biofilms of *P. aeruginosa* PAO1 or PDO300 ($p > 0.1$) (Figure 5A,C).

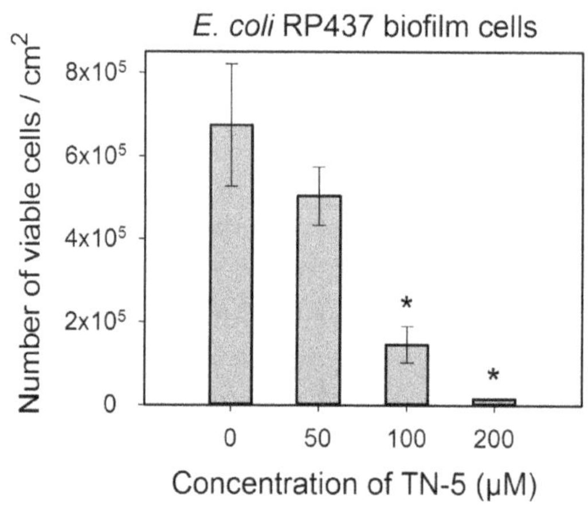

Figure 4. Effects of TN-5 on *E. coli* RP437 biofilms. The biofilms were cultured for 24 h in LB on stainless steel coupons prior to treatment with TN-5. All significant differences (compared to the TN-5 free control) with $p < 0.01$ are marked with an asterisk. Note: Data are plotted in linear scale.

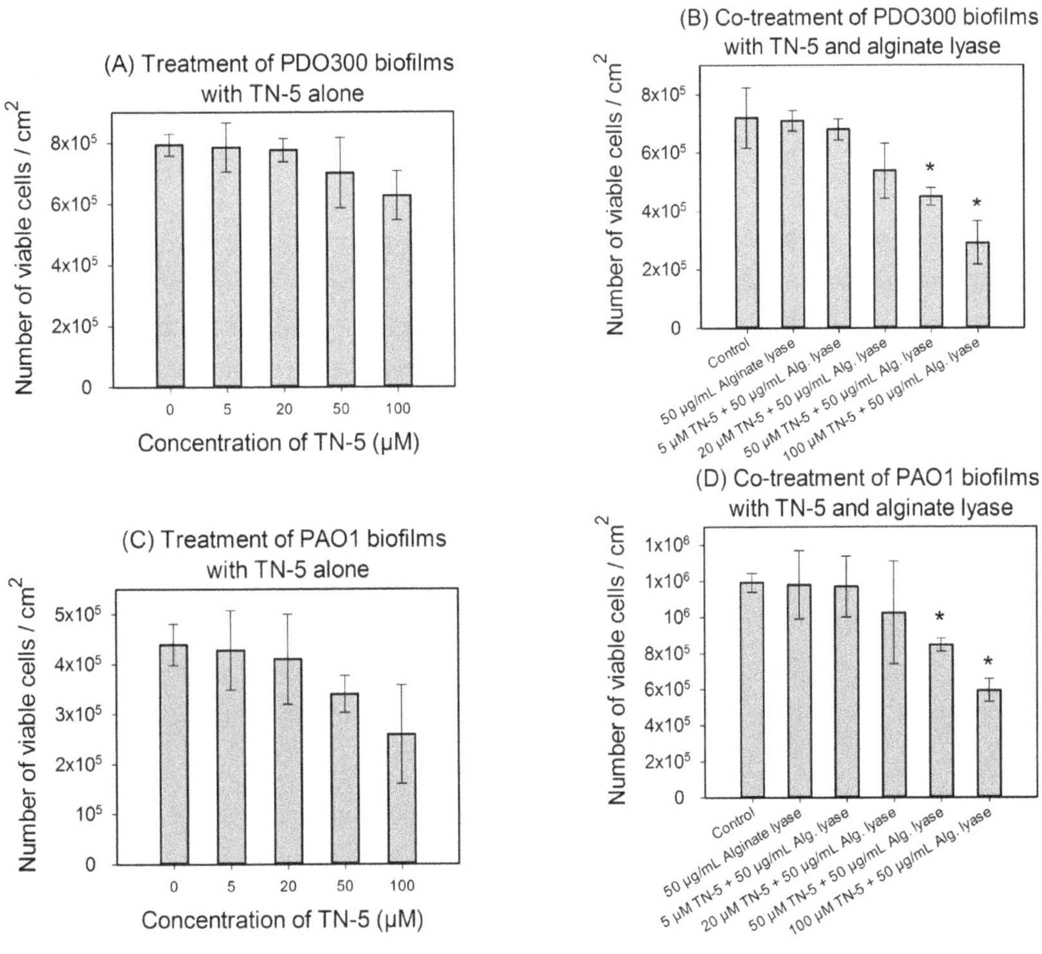

Figure 5. Effects of TN-5 alone and co-treatment with alginate lyase on *P. aeruginosa* PDO300 (**A** and **B**) and PAO1 (**C** and **D**) biofilms. The biofilms were grown for 24 h and treated with TN-5 alone or in combination with alginate lyase for 3.5 h. All significant differences (compared with the TN-5 free control) with $p < 0.01$ are marked with an asterisk. The co-treatment with TN-5 (at 50 or 100 μM) and 50 μg/mL alginate lyase caused significant killing compared to the control (with no TN-5 and alginate lyase) for both *P. aeruginosa* PDO300 and PAO1, but only significantly increased the killing by TN-5 alone for *P. aeruginosa* PDO300. Note: Data are plotted in linear scale.

We speculated that the lack of antimicrobial effects against *P. aeruginosa* biofilms is because of the presence of biofilm matrix. Alginate is a major component of biofilm matrix of the mucoid *P. aeruginosa* strains [13,33]. Thus we tested the concurrent treatment of *P. aeruginosa* biofilms with alginate lyase and TN-5. As shown in Figure 5B,D, alginate lyase itself showed no effect on the viability of biofilm cells ($p = 1$). However, the combination of TN-5 with alginate lyase increased the killing of PDO300 biofilms; e.g., addition of 50 μM alginate lyase increased the activity of 50 μM TN-5 on *P. aeruginosa* PDO300 biofilms from no significant killing ($p = 0.58$) to 36.5% ± 4.6% ($p < 0.005$) and that of 100 μM TN-5 from insignificant killing ($p = 0.12$) to 57% ± 9.6% ($p < 0.005$), respectively (Figure 5A *vs.* Figure 5B). The addition of alginate lyase did not show the same effects on *P. aeruginosa* PAO1 biofilm cells; e.g., 50 μM alginate lyase did not increase the effects of 50 and 100 μM

TN-5 ($p > 0.1$) (by comparing the % of killing based on corresponding data between Figure 5C,D). This is likely because alginate is not a major component of its biofilm matrix.

3. Discussion

AMPs have been proposed as a promising source of new antimicrobial agents [34–37]. Different strategies have been tested to achieve effective killing of microbes while maintaining low hemolytic activity. In some of these studies, Trp and Arg containing 1,3,5-triazine structures [32] and dendrimeric peptides [38] have been used as AMP templates and lead compounds with antimicrobial activities and low toxicity to red blood cells have been identified. In this study, one of the triazine-1,3,5 derivatives, TN-5, was tested on Gram-negative bacteria *E. coli* and *P. aeruginosa*. TN-5 was found to be effective against both species. The killing of persister and biofilm cells is of particular interest because these cells are difficult to eliminate and conventional antibiotics are generally ineffective. Our results show that TN-5 alone is effective against persister and biofilms of *E. coli* RP437 with up to 2-log killing achieved at 200 μM. It is also interesting that TN-5 caused more killing of biofilm cells than planktonic persister cells of *E. coli*. For example, 100 μM TN-5 killed 43.8% ± 2.8% and 78.5% ± 30.2% of persister and biofilm cells of *E. coli* RP437, respectively. This finding suggests that TN-5 can penetrate the biofilm matrix of *E. coli*.

Comparable effects were not observed for *P. aeruginosa* biofilms, however, possibly due to the difference in biofilm matrices. With a thick layer of alginate, biofilms of mucoid bacteria have high-level tolerance to some antimicrobials [13]. Thus, breaking down alginate in the biofilm matrix could be essential for certain antimicrobials to kill biofilm cells, especially the agents that can be absorbed or neutralized by the matrix components. Consistently, we found that the killing of *P. aeruginosa* PDO300 biofilm cells was enhanced by alginate lyase. It is worth noting that TN-5 is effective against *P. aeruginosa* PDO300 persister cells, but not its biofilm cells. Thus, it will be interesting to study if and how TN-5 interacts with alginate directly and if there is any difference in the amount and structure of alginate between biofilm matrix and the surface of mucoid cells. In comparison, the killing of *P. aeruginosa* PAO1 biofilms was not enhanced by alginate lyase. This is consistent with the report [39] that alginate is not the primary component of *P. aeruginosa* PAO1 biofilm matrix. It will be interesting to test other matrix degrading enzymes, such as DNase.

In a previous study performed in our labs, Trp- and Arg- containing synthetic AMPs were shown to inhibit 95.0% ± 1.1% of *E. coli* biofilm formation at 200 μM [40]. Intriguingly, TN-5 inhibited biofilm formation of both *E. coli* and *P. aeruginosa* completely at 20 μM (Figure S2). Therefore, TN-5 may also be an important antimicrobial agent for biofilm control.

The mechanism of bacterial killing by TN-5 or in fact any AMP deserves more study. The positive charge of the arginine (R) side chain can help an AMP to interact with negatively charged lipopolysaccharides in Gram-negative bacterial walls [41]. Tryptophan (W) residues favor location below the head groups of bilayers, and are assumed to provide lipophilic interaction sites to cause membrane disruption [42]. Thus, RW mimicking TN-5 might target the negatively charged bacterial cell membrane. Another observation is that TN-5 mode of action occurs within the first 3 h (Figure S1). This rapid mode of action is one of the major features of membrane acting AMPs.

E. coli is generally more susceptible to antibiotics than *P. aeruginosa*. Among 47 antimicrobial agents tested in an early study [43], only tobramycin showed the same MIC value (0.5 µg/mL) for *E. coli* (NCTC 10418 and ATCC 25922) and *P. aeruginosa* (NCTC 10662 and ATCC 27853) reference strains. The other antimicrobial agents tested all showed higher (up to 500 fold) MIC values for *P. aeruginosa* compared to *E. coli* [43]. Interestingly, TN-5 showed the same MIC value (12.8 µM) for both species in our study, which also indicates possible membrane targeting activities.

Overall, this study shows that the triazine derivative TN-5 is a promising lead compound for developing new synthetic AMPs. It is encouraging that TN-5 is effective against both *E. coli* and *P. aeruginosa* at different growth stages. Persister cells of *Pseudomonas* strains, especially those of the mucoid strains, are highly resistant to antibiotic treatments. Because most AMPs kill microbes through mechanisms that differ from those of conventional antibiotics, e.g. by targeting cell membranes, the killing effect of TN-5 on persister cells might be increased by synergy with some antibiotics if TN-5 can be effectively delivered to target cells. We assume that triazines are not susceptible to proteolytic degradation since there is no specific motif to trigger cleavage. Future experiments with mammalian cells and animal models are needed to evaluate the potential of using TN-5 and antibiotics together as a novel therapy for chronic infections involving biofilms and persister cells.

4. Experimental Section

4.1. Chemical Synthesis of TN-5

TN-5, *N*2-(4-Aminobutyl)-*N*4-benzyl-*N*6-naphthalenemethyl-2,4,6-triamino-1,3,5-triazine (Figure 1), white solid, was synthesized using an orthogonal synthetic approach based on cyanuric chloride stepwise reaction with naphthalenemethylamine, benzylamine and Boc-1,4-diaminobutane as documented previously [32]. In brief, (i) cyanuric chloride was mixed with naphthalenemethylamine (1.1 eq) in tetrahydrofuran (THF) at 0 °C, and N,N-diisopropylethylamine (DIEA) (2 eq). After 1 hour of stirring, the solution was extracted with water/ethyl acetate and product 2, 4-dichloro-*N*6-naphthalenemethylamino-1,3,5-triazine was purified using flash column chromatography. (ii) PAL aldehyde resin was mixed with benzylamine (5 eq.) in THF with 2% of acetic acid and shaked at room temperature for 1 h. At the end of the hour, NaB (OAc) 3H (7 eq.) was added and mixed at room temperature overnight. (iii) The product on resin (4 eq) was mixed with 2, 4-dichloro-*N*6-naphthalenemethylamino-1,3,5-triazine in THF and heated at 60 °C for 3 h. (iv) Resin from last step was mixed with Boc-1,4-Diaminobutane (10 eq) and DIEA (30 eq) in *N*-methylpyrrolidinone-*N*-butanol (NMP:*n*-BuOH) (1:1) and heated for 3 h at 120 °C. (v) Resin was cleaved with 10% TFA in dichloromethane (DCM) for 30 min. Final product was dried and purified with prep thin layer chromatography (TLC). The identity of TN-5 was confirmed by LC-MS (Agilent 6226 TOF LC/MS Mass Spectrometer, Agilent Technologies Inc., Santa Clara, CA, USA). HRMS *m/z*: calculated $[M + H]^+$ 428.2557, found 428.2566 (Figure S3). The composition of the product was verified by ^{1}H-NMR (300 MHz, CD_3OD): δ 8.10–7.86 (m, 1H), δ 7.8–7.65 (m, 2H), δ 7.55–7.30 (m, 4H), δ 7.30–7.15 (m, 3H), δ 7.18 (s, 2H), δ 5.05–4.92 (m, 2H), δ 4.55–4.40 (m, 2H), δ 3.40–3.25 (m, 2H), δ 2.90-2.61 (m, 2H), δ 1.65–1.40 (m, 4H) (Figure S4).

4.2. Bacterial Strains and Growth Media

E. coli RP437 [44] was provided by Dr. John S. Parkinson at the University of Utah. *P. aeruginosa* PAO1 [45] and *P. aeruginosa* PDO300 [12] were obtained from Dr. Matthew Parsek at the University of Washington. All strains were routinely grown in Lysogeny broth (LB) containing 10 g/L NaCl, 5 g/L yeast extract, and 10 g/L tryptone with pH 7.0. To ensure consistent experimental conditions throughout this study, all overnight cultures of a particular strain were started with single-use glycerol stocks originating from the same culture. Each experimental condition was tested with three independent cultures (three biological replicates).

4.3. MIC and MBC Values of TN-5

TN-5 was tested at different concentrations against *E. coli* and *P. aeruginosa* strains to determine MIC and MBC values following a previously described protocol [46] with slight modifications. MIC is referred to the concentration of an antimicrobial agent, which inhibits the visible growth of a given bacterium completely by checking with unaided eye. MBC is referred to the minimum concentration of an antimicrobial agent required to completely kill all the tested bacterial cells of a particular strain by checking colony formation on the plates after treatment. Briefly, exponential cultures of bacterial samples were used to inoculate test samples with a cell density of 5×10^5 colony forming unit (CFU)/mL. TN-5 was tested at concentrations from 0.2 μM to 96 μM, increasing logarithmically. Cultures grown without antimicrobial and sterile LB medium were used as positive and negative controls, respectively. All samples were adjusted to a final volume of 3 mL and incubated for 16–18 h at 37 °C. After incubation, the concentrations with no visible growth were assigned by visual check to determine MIC. MBC value of TN-5 was identified by spreading TN-5 (0.2 μM to 96 μM) treated cells (overnight in LB) on LB agar plates and checking for growth after 24 h. Three independent cultures for each concentration were tested for MIC and MBC tests.

4.4. Effects on Planktonic Cells

To examine the antimicrobial activity of TN-5, the planktonic growth of *E. coli* RP437, *P. aeruginosa* PAO1 and PDO300 with TN-5 added at different concentrations was examined. Overnight cultures of each strain were grown in 50 mL LB medium for 12–16 h at 37 °C. Subcultures were then prepared by inoculating LB medium with overnight cultures to an optical density at 600 nm (OD_{600}) of 0.01 and harvested when the OD_{600} reached 0.4–0.5. These exponential cultures were washed with phosphate buffered saline (PBS) three times. Cell pellets were washed with fresh PBS and re-suspended in 20 mL PBS buffer. For the TN-5 treatment, 3 mL of each sample was taken and mixed with TN-5 at different concentrations. The samples were incubated for 3, 6, 12 and 24 h at 37 °C with shaking at 200 rpm, and washed three times with PBS. Then a serial dilution of each sample was performed and the cells were spread on LB agar plates for counting CFU. The amount of DMSO (solvent to dissolve TN-5 in stock solutions) was adjusted to be the same for all samples to eliminate any solvent effect. Each condition was tested with three independent cultures.

4.5. Persister Isolation and Treatment

To isolate persister cells, overnight cultures of *P. aeruginosa* PAO1 and *P. aeruginosa* PDO300 strains were treated with 200 μg/mL ciprofloxacin while exponential cultures of *E. coli* RP437 were treated with 100 μg/mL ampicillin (all for 3 h at 37 °C) to kill normal cells as described previously [47–49]. Then the persister cells were harvested by centrifugation, washed with PBS three times to remove remaining antibiotic, and re-suspended in 20 mL PBS. Aliquots (1 mL) of each sample were supplemented with TN-5 at different concentrations. Three replicates were tested for each condition and all samples were incubated for 3 h at 37 °C with shaking at 200 rpm. No change in persister cell number was found before and after incubation for 3 and 6 h in PBS without TN-5 (regardless the presence of antibiotics; Figure S5). After TN-5 treatment, the samples were washed with PBS three times and plated to count CFU as described above. Each condition was tested with three independent cultures.

4.6. Biofilm Experiments

To study the effects of TN-5 on biofilm cells of *E. coli* and *P. aeruginosa*, biofilms were grown on 2 cm × 1 cm 316L stainless steel coupons. Each coupon was polished with 220 Grit sandpaper (3M, Sandblaster, St. Paul, MN, USA) on both sides and sterilized with 70% ethanol for at least 15 min. The coupons were then dried in a 50 °C oven for 15 min. Sterilization was confirmed by incubating three coupons in LB medium for 24 h at 37 °C and checking the turbidity of the cultures. To culture biofilms, sterilized coupons were transferred into new sterile plates containing 20 mL LB medium. To initiate biofilm formation, each sample was inoculated to OD_{600} of 0.01 with an overnight culture. The coupons were incubated for 24 h at 37 °C without shaking to grow biofilms. After 24 h incubation the coupons were washed gently with PBS three times to remove planktonic cells. Coupons were then placed in 12 well plates separately, each including a different concentration of TN-5 in 2 mL PBS buffer. The coupons were incubated for 3 h at 37 °C with no shaking followed by washing with PBS three times. Each coupon was then transferred into a 15 mL sterile conical test tube containing 3 mL PBS. Samples were gently sonicated for 4 min in a water sonication bath (Branson B200 Ultrasonic, Danbury, CT, USA) and then vortexed for 15 s. Three replicates were tested for each condition. The cells in the suspension were then spread on LB agar plates to count CFU as described above. Each condition was tested with three independent cultures.

Besides treating established biofilms, the capability of TN-5 to prevent biofilm formation was also evaluated by adding TN-5 prior to the inoculation of biofilm cultures. After incubation for 24 h, the coupons were washed gently and the cells were removed from coupon surface by sonication and vortex for CFU count as described above.

4.7. Alginate Lyase Enhanced the Effects of TN-5 on P. aeruginosa Biofilm Cells

To determine the contribution of a biofilm degrading agent in bacterial killing by TN-5, we repeated biofilm killing experiments in the presence of 50 μg/mL alginate lyase during TN-5 treatments. The same protocols were used for *P. aeruginosa* PAO1 and PDO300 biofilms. Each condition was tested with three independent cultures.

4.8 Statistical Analysis

The data from CFU experiments were analyzed with one-way ANOVA followed by Tukey test using SAS version 9.4 (SAS Institute, Cary, NC, USA). Differences with $p < 0.01$ were considered as statistically significant.

Acknowledgements

We thank John S. Parkinson at University of Utah for sharing the strain *E. coli* RP437 and Matthew Parsek at University of Washington for providing the *P. aeruginosa* PAO1 and PDO300 strains.

Author Contributions

A.A.B., Z.L., N.K., and D.R. conceived and designed the experiments; A.A.B., Z.L., and M.G. performed the experiments; A.A.B. and Z.L. analyzed the data; A.A.B and D.R. led the preparation of the manuscript with participation of all authors.

References

1. Fleming, A. On the antibacterial action of cultures of a *Penicillium*, with special reference to their use in the isolation of *B. influenzae*. *Brit. J. Exp. Pathol.* **1929**, *10*, 226–236.
2. Myers, R.S.; Aldrich, R.H.; Howard, R.W.; Walsh, R.A. The use of gauze inoculated with *Penicillium notatum* or impregnated with crude penicillin in the treatment of surface infections. *New Engl. J. Med.* **1944**, *231*, 761–764.
3. Kardos, N.; Demain, A.L. Penicillin: The medicine with the greatest impact on therapeutic outcomes. *Appl. Microbiol. Biotechnol.* **2011**, *92*, 677–687.
4. Witte, W.; Cuny, C.; Klare, I.; Nubel, U.; Strommenger, B.; Werner, G. Emergence and spread of antibiotic-resistant Gram-positive bacterial pathogens. *Int. J. Med. Microbiol.* **2008**, *298*, 365–377.
5. O'Connell, K.M.; Hodgkinson, J.T.; Sore, H.F.; Welch, M.; Salmond, G.P.; Spring, D.R. Combating multidrug-resistant bacteria: Current strategies for the discovery of novel antibacterials. *Angew. Chem. Int. Ed.* **2013**, *52*, 10706–10733.
6. Davies, J. Inactivation of antibiotics and the dissemination of resistance genes. *Science* **1994**, *264*, 375–382.
7. Lewis, K. Persister cells: Molecular mechanisms related to antibiotic tolerance. In *Antibiotic Resistance*; Springer Berlin Heidelberg: Berlin, Germany, 2012; Volume 211, pp. 121–133.
8. Lewis, K. Riddle of biofilm resistance. *Antimicrob. Agents Chemother.* **2001**, *45*, 999–1007.
9. Barlow, M. What antimicrobial resistance has taught us about horizontal gene transfer. *Methods Mol. Biol.* **2009**, *532*, 397–411.
10. Lewis, K. Persister cells, dormancy and infectious disease. *Nat. Rev. Microbiol.* **2007**, *5*, 48–56.

11. Hall-Stoodley, L.; Costerton, J.W.; Stoodley, P. Bacterial biofilms: From the natural environment to infectious diseases. *Nat. Rev. Microbiol.* **2004**, *2*, 95–108.
12. Mathee, K.; Ciofu, O.; Sternberg, C.; Lindum, P.W.; Campbell, J.I.; Jensen, P.; Johnsen, A.H.; Givskov, M.; Ohman, D.E.; Molin, S.; *et al.* Mucoid conversion of *Pseudomonas aeruginosa* by hydrogen peroxide: A mechanism for virulence activation in the cystic fibrosis lung. *Microbiology* **1999**, *145*, 1349–1357.
13. Pritt, B.; O'Brien, L.; Winn, W. Mucoid *Pseudomonas* in cystic fibrosis. *Am. J. Clin. Pathol.* **2007**, *128*, 32–34.
14. Cabral, D.A.; Loh, B.A.; Speert, D.P. Mucoid *Pseudomonas aeruginosa* resists nonopsonic phagocytosis by human neutrophils and macrophages. *Pediatr. Res.* **1987**, *22*, 429–431.
15. Govan, J.R.; Harris, G.S. *Pseudomonas aeruginosa* and cystic fibrosis: Unusual bacterial adaptation and pathogenesis. *Microbiol. Sci.* **1986**, *3*, 302–308.
16. Torcato, I.M.; Huang, Y.H.; Franquelim, H.G.; Gaspar, D.; Craik, D.J.; Castanho, M.A.; Henriques, S.T. Design and characterization of novel antimicrobial peptides, R-BP100 and RW-BP100, with activity against Gram-negative and Gram-positive bacteria. *Biomembranes* **2013**, *1828*, 944–955.
17. Chongsiriwatana, N.P.; Wetzler, M.; Barron, A.E. Functional synergy between antimicrobial peptoids and peptides against Gram-negative bacteria. *Antimicrob. Agents Ch.* **2011**, *55*, 5399–5402.
18. Wiesner, J.; Vilcinskas, A. Antimicrobial peptides the ancient arm of the human immune system. *Virulence* **2010**, *1*, 440–464.
19. Wang, G. Human antimicrobial peptides and proteins. *Pharmaceuticals* **2014**, *7*, 545–594.
20. Izadpanah, A.; Gallo, R.L. Antimicrobial peptides. *J. Am. Acad. Dermatol.* **2005**, *52*, 381–390.
21. Jenssen, H.; Hamill, P.; Hancock, R.E. Peptide antimicrobial agents. *Clin. Microbiol. Rev.* **2006**, *19*, 491–511.
22. Bahar, A.A.; Ren, D. Antimicrobial peptides. *Pharmaceuticals* **2013**, *6*, 1543–1575.
23. Madani, F.; Lindberg, S.; Langel, U.; Futaki, S.; Graslund, A. Mechanisms of cellular uptake of cell-penetrating peptides. *J. Biophys.* **2011**, *2011*, 414729.
24. Gordon, Y.J.; Romanowski, E.G.; McDermott, A.M. A review of antimicrobial peptides and their therapeutic potential as anti-infective drugs. *Curr. Eye Res.* **2005**, *30*, 505–515.
25. Donnelly, J.P.; Bellm, L.A.; Epstein, J.B.; Sonis, S.T.; Symonds, R.P. Antimicrobial therapy to prevent or treat oral mucositis. *Lancet Infect. Dis.* **2003**, *3*, 405–412.
26. Zhang, L.; Parente, J.; Harris, S.M.; Woods, D.E.; Hancock, R.E.; Falla, T.J. Antimicrobial peptide therapeutics for cystic fibrosis. *Antimicrob. Agents Chemother.* **2005**, *49*, 2921–2927.
27. Yedery, R.D.; Reddy, K.V. Antimicrobial peptides as microbicidal contraceptives: Prophecies for prophylactics—a mini review. *Eur. J. Contracep. Repr.* **2005**, *10*, 32–42.
28. Laverty, G.; Gorman, S.P.; Gilmore, B.F. The potential of antimicrobial peptides as biocides. *Int. J. Mol. Sci.* **2011**, *12*, 6566–6596.
29. Noto, P.B.; Abbadessa, G.; Cassone, M.; Mateo, G.D.; Agelan, A.; Wade, J.D.; Szabo, D.; Kocsis, B.; Nagy, K.; Rozgonyi, F.; *et al.* Alternative stabilities of a proline-rich antibacterial peptide *in vitro* and *in vivo*. *Protein Sci.* **2008**, *17*, 1249–1255.

30. Bruhn, O.; Grotzinger, J.; Cascorbi, I.; Jung, S. Antimicrobial peptides and proteins of the horse-insights into a well-armed organism. *Vet. Res.* **2011**, *42*, 98.
31. Lehrer, R.I.; Lichtenstein, A.K.; Ganz, T. Defensins: Antimicrobial and cytotoxic peptides of mammalian cells. *Annu. Rev. Immunol.* **1993**, *11*, 105–128.
32. Zhou, C.H.; Min, J.K.; Liu, Z.G.; Young, A.; Deshazer, H.; Gao, T.; Chang, Y.T.; Kallenbach, N.R. Synthesis and biological evaluation of novel 1,3,5-triazine derivatives as antimicrobial agents. *Bioorg. Med. Chem. Lett.* **2008**, *18*, 1308–1311.
33. Hay, I.D.; Remminghorst, U.; Rehm, B.H. MucR, a novel membrane-associated regulator of alginate biosynthesis in *Pseudomonas aeruginosa*. *Appl. Environ. Microbiol.* **2009**, *75*, 1110–1120.
34. Som, A.; Vemparala, S.; Ivanov, I.; Tew, G.N. Synthetic mimics of antimicrobial peptides. *Biopolymers* **2008**, *90*, 83–93.
35. Thaker, H.D.; Som, A.; Ayaz, F.; Lui, D.; Pan, W.; Scott, R.W.; Anguita, J.; Tew, G.N. Synthetic mimics of antimicrobial peptides with immunomodulatory responses. *J. Am. Chem. Soc.* **2012**, *134*, 11088–11091.
36. He, J.; Eckert, R.; Pharm, T.; Simanian, M.D.; Hu, C.; Yarbrough, D.K.; Qi, F.; Anderson, M.H.; Shi, W. Novel synthetic antimicrobial peptides against *Streptococcus mutans*. *Antimicrob. Agents Chemother.* **2007**, *51*, 1351–1358.
37. Diehnelt, C.W. Peptide array based discovery of synthetic antimicrobial peptides. *Front. Microbiol.* **2013**, *4*, 402.
38. Bahar, A.A.; Liu, Z.; Totsingan, F.; Buitrago, C.; Kallenbach, N.; Ren, D. Synthetic dendrimeric peptide active against biofilm and persister cells of *Pseudomonas aeruginosa*. *Appl. Microbiol. Biotechnol.* **2015**, *99*, 8125–8135.
39. Wozniak, D.J.; Wyckoff, T.J.; Starkey, M.; Keyser, R.; Azadi, P.; O'Toole, G.A.; Parsek, M.R. Alginate is not a significant component of the extracellular polysaccharide matrix of PA14 and PAO1 *Pseudomonas aeruginosa* biofilms. *Proc. Natl. Acad. Sci. USA* **2003**, *100*, 7907–7912.
40. Hou, S.Y.; Liu, Z.G.; Young, A.W.; Mark, S.L.; Kallenbach, N.R.; Ren, D.C. Effects of Trp- and Arg-containing antimicrobial peptide structure on inhibition of *Escherichia coli* planktonic growth and biofilm formation. *Appl. Environ. Microbiol.* **2010**, *76*, 1967–1974.
41. Brown, K.L.; Hancock, R.E. Cationic host defense (antimicrobial) peptides. *Curr. Opin. Immunol.* **2006**, *18*, 24–30.
42. Tam, J.P.; Lu, Y.A.; Yang, J.L. Antimicrobial dendrimeric peptides. *Eur. J. Biochem.* **2002**, *269*, 923–932.
43. Andrews, J.M. Determination of minimum inhibitory concentrations. *J. Antimicrob. Chemother.* **2001**, *48*, 5–16.
44. Parkinson, J.S.; Houts, S.E. Isolation and behavior of *Escherichia coli* deletion mutants lacking chemotaxis functions. *J. Bacteriol.* **1982**, *151*, 106–113.
45. DeVries, C.A.; Ohman, D.E. Mucoid-to-nonmucoid conversion in alginate-producing *Pseudomonas aeruginosa* often results from spontaneous mutations in algt, encoding a putative alternate sigma factor, and shows evidence for autoregulation. *J. Bacteriol.* **1994**, *176*, 6677–6687.
46. Wiegand, I.; Hilpert, K.; Hancock, R.E. Agar and broth dilution methods to determine the minimal inhibitory concentration (MIC) of antimicrobial substances. *Nat. Protoc.* **2008**, *3*, 163–175.

47. Niepa, T.H.R.; Gilbert, J.L.; Ren, D.C. Controlling *Pseudomonas aeruginosa* persister cells by weak electrochemical currents and synergistic effects with tobramycin. *Biomaterials* **2012**, *33*, 7356–7365.
48. Pan, J.C.; Bahar, A.A.; Syed, H.; Ren, D.C. Reverting antibiotic tolerance of *Pseudomonas aeruginosa* PAO1 persister cells by (*Z*)-4-bromo-5-(bromomethylene)-3-methylfuran-2(5H)-one. *PloS ONE* **2012**, *7*, e45778.
49. Chen, X.; Zhang, M.; Zhou, C.H.; Kallenbach, N.R.; Ren, D.C. Control of bacterial persister cells by Trp/Arg-containing antimicrobial peptides. *Appl. Environ. Microbiol.* **2011**, *77*, 4878–4885.

2

Control of Biofilms with the Fatty Acid Signaling Molecule *cis*-2-Decenoic Acid

Cláudia N. H. Marques [1,2,*], David G. Davies [1,2] and Karin Sauer [1,2]

[1] Department of Biological Sciences, Binghamton University, Binghamton, NY 13902, USA; E-Mails: dgdavies@binghamton.edu (D.G.D.); ksauer@binghamton.edu (K.S.)

[2] Binghamton Biofilm Research Center (BBRC), Binghamton University, Binghamton, NY 13902, USA

* Author to whom correspondence should be addressed; E-Mail: cmarques@binghamton.edu;

Academic Editor: Dacheng Ren

Abstract: Biofilms are complex communities of microorganisms in organized structures attached to surfaces. Importantly, biofilms are a major cause of bacterial infections in humans, and remain one of the most significant challenges to modern medical practice. Unfortunately, conventional therapies have shown to be inadequate in the treatment of most chronic biofilm infections based on the extraordinary innate tolerance of biofilms to antibiotics. Antagonists of quorum sensing signaling molecules have been used as means to control biofilms. QS and other cell-cell communication molecules are able to revert biofilm tolerance, prevent biofilm formation and disrupt fully developed biofilms, albeit with restricted effectiveness. Recently however, it has been demonstrated that *Pseudomonas aeruginosa* produces a small messenger molecule *cis*-2-decenoic acid (*cis*-DA) that shows significant promise as an effective adjunctive to antimicrobial treatment of biofilms. This molecule is responsible for induction of the native biofilm dispersion response in a range of Gram-negative and Gram-positive bacteria and in yeast, and has been shown to reverse persistence, increase microbial metabolic activity and significantly enhance the cidal effects of conventional antimicrobial agents. In this manuscript, the use of *cis*-2-decenoic acid as a novel agent for biofilm control is discussed. Stimulating the biofilm dispersion response as a novel antimicrobial strategy holds significant promise for enhanced treatment of infections and in the prevention of biofilm formation.

Keywords: biofilms; biofilm dispersion; persister cells; signaling molecules; bacterial physiology; *cis*-2-decenoic acid

---

1. Introduction

Biofilms are dynamic communities of closely associated microbial cells embedded within a hydrated extracellular polymeric substance (EPS) on an air-liquid interface, or adherent to inert (abiotic) or living surfaces and constitute the major proportion of bacterial biomass in nature [1]. The EPS of biofilm communities is composed of any combination of polysaccharides, nucleic acids, proteins, lipids, and under certain conditions, humic substances [2–6]. Biofilms are formed in a sequential manner [7–10], initiated by the reversible attachment of free-floating cells to a surface [11]. A transition to irreversible attachment is mediated by a variety of direct interactions with the substratum and is generally associated with the onset of the production of EPS which entraps bacteria and results in cell-cell aggregation and the formation of microcolonies [12–16]. As microcolonies grow, their resident bacteria alter their physiological state toward a biofilm-specific phenotype, described as maturation stages I and II [7–9,17,18]. When biofilm microcolonies begin to experience limited transport with the bulk liquid, the bacteria within the microcolony may undergo a final switch in phenotype from a sessile to a mobile state, releasing cells through breaches formed in the microcolony wall in a process called dispersion [7,19].

As three dimensionally structured communities, biofilms encompass a variety of environmental microniches characterized by bacteria with differences in metabolic activity, growth rate, nutrient utilization, and the expression of surface molecules and virulence factors [20–29]. The behavior of bacteria within a biofilm has been shown in many cases to be governed by secreted chemical signals that are used in cell-to-cell communication. An example of such behavioral modification is demonstrated by *Pseudomonas aeruginosa*, which as a wild-type forms highly structured heterogeneous biofilms, but in strains lacking the ability to produce the quorum sensing (QS) inducer 3-oxododecanoyl-HSL, forms biofilms displaying a flat, densely packed and homogeneous architecture [17]. Likewise, a *P. aeruginosa pqs* mutant, deficient in the production of 2-heptyl-3-hydroxy-4-quinolone (*Pseudomonas* quinolone signal—PQS) has been observed to form thin biofilms containing reduced levels of extracellular DNA, compared to the wild-type strain [6,30–32]. In addition to influencing biofilm architecture, bacterial cell-to-cell signaling within a biofilm can provide cues about local environmental conditions, modulate inter-species and intra-species interactions with other individuals within the community, and allow coordinated responses to enable group-level expression of virulence and bacterial persistence. Cell-cell signaling in bacteria is accomplished through the production and detection of small diffusible signaling molecules of different chemical classes, that can be produced by Gram-negative bacteria, Gram-positive bacteria, or both Gram-positive and Gram-negative. To date, more than 50 such molecules have been identified and include: autoinducer-1 (AI-1) also known as *N*-acylhomoserine lactones (*N*-AHL), autoinducer-2 (AI-2) a furanosyl borate, PQS, oligopeptides (5–10 amino acid cyclic thiolactone) known as autoinducer peptides (AIP), and short-chain fatty acids which are typically unsaturated at the number 2 carbon in a *cis* configuration [33–37]. Secondary messengers provide a further level of regulation and include c-di-GMP, known to regulate biofilm development, with high intercellular levels of this molecule inducing the production of adhesins and EPS, and low levels having the opposite effect [38–40]. In *P. aeruginosa*, biofilm formation has been linked to elevated c-di-GMP levels, while dispersion is coincident with significantly reduced c-di-GMP [38,41].

One of the hallmark characteristics of biofilms is their profound tolerance to antimicrobial agents [42–44], which results in persistent infections and renders biofilms difficult to control [42,45]. Biofilms are typically 10 to 1000 times less susceptible to antimicrobials when compared to their planktonic counterparts [22,46–48]. This resistance to killing, referred to here as "biofilm tolerance" is distinct from specific resistance conferred by mutation of an antimicrobial target or the acquisition of a plasmid-borne resistance marker [46,49–52]. Instead, biofilm tolerance is observed as a generalized reduction in susceptibility to killing by essentially any antimicrobial agent. This tolerance can be overcome, in part, by significantly increasing the concentration of antimicrobial, suggesting that specific resistance is not a component of biofilm tolerance, and that cells derived from biofilms do not exhibit specific tolerance. Furthermore, biofilm tolerance may be overcome by inducing biofilm bacteria to mount a dispersion response, resulting in the evacuation of bacterial cells from the biofilm; a behavior that re-establishes susceptibility of the participating bacteria to antimicrobial agents [53–55].

Biofilm tolerance arises from a combination of factors that are associated with the unique environment bacteria experience within the confines of a biofilm, including: failure of the antimicrobial to penetrate throughout the biofilm matrix [21,56–58], reduced growth rates by the resident bacteria, a wide range of non-specific protective adaptations (such as enhanced efflux) associated with the biofilm phenotype [59–63], and the formation of a persister cell sub-population [64–67]. Biofilm tolerance further extends to resistance to killing by the immune system, with bacteria deriving protection against immune cells by being embedded within the mesh-like network of EPS [68].

Considering the multifactorial nature of biofilm resistance, it is apparent that strategies to eradicate biofilms have to address all the above mechanisms simultaneously [21]. While there are numerous ongoing efforts to address the biofilm problem, the most promising strategies include those aimed at manipulating the mode of growth, by either preventing biofilms from forming, or by disrupting existing biofilms. This is further supported by the finding that dispersion, the evacuation of bacterial cells from the biofilm, coincides with the reinstatement of antibiotic sensitivity [53–55]; suggesting that the potential to overcome the biofilm resistance mechanisms or to induce the transition of biofilm bacteria from a resistant to a susceptible phenotype would likely result in enhanced treatment options in fighting biofilm infections.

In this review, we will discuss a novel strategy to enhance the susceptibility of biofilm cells by subverting the normal biological processes that maintain biofilm integrity and promote biofilm tolerance. This strategy takes advantage of the regulatory networks that are controlled by inter and intra-species signaling and will focus primarily upon the cross-kingdom signaling molecule *cis*-2-decenoic acid, a fatty acid signal that has been shown to prevent biofilm development, induce biofilm dispersion, cause global changes in cellular phenotype, modulate bacterial virulence and override microbial persistence.

2. Jamming Bacterial Communication

Intercellular signaling regulates functions contributing to persistence and virulence of many bacterial pathogens. Thus, signaling interference holds significant promise as the basis of novel therapeutic strategies to improve the outcome of bacterial, and in particular biofilm infections [69–71]. QS plays a vital role in the regulation of virulence and biofilm-related behaviors both in the natural environment and in persistent infections [72,73], and is currently considered to be the main target as means of control of biofilm infections in the lung infections of cystic fibrosis patients, as well as other infections. Disruption of QS can be achieved by reducing the *N*-acylhomoserine lactone (*N*-AHL) synthase activity, inhibiting the production of the QS molecules, degrading *N*-AHLs (the most used), and using synthetic compounds that mimic QS inducer molecules [33]. Below, we provide several examples of quorum quenching, where interference with QS signalling occurs, leading to the arrest of biofilm development, disaggregation of established biofilms, and reduction of virulence.

P. aeruginosa, is an opportunistic microorganism found in a variety of niches and able to cause acute and chronic infections such as those found in the lungs of cystic fibrosis patients [74–77]. It is thus, a principal model microorganism used in the study of biofilms and biofilm cell behavior, has one of the most extensively investigated QS systems, and has been found to have approximately 10% of its genome dedicated to signalling [78]. *P. aeruginosa* contains one of the most complex QS systems known in bacteria, where an interconnected signaling cascade coordinates virulence, persistence, the transition to sessile growth, and biofilm development [79]. Three separate and inter-related QS systems are used by *P. aeruginosa*. These systems are regulated in a hierarchical manner with LasI/LasR positively regulating the RhlI/RhlR system, while the PQS system is regulated by both the LasI/LasR and RhlI/RhlR systems [30,73,80]. Considering that establishment of infections by *P. aeruginosa* is dependent on the LasI/LasR system, with LasR being required for QS signal binding and activation of the regulatory cascade, it is not surprising that the identification of LasR inhibitors has become a major focus in *P. aeruginosa* research [34]. The LasR inhibitors comprise N-AHL signaling molecules that have been chemically altered with respect to the acyl side chain, and the lactone ring [34]. In addition to inhibiting LasR, some of these molecules have been found to be potent inhibitors of RhlR [81,82]. For instance, *meta*-bromothiolactone was shown to inhibit pyocyanin production (a virulence factor) and biofilm formation, and to protect *Caenorhabditis elegans* and human lung epithelial cells from killing by *P. aeruginosa* [83]. Additional N-AHL inhibitors comprise various naturally occurring products such as furanones and patulin and their derivatives [83–88], with synthetic furanones having been demonstrated to significantly attenuate *P. aeruginosa* lung infections [83].

Modulators of the secondary messenger molecule c-di-GMP have also previously been described. The molecules LP 3134, LP 3145, LP 4010 and LP 1062 were found to inhibit the diguanylate cyclases (DCG), WspR from *P. aeruginosa* and tDGC from *Thermotoga maritime*; both required for c-di-GMP synthesis [89]. The inactivation of these DGCs resulted in biofilm inhibition of *P. aeruginosa* and *Acinetobacter baumannii* [89]. In addition, the compounds ebselen and ebselen oxide were also found to inhibit c-di-GMP by covalently modifying cysteine residues and inhibiting the binding of c-di-GMP to particular receptors, ultimately regulating biofilm development [90].

Most of these QS inhibitors are naturally occurring and show promise as therapeutic agents due to their cross species activity. For example, the QS modulators curcumin and its derivatives have been shown to reduce biofilm growth of *P. aeruginosa*, *Escherichia coli* and *Vibrio harveyi*, while also reducing pathogenicity and the accumulation of microorganisms on wet surfaces (biofouling) [85]. A combination of quorum quenching metabolites (inhibitors of QS) with antimicrobials has been demonstrated to enhance the susceptibility of *P. aeruginosa* and other microorganisms [86,91]. In *Staphylococcus aureus*, a Gram-positive organism, an RNAIII-inhibiting peptide suppressed the staphylococcal TRAP/agr two-component regulatory systems [involved in biofilm formation and virulence), reduced biofilm formation *in vivo*, and prevented infections by methicillin-resistant *S. aureus* (MRSA) in rats [92]. Several anti-QS peptides have also led to the inhibition of biofilm formation in Gram-positive oral bacteria [93].

These examples underscore the importance of QS quenching in the control of biofilm infections. However, molecules involved in QS have different chemical compositions, different targets, and vary significantly from species to species, and to date, no universal molecule that is able to quench (or inhibit) all different QS systems has been found. While QS quenching shows promise as a strategy for combatting biofilms [94], this approach is limited due to the complex interactions between the signals of the various species, and the various mechanisms that control QS sensing [95].

3. Fatty Acid Signaling Systems

Fatty acid signals comprise a growing group of recently identified structurally related inducer molecules, which regulate a wide range of cellular functions. These signals have been identified in a range of Gram-positive and Gram-negative bacteria, as well as in the yeast *Candida albicans* (Table 1), and are now recognized as a unique chemical class of QS signals [96]. To readily distinguish these molecules from one another, a new nomenclature has been proposed where the methyl (Me) substitution, if present, and its position are indicated first, the number of carbons in the fatty acyl side-chain is indicated second, and the position of the double bond (Δ) is indicated last, for example: *cis*-11-methyl-2-dodecenoic acid is named 11-Me-C_{12}:Δ^2, where the methyl group is on carbon 11, in a 12 carbon fatty acid, with a double bond in the *cis* configuration between the number 2 and number 3 carbons (Table 1) [97].

Akin to *N*-AHLs and other cell-cell signaling molecules, fatty acid signals are involved in intra-species, inter-species, and cross-kingdom communication where they regulate bacterial growth, virulence, motility, polymer production, biofilm development, biofilm dispersion and persistence. The fatty acid signal 11-Me-C_{12}:Δ^2 in *Xanthomonas campestris* was the first member of this sub-class of *cis*-2-unsaturated fatty acids to be characterized and was named DSF for diffusible signal factor [98,99]. Interspecies signaling using fatty acids has been shown to occur between members of the *Burkholderia cepacia* complex [97], and between the members of *Xanthomonads* [96]. The *Xyllela fastidiosa* signal C14:Δ^2, and the *Burkholderia cenocepacia* signals C_{12}:Δ^2 and 11-Me-C_{12}:$\Delta^{2,5}$ are functional homologues of 11-Me-C_{12}:Δ^2 and both regulate biofilm formation and production of virulence factors [97,100]. Additionally, both signaling molecules contribute to the regulation of virulence factors, in response to elevated concentrations of 11-Me-C_{12}:Δ^2 produced by *Xanthomonas campestris*, *Stenotrophomonas maltophilia* and *Xylella* sp., leading to the disaggregation of flocs (clumps) present in planktonic cultures, acriflavin resistance and detoxification [101–103].

Table 1. Fatty acid signaling molecules with known functions in the various microorganisms.

Compound	New Nomenclature	Structure	Bacterial Species	Function	Reference
cis-11-Methyl-2-dodecenoic acid (DSF)	11-Me-C_{12}:Δ^2	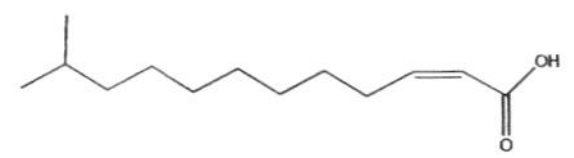	*Xanthomonas campestris*, *Xanthomonas oryzae*, *Stenotrophomonas maltophilia*, *Burkholderia multivorans*	Virulence, biofilm formation, floc disaggregation, microcolony formation, tolerance to antibiotics, detoxification, hyphal growth inhibition	[96–99,104,105]
cis-2-Dodecenoic acid (BDSF)	C_{12}:Δ^2	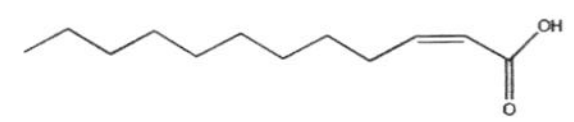	*Burkholderia cenocepacia*, *Burkholderia lata* *Burkholderia stabilis* *Burkholderia vietnamiensis* *Burkholderia dolorosa* *Burkholderia ambifaria* *Burkholderia anthina* *Burkholderia pyrrocinia* *B. multivorans*, *X. oryzae*	Virulence, hyphal growth inhibition	[97,104,106,107]
cis-2-Decenoic acid (*cis*-DA)	C_{10}:Δ^2	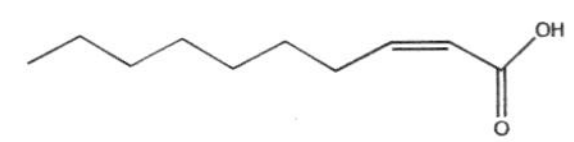	*Pseudomonas aeruginosa*	Biofilm formation, biofilm dispersion, persister cell formation, persister cell awakening, tolerance to antimicrobials.	[19,65,108–110]
cis-2-Tetradecenoic acid	C_{14}:Δ^2	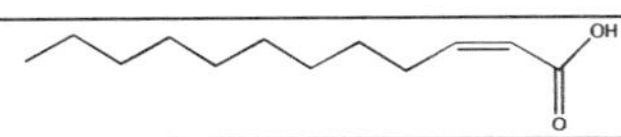	*Xylella fastidiosa*	Virulence and aggregation	[100]
trans-2-Decenoic acid (SDSF)	C_{10}:Δ^{2t}	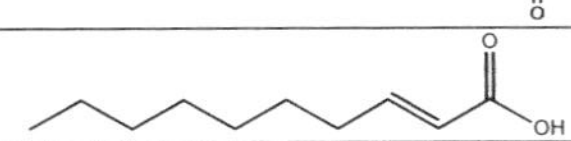	*Streptococcus mutans*	Hyphal growth inhibition	[111]
cis-11-Methyldodeca-2,5-dienoic acid (CDSF)	11-Me-C_{12}:$\Delta^{2,5}$	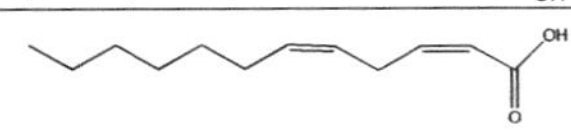	*B. multivorans*, *B. stabilis* *B. anthina*, *B. pyrrocinia*, *X. oryzae*	Hyphal growth inhibition	[97,104]
10-Methyldodecanoic acid	10-Me-C_{12}	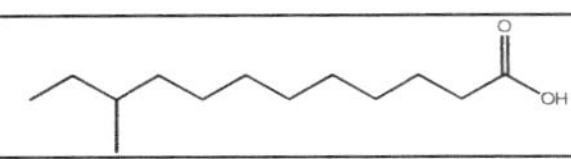	*S. maltophilia*	Stress tolerance and antibiotic tolerance	[105]
11-Methyldodecanoic acid	11-Me-C_{12}	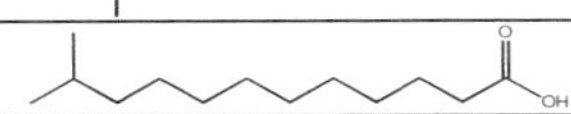	*S. maltophilia*	Stress tolerance and antibiotic tolerance	[105]
12-Methyltetradecanoic acid	12-Me-C_{14}	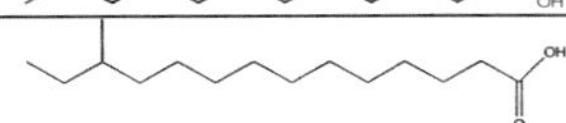	*Xylella fastidiosa*	Virulence, biofilm formation, motility	[112,113]
3-Hydroxypalmitic acid	3OH-PAME	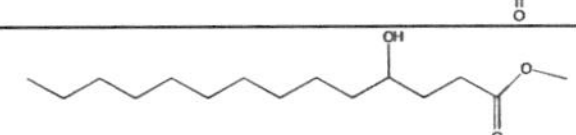	*Ralstonia solanacearum*	Virulence	[114]
Farnesoic acid	3,7,11-Me-C_{12}:$\Delta^{2t,6t,10t}$	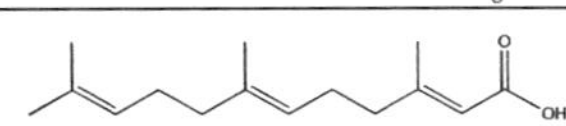	*Candida albicans*	Inhibition of germ tube formation	[115]

In *S. maltophilia*, 11-Me-C_{12}:Δ^2 production has been linked to changes in the lipopolysaccharide composition of LPS, increased protease synthesis, enhanced cell aggregation, biofilm microcolony formation and increased tolerance to antibiotics and heavy metals [116]. However, fatty acid signals do not only increase aggregative behavior or biofilm formation capability. For instance, 12-Me-C_{14} which was originally isolated from *Xylella fastidiosa*, has been found to abolish swarming motility and to reduce biofilm formation by *X. fastidiosa* and in *P. aeruginosa* [117]. The various fatty acids produced by *S. maltophilia* (Table 1) have also been demonstrated to have interspecies activity, and have been found to increase the stress tolerance and resistance of *P. aeruginosa* to antimicrobial peptides [118]. Production of fatty acid signals is not limited to bacteria, but is also found within the fungi where *C. albicans* has been shown to produce farnesoic acid which regulates the morphological transition from a yeast-form to a hyphal-form by inhibiting germ tube formation [96,115]. In addition to responding to its own fatty acid signal, *C. albicans* has been shown to respond to a variety of signals from other organisms, including C_{12}:Δ^2 from *B. cenocepacia*, C_{10}:Δ^{2t} from *S. mutans*, 11-Me-C_{12}:Δ^2 from *X. campestris* and 11-Me-C_{12}:$\Delta^{2,5}$ from *Burkholderia multivorans*, demonstrating the cross-kingdom capabilities of fatty acid signals [96,97,106,111].

4. The Fatty Acid System in *P. aeruginosa* and Its Role in Biofilm Control

The fatty acid signal *cis*-2-decenoic acid (C_{10}:Δ^2), which will be referred to subsequently as *cis*-DA, was recently identified in *P. aeruginosa* and found to act as a biofilm dispersion autoinducer [19]. The *cis*-DA signaling system has been shown to regulate 666 genes encoding proteins involved in motility, chemotaxis, cell attachment, TCA cycle, EPS and LPS synthesis and secretion, virulence, iron uptake, and respiration [119]. Current findings have shown the *dspI* gene (PA14_54640), a *PA0745* ortholog, to be required for the production of *cis*-DA [120]. Mutation of *dspI* in *P. aeruginosa* leads to a reduction of pyoverdine (a virulence factor) production, and to defective swarming motility [121]. However, the complete signaling transduction mechanism of the *cis*-DA system has yet to be fully elucidated. Most importantly and unlike other fatty acids and other signaling systems, *cis*-DA has been demonstrated to control biofilms by three different mechanisms: by impairing biofilm formation, by inducing biofilm dispersion [19], and by modulating the persister cell sub-population [65]. Additionally, *cis*-DA shows interspecies activity and also induces a dispersion response similar to that observed by *P. aeruginosa* in biofilms formed by *C. albicans* and various Gram-negative and Gram-positive bacteria, including other *Pseudomonads*, *E. coli*, *Streptococcus pyogenes*, *S. aureus*, *Klebsiella pneumoniae*, *Proteus mirabilis* and *Bacillus subtilis* [19]. Furthermore, *cis*-DA has been shown to induce biofilm dispersion in *Propionibacterium acnes* [122], *Actimomyces naeslundii*, *Lactobacillus casei* and *Streptococcus mutans* [123], when grown as single or mixed species biofilms. To date no other fatty acid has been shown to induce biofilm dispersion. However, 11-Me-C_{12}:Δ^2 has been demonstrated to disaggregate bacterial flocs of *X. campestris* [101].

4.1. Dispersion Induced by cis-2-Decenoic Acid as Means of Biofilm Control

Dispersion is one of the only bacterial behaviors known to be unique to biofilms, and is modulated by cell-to-cell communication. Biofilm dispersion is characterized by the release of bacteria from established biofilms, presumably to escape overcrowding or degraded environmental conditions.

Autoinduction of dispersion occurs when microcolonies within a biofilm have attained a critical maximum size, and provides a mechanism by which bacteria can evacuate a microcolony, enter the bulk liquid environment, and disseminate to new sites for attachment [19]. The biofilm-specific nature of the dispersion response implies that drug development focusing on this behavior, could lead to an effective treatment of biofilm infections. The principal advantage to using drugs that target biofilm dispersion is that as bacteria (and fungi) transition from a biofilm to planktonic phenotype, the biofilm tolerance is reversed, enhancing susceptibility to antimicrobial agents and eliminating problems associated with antimicrobial penetration, the biofilm-specific phenotype and reduced growth rates [19,124]. Hence, exposure of biofilms to a dispersion inducer before and/or in combination with conventional antimicrobials would provide a novel mechanism for enhancing the activity of the treatments, through the disruption of existing biofilms [19,124]. Recent studies have shown that when compared to treatment with antimicrobials alone, targeting mature biofilms with antimicrobials in combination with *cis*-DA significantly reduces and removes the overall biofilm biomass in bacteria that are associated with biofilm infections and contamination of foods [108–110,125]. Exposure of pre-formed biofilms to different antimicrobials in the presence of *cis*-DA has been shown to lead to a reduction of the viability of MRSA *S. aureus* biofilm cells [109], and to significantly remove the biofilm biomass of *B. subtilis*, *Salmonella enterica*, *S. aureus* and *E. coli* from stainless steel and polystyrene surfaces, where they are known to cause food contamination [108]. In addition, co-exposure of *P. aeruginosa* PAO1 biofilm cells to antibiotics with *cis*-DA enhanced the killing efficacy of tobramycin or ciprofloxacin by greater than 1.5 Log, compared to treatment with the antibiotics alone [125]. It is notable that exposure to *cis*-DA alone shows no reduction in viability of bacteria [126], and has no cytotoxic effects on fibroblasts [109].

4.2. Persister Cell Control by Exposure to cis-DA

Persister cells are a subgroup (typically 0.0001%–1%) of a bacterial clonal population, consisting of dormant/semi-dormant specialised survivor cells that fail to succumb to treatment with antimicrobial agents, when the majority of the population shows susceptibility [64–66,127,128]. Various mechanisms have been proposed to explain persister cell formation, including down-regulation of genes involved in energy generation and cell maintenance [129,130], activation of the stringent response [131], and increased expression of toxin-antitoxin (TA) modules [64,132,133]. The growth rate distribution of cells within the differing microniches of a biofilm likely accounts for persister cell survival, with dormant or semi-dormant sub-populations existing at locations within the biofilm that are nutrient-limited, inducing a bacterial stress response [83,84].

The control of persister cells could be accomplished if this population were to be resuscitated to a metabolically active state, rendering them susceptible to challenge with antimicrobials. Exposure to metabolic stimuli such as—mannitol, glucose, fructose and pyruvate—have been shown to increase bacterial growth of *S. aureus* persister cells by more than 600 fold [134]. In addition to experiencing increased growth, exposure of *S. aureus* and *E. coli* persister cells to mannitol, glucose, fructose and pyruvate, also increased central metabolic activity as measured by increased respiration rates and membrane permeability [134]. This finding suggests that metabolic stimuli have the potential to resuscitate persister cells [134]. This is further supported by the finding that exposure to metabolic stimuli coincided with the eradication of *E. coli* persister cells by gentamicin [135]. In addition to metabolic stimuli, resuscitation of persister cells has been accomplished using QS inhibitors. The QS inhibitor (*Z*)-4-bromo-5-(bromomethylene)-3-methylfuran-2(5*H*)-one (BF8), known to inhibit AI-1 mediated QS through the inhibition of *lasB* expression, was shown to sensitize *P. aeruginosa* persister cells to ciprofloxacin and tobramycin, however the efficacy was reduced compared to exposure to metabolic stimuli has it did not result in complete cell eradication [136]. Although not used as a carbon source [65], *cis*-DA was found to significantly increase the number of cells recovered in agar plates [126], suggesting *cis*-DA to somehow trigger a transition from a dormant to an awake, likely more metabolically active susceptible state. This is supported by the finding that *P. aeruginosa* and *E. coli* persister cells obtained from planktonic and biofilm populations experience an additional decrease in viability (greater than 2 Log) to the point of eradication upon exposure to antimicrobials in combination with *cis*-DA compared to treatment with antimicrobials alone [65]. The effect of *cis*-DA is not limited to persister cells formed by Gram-negative organisms, as exposure of persister cells by the Gram-positive *S. aureus* to antibiotics in the presence of *cis*-DA also led to a significant decrease in cell viability, while antibiotics alone had no effect on the persister population [137]. In addition to increasing the metabolic status of persister cells and reverting their tolerant state, *cis*-DA also prevents persister cell formation. This is supported by the finding that the presence of *cis*-DA during persister cell isolation resulted in a 2-Log decrease in persister cell numbers in both *P. aeruginosa* and *E. coli* [65]. Thus, under the conditions tested challenge with *cis*-DA leads to a change in the cell's metabolic status, resulting in a reduction of persister cell formation and a reversion of the persister cell's tolerant state.

4.3. Prevention of Biofilm Formation by Exposure to cis-DA

It has been shown that certain fatty acid signals can act as agents to reduce, or negatively impact biofilm formation in several bacteria as demonstrated with 11-Me-C_{12}:Δ^2, 11-Me-C_{12}:$\Delta^{2,5}$, and 12-Me-C_{14} [97,99,117]. It is therefore not surprising that *cis*-DA can also prevent biofilm formation, likely by keeping the cells in a constant metabolically active and/or dispersive growth state, as *cis*-DA also induces dispersion and increases the persister cells' metabolic status. In a *S. aureus* MRSA strain, inhibition of bacterial growth was achieved in the presence of 2.94 mM of *cis*-DA, while as little as 734 μM *cis*-DA prevented biofilm formation [109]. Similarly, *cis*-DA prevented biofilm formation by single and dual-species biofilms of *E. coli* and *K. pneumoniae* in catheters [110], and in biofilms of *P. aeruginosa* PAO1 grown in flow cell reactors [19], when used at concentrations of 310 nM and 2.5 nM respectively. Continuous exposure of *P. aeruginosa* PAO1 expressing green fluorescent protein (GFP), in flow cell reactors to Electric Power Research Institute (EPRI) medium supplemented with 1 nM, 1 μM or 1 mM of *cis*-DA (Figure 1B–D, Table 2) under the conditions described by Davies and Marques [19], resulted in a significant reduction of maximum biofilm thickness (2-fold), average biofilm thickness (2.2-, 1.4- and 8-fold lower, respectively), roughness coefficient (1.5-fold lower, 6.4-fold lower and no-fold change, respectively), and total biomass (2.4-, 1.7- and 8.5-fold lower respectively), when compared to the absence of *cis*-DA (Figure 1, Table 2). Interestingly, *cis*-DA effectiveness does not increase with increasing concentration. Challenge of bacteria with 1 μM *cis*-DA (Figure 1C) actually resulted in higher biomass and increased microcolony average thickness than

1 nM *cis*-DA (Figure 1B), demonstrating that the effects of this fatty acid are due to its activity as a signaling molecule and not due to direct effects of the fatty acid as a detergent. However, increasing the *cis*-DA concentration to the mM range (Figure 1D), its effectiveness is once again present, indicating that at high concentrations *cis*-DA might have a detergent effect.

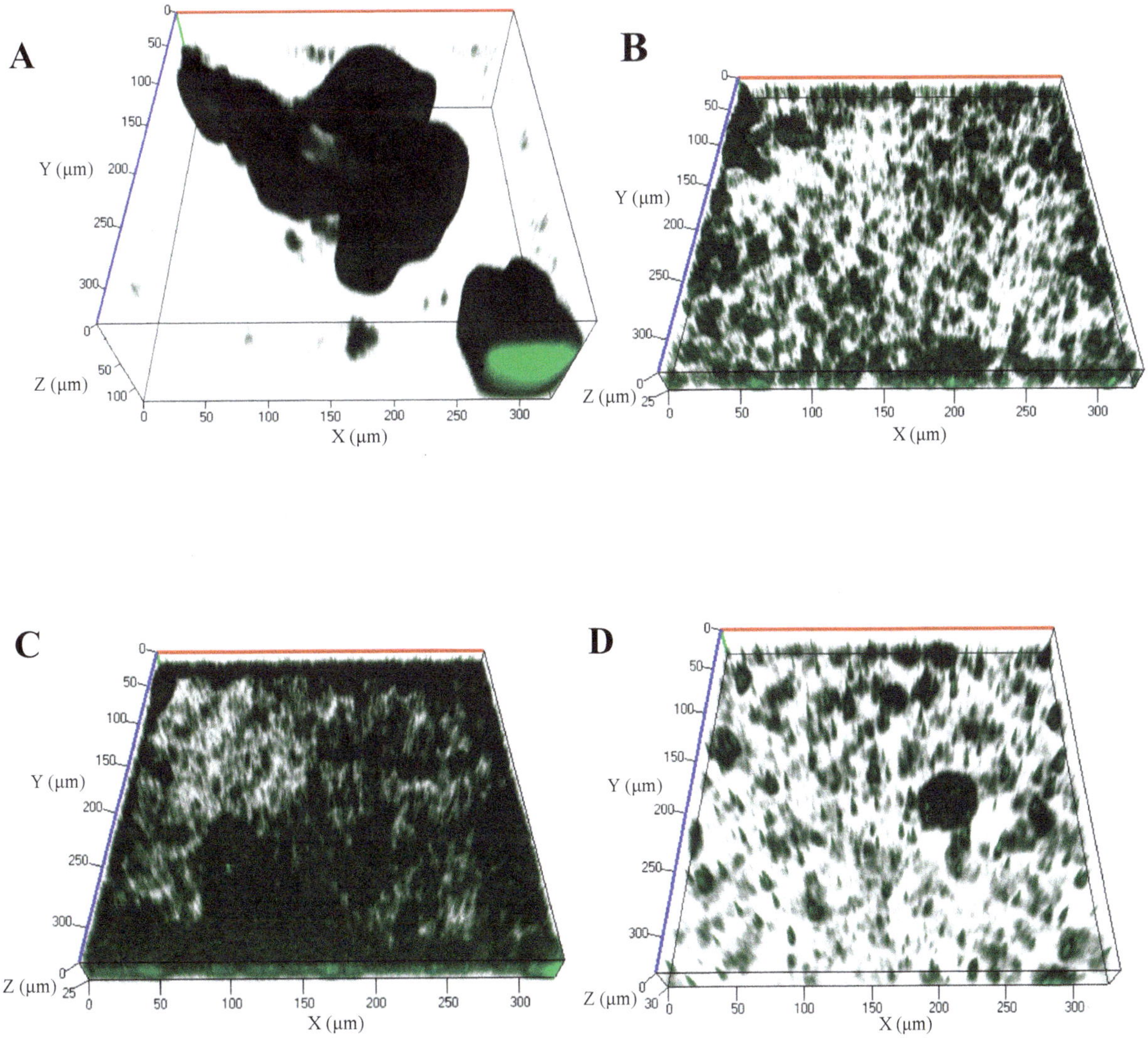

Figure 1. Biofilms of *P. aeruginosa* PA01 harboring the green fluorescence protein (GFP) expressing vector pMRP9-1 were grown in flow cell reactors in EPRI media, in the continuous presence of different concentrations of *cis*-2-decenoic acid (*cis*-DA). (**A**) Control (media); (**B**) 1 nM of *cis*-DA; (**C**) 1 μM of *cis*-DA; (**D**) 1 mM *cis*-DA. Following 96 h of culture, biofilms were observed using a confocal microscope at a 400× magnification.

Table 2. COMSTAT analysis of *P. aeruginosa* PAO1 harboring the green fluorescence protein expressing vector pMRP9-1, grown in flow cell reactors in EPRI media, in the continuous presence of different concentrations of *cis*-DA. Results are the average of at least 4 replicates.

Biofilm Structure Quantification	**Control**	***cis*-DA 1 nM**	***cis*-DA 1 µM**	***cis*-DA 1 mM**
	Average ± SD	**Average ± SD**	**Average ± SD**	**Average ± SD**
Total biomass ($\mu m^3/\mu m^2$)	20.66 ± 7.02	8.49 ± 2.38	12.29 ± 3.5	2.42 ± 0.86
Average thickness (µm)	26.46 ± 14.99	12.21 ± 3.73	19.37 ± 1.92	3.31 ± 1.19
Maximum thickness (µm)	146.13 ± 27.69	61.38 ± 13.22	50.80 ± 6.35	59.27 ± 15.98
Roughness coefficient (dimensionless, range: zero-infinity)	1.40 ± 0.38	0.90 ± 0.27	0.22 ± 0.05	1.86 ± 0.07

5. Conclusions

During the past 15 years, efforts to develop new and effective anti-biofilm chemotherapeutics has focused to a large degree on blocking the mechanism of signal transduction involved in cell-to-cell communication via quorum sensing, and has led to the development of several QS antagonists. The potential chemotherapeutic use of *cis*-DA differs significantly from existing anti-biofilm strategies, all of which exert a negative effect upon biofilm bacteria and are likely to be both limited in the spectrum of bacteria they act upon and likely to lead to some form of resistance. The biofilm dispersion response is undoubtedly necessary for the long-term survival of bacterial biofilm populations/communities. Thus, the development of resistance to autoinduction of biofilm dispersion is not likely to be an advantage to a bacterial population. This approach to infection management, therefore, presents a new direction in microbial control. One of the key advantages to the potential use of *cis*-DA in therapeutic contexts is the multiple desirable effects it induces in bacteria, inhibiting biofilm development, altering virulence, inducing the disaggregation of the biofilm by inducing a dispersion response, and reverting persister cells to a susceptible state. Studies with this signal have also shown that *cis*-DA has no bactericidal activity nor does it induce cytotoxic effects in fibroblasts. Instead, *cis*-DA increases cell recovery on agar plates while reducing the number of persister cells formed in a population. Using *cis*-DA as an adjunctive to antimicrobial agents enhances killing of biofilm bacteria and persister cells, in some instances to the point of eradication. This effect is achieved by overcoming the mechanisms that lead to biofilm tolerance including: penetration failure, reduced metabolic rates and slow growth, biofilm-specific resistance (such as is mediated by the activation of efflux pumps), and the existence of a persister cell sub-population. Considering the urgent need to find new therapies to effectively treat chronic and unresolved infections, *cis*-DA provides a promising new potential chemotherapeutic to treat biofilm infections. Further research on this novel cell-to-cell signal is likely to elucidate the mechanism of signal transduction and reversion of persistence in cells challenged with *cis*-DA, as well as, reveal other molecules with similar activity and extend the list of microorganisms that use cell-to-cell signaling to induce biofilm dispersion.

Acknowledgments

The authors would like to acknowledge NIH grant 1 R15 AI094485-01 for the funding support and all the students that worked in their laboratories throughout the years.

Author Contributions

C.N.H.M., D.G.D. and K.S. wrote the paper.

References

1. De Kievit, T.R.; Gillis, R.; Marx, S.; Brown, C.; Iglewski, B.H.; Kievit, T.R.D.E. Quorum-sensing genes in *Pseudomonas aeruginosa* biofilms: Their role and expression patterns. *Appl. Environ. Microbiol.* **2001**, *67*, 1865–1873.
2. Costerton, J.W. Introduction to biofilm. *Int. J. Antimicrob. Agents* **1999**, *11*, 217–219.
3. Donlan, R.M. Biofilms: Microbial life on surfaces. *Emerg. Infect. Dis.* **2002**, *8*, 881–890.
4. Wimpenny, J.W. An overview of biofilms as functional communities. In *Community Structure and Cooperation in Biofilms*; Allison Gilbert, P., Lappin-Scott, H., Wilson, M.D., Eds.; Cambridge University Press: Cambridge, UK, 2000; pp. 1–24.
5. Flemming, H.-C.; Neu, T.R.; Wozniak, D.J. The EPS matrix: The "house of biofilm cells". *J. Bacteriol.* **2007**, *189*, 7945–7947.
6. Allesen-Holm, M.; Barken, K.B.; Yang, L.; Klausen, M.; Webb, J.S.; Kjelleberg, S.; Molin, S.; Givskov, M.; Tolker-Nielsen, T. A characterization of DNA release in *Pseudomonas aeruginosa* cultures and biofilms. *Mol. Microbiol.* **2006**, *59*, 1114–1128.
7. Sauer, K.; Camper, A.; Ehrlich, G.; Costerton, J.; Davies, D. *Pseudomonas aeruginosa* displays multiple phenotypes during development as a biofilm. *J. Bacteriol.* **2002**, *184*, 1140–1154.
8. Stoodley, P.; Sauer, K.; Davies, D.G.; Costerton, J.W. Biofilms as complex differentiated communities. *Annu. Rev. Microbiol.* **2002**, *56*, 187–209.
9. Allegrucci, M.; Hu, F.Z.; Shen, K.; Hayes, J.; Ehrlich, G.D.; Post, J.C.; Sauer, K. Phenotypic characterization of *Streptococcus pneumoniae* biofilm development. *J. Bacteriol.* **2006**, *188*, 2325–2335.
10. O'Toole, G.; Kaplan, H.B.; Kolter, R. Biofilm formation as microbial development. *Annu. Rev. Microbiol.* **2000**, *54*, 49–79.
11. Watnick, P.; Kolter, R. Biofilm, city of microbes. *J. Bacteriol.* **2000**, *182*, 2675–2679.
12. Hinsa, S.M.; Espinosa-Urgel, M.; Ramos, J.L.; O'Toole, G.A. Transition from reversible to irreversible attachment during biofilm formation by *Pseudomonas fluorescens* WCS365 requires an ABC transporter and a large secreted protein. *Mol. Microbiol.* **2003**, *49*, 905–918.
13. Caiazza, N.C.; O'Toole, G.A. SadB is required for the transition from reversible to irreversible attachment during biofilm formation by *Pseudomonas aeruginosa* PA14. *J. Bacteriol.* **2004**, *186*, 4476–4485.

14. Chiang, P.; Burrows, L.L. Biofilm formation by hyperpiliated mutants of *Pseudomonas aeruginosa*. *J. Bacteriol.* **2003**, *185*, 2374–2378.
15. Sutherland, I.W. The biofilm matrix—An immobilized but dynamic microbial environment. *Trends Microbiol.* **2001**, *9*, 222–227.
16. Gómez-Suárez, C.; Pasma, J.; van der Borden, A.J.; Wingender, J.; Flemming, H.-C.; Busscher, H.J.; van der Mei, H.C. Influence of extracellular polymeric substances on deposition and redeposition of *Pseudomonas aeruginosa* to surfaces. *Microbiology* **2002**, *148*, 1161–1169.
17. Davies, D.G.; Parsek, M.R.; Pearson, J.P.; Iglewski, B.H.; Costerton, J.W.; Greenberg, E.P. The involvement of cell-to-cell signals in the development of a bacterial biofilm. *Science* **1998**, *280*, 295–298.
18. Harmsen, M.; Yang, L.; Pamp, S.J.; Tolker-Nielsen, T. An update on *Pseudomonas aeruginosa* biofilm formation, tolerance, and dispersal. *FEMS Immunol. Med. Microbiol.* **2010**, *59*, 253–268.
19. Davies, D.G.; Marques, C.N.H. A fatty acid messenger is responsible for inducing dispersion in microbial biofilms. *J. Bacteriol.* **2009**, *191*, 1393–1403.
20. Kinniment, S.L.; Wimpenny, J.W. Measurements of the distribution of adenylate concentrations and adenylate energy charge across *Pseudomonas aeruginosa* biofilms. *Appl. Environ. Microbiol.* **1992**, *58*, 1629–1635.
21. Schierholz, J.M.; Beuth, J.; Konig, D.; Nurnberger, A.; Pulverer, G. Antimicrobial substances and effects on sessile bacteria. *Zentralblatt Bakteriol.* **1999**, *289*, 165–177.
22. Gilbert, P.; Collier, P.J.; Brown, M.R. Influence of growth rate on susceptibility to antimicrobial agents: Biofilms, cell cycle, dormancy, and stringent response. *Antimicrob. Agents Chemother.* **1990**, *34*, 1865–1868.
23. Bagge, N.; Hentzer, M.; Andersen, J.B.; Ciofu, O.; Givskov, M.; Høiby, N. Dynamics and spatial distribution of b-lactamase expression in *Pseudomonas aeruginosa* biofilms. *Antimicrob. Agents Chemother.* **2004**, *48*, 1168–1174.
24. Lenz, A.P.; Williamson, K.S.; Pitts, B.; Stewart, P.S.; Franklin, M.J. Localized gene expression in *Pseudomonas aeruginosa* biofilms. *Appl. Environ. Microbiol.* **2008**, *74*, 4463–4471.
25. Vuong, C.; Whitney, A.R.; Deleo, F.R. Polysaccharide intercellular adhesin (PIA) protects *Staphylococcus epidermidis* against major components of the human innate immune system. *Cell. Microbiol.* **2004**, *6*, 269–275.
26. Pearson, M.M.; Laurence, C.A.; Guinn, S.E.; Hansen, E.J. Biofilm formation by *Moraxella catarrhalis in vitro*: Roles of the UspA1 adhesin and the Hag hemagglutinin. *Infect. Immun.* **2006**, *74*, 1588–1596.
27. Jurcisek, J.A.; Bakaletz, L.O. Biofilms formed by nontypeable *Haemophilus influenzae in vivo* contain both double-stranded DNA and type IV pilin protein. *J. Bacteriol.* **2007**, *189*, 3868–3875.
28. Zhang, L.; Fritsch, M.; Hammond, L.; Landreville, R.; Slatculescu, C.; Colavita, A.; Mah, T.-F. Identification of genes involved in *Pseudomonas aeruginosa* biofilm-specific resistance to antibiotics. *PLoS ONE* **2013**, *8*, e61625.
29. Zhang, L.; Mah, T.-F. Involvement of a novel efflux system in biofilm-specific resistance to antibiotics. *J. Bacteriol.* **2008**, *190*, 4447–4452.

30. Diggle, S.P.; Winzer, K.; Chhabra, S.R.; Worrall, K.E.; Camara, M.; Williams, P. The *Pseudomonas aeruginosa* quinolone signal molecule overcomes the cell density-dependency of the quorum sensing hierarchy, regulates *rhl*-dependent genes at the onset of stationary phase and can be produced in the absence of LasR. *Mol. Microbiol.* **2003**, *50*, 29–43.
31. Yang, L.; Barken, K.B.; Skindersoe, M.E.; Christensen, A.B.; Givskov, M.; Tolker-Nielsen, T. Effects of iron on DNA release and biofilm development by *Pseudomonas aeruginosa*. *Microbiology* **2007**, *153*, 1318–1328.
32. Yang, L.; Nilsson, M.; Gjermansen, M.; Givskov, M.; Tolker-Nielsen, T. Pyoverdine and PQS mediated subpopulation interactions involved in *Pseudomonas aeruginosa* biofilm formation. *Mol. Microbiol.* **2009**, *74*, 1380–1392.
33. Kalia, V.C. Quorum sensing inhibitors: An overview. *Biotechnol. Adv.* **2013**, *31*, 224–245.
34. Rutherford, S.T.; Bassler, B.L. Bacterial quorum sensing: Its role in virulence and possibilities for its control. *Cold Spring Harb. Perspect. Med.* **2012**, *2*, 1–26.
35. Bassler, B.L. How bacteria talk to each other: Regulation of gene expression by quorum sensing. *Curr. Opin. Microbiol.* **1999**, *2*, 582–587.
36. Parsek, M.R.; Greenberg, E.P. Sociomicrobiology: The connections between quorum sensing and biofilms. *Trends Microbiol.* **2005**, *13*, 27–33.
37. Ryan, R.P.; Dow, J.M. Communication with a growing family: Diffusible signal factor (DSF) signaling in bacteria. *Trends Microbiol.* **2011**, *19*, 145–152.
38. Hengge, R. Principles of c-di-GMP signalling in bacteria. *Nat. Rev. Microbiol.* **2009**, *7*, 263–273.
39. Romling, U.; Amikam, D. Cyclic di-GMP as a second messenger. *Curr. Opin. Microbiol.* **2006**, *9*, 218–228.
40. Ryan, R.P.; Fouhy, Y.; Lucey, J.F.; Dow, J.M. Cyclic di-GMP signaling in bacteria: Recent advances and new puzzles. *J. Bacteriol.* **2006**, *188*, 8327–8334.
41. Simm, R.; Morr, M.; Kader, A.; Nimtz, M.; Romling, U. GGDEF and EAL domains inversely regulate cyclic di-GMP levels and transition from sessility to motility. *Mol. Microbiol.* **2004**, *53*, 1123–1134.
42. Costerton, J.W.; Stewart, P.S.; Greenberg, E.P. Bacterial biofilms: A common cause of persistent infections. *Science* **1999**, *284*, 1318–1322.
43. Lewis, K. Multidrug tolerance of biofilms and persister cells. *Curr. Top. Microbiol. Immunol.* **2008**, *322*, 107–131.
44. Spoering, A.; Lewis, K. Biofilms and planktonic cells of *Pseudomonas aeruginosa* have similar resistance to killing by antimicrobials. *J. Bacteriol.* **2001**, *183*, 6746–6751.
45. O'Toole, G.A. Microbiology: A resistance switch. *Nature* **2002**, *416*, 695–696.
46. Gilbert, P.; Maira-Litran, T.; McBain, A.J.; Rickard, A.H.; Whyte, F.W. The physiology and collective recalcitrance of microbial biofilm communities. *Adv. Microb. Physiol.* **2002**, *46*, 202–256.
47. Prosser, B.L.; Taylor, D.; Dix, B.A.; Cleeland, R. Method of evaluating effects of antibiotics on bacterial biofilm. *Antimicrob. Agents Chemother.* **1987**, *31*, 1502–1506.
48. Gristina, A.G.; Hobgood, C.D.; Webb, L.X.; Myrvik, Q.N. Adhesive colonization of biomaterials and antibiotic resistance. *Biomaterials* **1987**, *8*, 423–426.
49. Donlan, R.M.; Costerton, J.W. Biofilms: Survival mechanisms of clinically relevant microorganisms. *Clin. Microbiol. Rev.* **2002**, *15*, 167–193.

50. Lewis, K. Riddle of biofilm resistance. *Antimicrob. Agents Chemother.* **2001**, *45*, 999–1007.
51. Mah, T.F.; O'Toole, G.A. Mechanisms of biofilm resistance to antimicrobial agents. *Trends Microbiol.* **2001**, *9*, 34–39.
52. Stewart, P.S.; William Costerton, J. Antibiotic resistance of bacteria in biofilms. *Lancet* **2001**, *358*, 135–138.
53. Stewart, P.S. Mechanisms of antibiotic resistance in bacterial biofilms. *Int. J. Med. Microbiol.* **2002**, *292*, 107–113.
54. Gupta, K.; Marques, C.N.H.; Petrova, O.E.; Sauer, K. Antimicrobial tolerance of *Pseudomonas aeruginosa* biofilms is activated during an early developmental stage and requires the two-component hybrid SagS. *J. Bacteriol.* **2013**, *195*, 4975–4987.
55. Davies, D. Understanding biofilm resistance to antibacterial agents. *Nat. Rev. Drug Discov.* **2003**, *2*, 114–122.
56. Yassien, M.; Khardori, N.; Ahmedy, A.; Toama, M. Modulation of biofilms of *Pseudomonas aeruginosa* by quinolones. *Antimicrob. Agents Chemother.* **1995**, *39*, 2262–2268.
57. Stewart, P.S. Theoretical aspects of antibiotic diffusion into microbial biofilms. *Antimicrob. Agents Chemother.* **1996**, *40*, 2517–2522.
58. Allison, D.G. The biofilm matrix. *Biofouling* **2003**, *19*, 139–150.
59. Liao, J.; Sauer, K. The MerR-like transcriptional regulator BrlR contributes to *Pseudomonas aeruginosa* biofilm tolerance. *J. Bacteriol.* **2012**, *194*, 4823–4836.
60. Newberry, K.J.; Brennan, R.G. The structural mechanism for transcription activation by MerR family member multidrug transporter activation, N terminus. *J. Biol. Chem.* **2004**, *279*, 20356–20362.
61. Heldwein, E.E.; Brennan, R.G. Crystal structure of the transcription activator BmrR bound to DNA and a drug. *Nature* **2001**, *409*, 378–382.
62. Zheleznova, E.E.; Markham, P.N.; Neyfakh, A.A.; Brennan, R.G. Structural basis of multidrug recognition by BmrR, a transcription activator of a multidrug transporter. *Cell* **1999**, *96*, 353–362.
63. Drenkard, E.; Ausubel, F.M. *Pseudomonas* biofilm formation and antibiotic resistance are linked to phenotypic variation. *Nature* **2002**, *416*, 740–743.
64. Keren, I.; Kaldalu, N.; Spoering, A.; Wang, Y.; Lewis, K. Persister cells and tolerance to antimicrobials. *FEMS Microbiol. Lett.* **2004**, *230*, 13–18.
65. Marques, C.N.H.; Morozov, A.; Planzos, P.; Zelaya, H.M. The fatty acid signaling molecule *cis*-2-decenoic acid increases metabolic activity and reverts persister cells to an antimicrobial susceptible state. *Appl. Environ. Microbiol.* **2014**, *80*, 6976–6991.
66. Keren, I.; Shah, D.; Spoering, A. Specialized persister cells and the mechanism of multidrug tolerance in *Escherichia coli*. *J. Bacteriol.* **2004**, *186*, 8172–8180.
67. Sufya, N.; Allison, D.G.; Gilbert, P. Clonal variation in maximum specific growth rate and susceptibility towards antimicrobials. *J. Appl. Microbiol.* **2003**, *95*, 1261–1267.
68. Hoyle, B.D.; Jass, J.; Costerton, J.W. The biofilm glycocalyx as a resistance factor. *J. Antimicrob. Chemother.* **1990**, *26*, 1–5.
69. McCarthy, Y.; Yang, L.; Twomey, K.B.; Sass, A.; Tolker-Nielsen, T.; Mahenthiralingam, E.; Dow, J.M.; Ryan, R.P. A sensor kinase recognizing the cell-cell signal BDSF (*cis*-2-dodecenoic acid) regulates virulence in *Burkholderia cenocepacia*. *Mol. Microbiol.* **2010**, *77*, 1220–1236.

70. Rasko, D.A.; Sperandio, V. Novel approaches to bacterial infection therapy by interfering with cell-to-cell signaling. *Curr. Protoc. Microbiol.* **2009**, *12*, doi: 10.1002/9780471729259.mc1703s12.
71. Rasko, D.A.; Sperandio, V. Anti-virulence strategies to combat bacteria-mediated disease. *Nat. Rev. Drug Discov.* **2010**, *9*, 117–128.
72. Singh, P.K.; Schaefer, A.L.; Parsek, M.R.; Moninger, T.O.; Welsh, M.J.; Greenberg, E.P. Quorum-sensing signals indicate that cystic fibrosis lungs are infected with bacterial biofilms. *Nature* **2000**, *407*, 762–764.
73. Chugani, S.A.; Whiteley, M.; Lee, K.M.; D'Argenio, D.; Manoil, C.; Greenberg, E.P. QscR, a modulator of quorum-sensing signal synthesis and virulence in *Pseudomonas aeruginosa*. *Proc. Natl. Acad. Sci. USA* **2001**, *98*, 2752–2757.
74. Green, S.K.; Schroth, M.N.; Cho, J.J.; Kominos, S.K.; Vitanza-jack, V.B. Agricultural plants and soil as a reservoir for *Pseudomonas aeruginosa*. *Appl. Microbiol.* **1974**, *28*, 987–991.
75. Zimakoff, J.; Høiby, N.; Rosendal, K.; Guilbert, J.P. Epidemiology of *Pseudomonas aeruginosa* infection and the role of contamination of the environment in a cystic fibrosis clinic. *J. Hosp. Infect.* **1983**, *4*, 31–40.
76. Moreau-Marquis, S.; Stanton, B.A.; O'Toole, G.A. *Pseudomonas aeruginosa* biofilm formation in the cystic fibrosis airway. *Pulm. Pharmacol. Ther.* **2008**, *21*, 595–599.
77. Van Delden, C.; Iglewski, B.H. Cell-to-cell signaling and *Pseudomonas aeruginosa* infections. *Emerg. Infect. Dis.* **1998**, *4*, 551–560.
78. Williams, P.; Cámara, M. Quorum sensing and environmental adaptation in *Pseudomonas aeruginosa*: A tale of regulatory networks and multifunctional signal molecules. *Curr. Opin. Microbiol.* **2009**, *12*, 182–191.
79. Jimenez, P.N.; Koch, G.; Thompson, J.A.; Xavier, K.B.; Cool, R.H.; Quax, W.J. The multiple signaling systems regulating virulence in *Pseudomonas aeruginosa*. *Microbiol. Mol. Biol. Rev.* **2012**, *76*, 46–65.
80. Pearson, J.P.; Pesci, E.C.; Iglewski, B.H. Roles of *Pseudomonas aeruginosa las* and *rhl* quorum-sensing systems in control of elastase and rhamnolipid biosynthesis genes. *J. Bacteriol.* **1997**, *179*, 5756–5767.
81. Smith, R.S.; Iglewski, B.H. *Pseudomonas aeruginosa* quorum sensing as a potential antimicrobial target. *J. Clin. Investig.* **2003**, *112*, 1460–1465.
82. Ishida, T.; Ikeda, T.; Takiguchi, N.; Kuroda, A.; Kato, J.; Ohtake, H. Inhibition of quorum sensing in *Pseudomonas aeruginosa* by *N*-acyl cyclopentylamides. *Appl. Environ. Microbiol.* **2007**, *73*, 3183–3188.
83. Wu, H.; Song, Z.; Hentzer, M.; Andersen, J.B.; Molin, S.; Givskov, M.; Høiby, N. Synthetic furanones inhibit quorum-sensing and enhance bacterial clearance in *Pseudomonas aeruginosa* lung infection in mice. *J. Antimicrob. Chemother.* **2004**, *53*, 1054–1061.
84. Hentzer, M.; Riedel, K.; Rasmussen, T.B.; Heydorn, A.; Andersen, J.B.; Parsek, M.R.; Rice, S.A; Eberl, L.; Molin, S.; Høiby, N.; *et al.* Inhibition of quorum sensing in *Pseudomonas aeruginosa* biofilm bacteria by a halogenated furanone compound. *Microbiology* **2002**, *148*, 87–102.

85. Bjarnsholt, T.; Jensen, P.Ø.; Rasmussen, T.B.; Christophersen, L.; Calum, H.; Hentzer, M.; Hougen, H.-P.; Rygaard, J.; Moser, C.; Eberl, L.; *et al.* Garlic blocks quorum sensing and promotes rapid clearing of pulmonary *Pseudomonas aeruginosa* infections. *Microbiology* **2005**, *151*, 3873–3880.

86. Ren, D.; Zuo, R.; Barrios, A.F.G.; Bedzyk, L.A; Eldridge, G.R.; Pasmore, M.E.; Wood, T.K. Differential gene expression for investigation of *Escherichia coli* biofilm inhibition by plant extract ursolic acid. *Appl. Environ. Microbiol.* **2005**, *71*, 4022–4034.

87. Kim, C.; Kim, J.; Park, H.-Y.; Park, H.-J.; Lee, J.H.; Kim, C.K.; Yoon, J. Furanone derivatives as quorum-sensing antagonists of *Pseudomonas aeruginosa*. *Appl. Microbiol. Biotechnol.* **2008**, *80*, 37–47.

88. Borlee, B.R.; Goldman, A.D.; Murakami, K.; Samudrala, R.; Wozniak, D.J.; Parsek, M.R. *Pseudomonas aeruginosa* uses a cyclic-di-GMP-regulated adhesin to reinforce the biofilm extracellular matrix. *Mol. Microbiol.* **2010**, *75*, 827–842.

89. Sambanthamoorthy, K.; Luo, C.; Pattabiraman, N.; Feng, X.; Koestler, B.; Waters, C.M.; Palys, T.J. Identification of small molecules inhibiting diguanylate cyclases to control bacterial biofilm development. *Biofouling* **2014**, *30*, 17–28.

90. Lieberman, O.O.J.; Orr, M.M.W.; Wang, Y.; Lee, V.T. High-throughput screening using the differential radial capillary action of ligand assay identifies ebselen as an inhibitor of diguanylate cyclases. *ACS Chem. Biol.* **2014**, *9*, 183–192.

91. Garo, E.; Eldridge, G.R.; Goering, M.G.; Pulcini, E.D.; Hamilton, M.A.; Costerton, J.W.; James, G.A. Asiatic acid and corosolic acid enhance the susceptibility of *Pseudomonas aeruginosa* biofilms to tobramycin. *Antimicrob. Agents Chemother.* **2007**, *51*, 1813–1817.

92. Balaban, N.; Cirioni, O.; Giacometti, A. Treatment of *Staphylococcus aureus* biofilm infection by the quorum-sensing inhibitor RIP. *Antimicrob. Agents Chemother.* **2007**, *51*, 2226–2229.

93. LoVetri, K.; Madhyastha, S. Antimicrobial and antibiofilm activity of quorum sensing peptides and Peptide analogues against oral biofilm bacteria. *Methods Mol. Biol.* **2010**, *618*, 383–392.

94. Kjelleberg, S.; Molin, S. Is there a role for quorum sensing signals in bacterial biofilms? *Curr. Opin. Microbiol.* **2002**, *5*, 254–258.

95. Estrela, A.B.; Abraham, W.-R. Combining Biofilm-Controlling Compounds and Antibiotics as a Promising New Way to Control Biofilm Infections. *Pharmaceuticals* **2010**, *3*, 1374–1393.

96. Wang, L.H.; He, Y.; Gao, Y.; Wu, J.E.; Dong, Y.H.; He, C.; Wang, S.X.; Weng, L.X.; Xu, J.L.; Tay, L.; *et al.* A bacterial cell-cell communication signal with cross-kingdom structural analogues. *Mol. Microbiol.* **2004**, *51*, 903–912.

97. Deng, Y.; Wu, J.; Eberl, L.; Zhang, L.H. Structural and functional characterization of diffusible signal factor family quorum-sensing signals produced by members of the *Burkholderia cepacia* complex. *Appl. Environ. Microbiol.* **2010**, *76*, 4675–4683.

98. Tang, J.L.; Liu, Y.N.; Barber, C.E.; Dow, J.M.; Wootton, J.C.; Daniels, M.J. Genetic and molecular analysis of a cluster of rpf genes involved in positive regulation of synthesis of extracellular enzymes and polysaccharide in *Xanthomonas campestris* pathovar *campestris*. *Mol. Gen. Genet.* **1991**, *226*, 409–417.

99. Barber, C.E.; Tang, J.L.; Feng, J.X.; Pan, M.Q.; Wilson, T.J.G.; Slater, H.; Dow, J.M.; Williams, P.; Daniels, M.J. A novel regulatory system required for pathogenicity of *Xanthomonas campestris* is mediated by a small diffusible signal molecule. *Mol. Microbiol.* **1997**, *24*, 555–566.
100. Beaulieu, E.D.; Ionescu, M.; Chatterjee, S.; Yokota, K.; Trauner, D.; Lindow, S. Characterization of a diffusible signaling factor from *Xylella fastidiosa*. *MBio* **2013**, *4*, 9–14.
101. Dow, J.M.; Crossman, L.; Findlay, K.; He, Y.Q.; Feng, J.X.; Tang, J.L. Biofilm dispersal in *Xanthomonas campestris* is controlled by cell-cell signaling and is required for full virulence to plants. *Proc. Natl. Acad. Sci. USA* **2003**, *100*, 10995–11000.
102. He, Y.W.; Xu, M.; Lin, K.; Ng, Y.J.A.; Wen, C.M.; Wang, L.H.; Liu, Z.D.; Zhang, H.B.; Dong, Y.H.; Dow, J.M.; *et al.* Genome scale analysis of diffusible signal factor regulon in *Xanthomonas campestris* pv. *campestris*: Identification of novel cell-cell communication-dependent genes and functions. *Mol. Microbiol.* **2006**, *59*, 610–622.
103. Minkwitz, A.; Berg, G. Comparison of antifungal activities and 16S ribosomal DNA sequences of clinical and environmental isolates of *Stenotrophomonas maltophilia*. *J. Clin. Microbiol.* **2001**, *39*, 139–145.
104. He, Y.-W.; Wu, J.; Cha, J.-S.; Zhang, L.-H. Rice bacterial blight pathogen Xanthomonas oryzae pv. oryzae produces multiple DSF-family signals in regulation of virulence factor production. *BMC Microbiol.* **2010**, *10*, 187, doi:10.1186/1471-2180-10-187.
105. Huang, T.-P.; Lee Wong, A.C. Extracellular fatty acids facilitate flagella-independent translocation by *Stenotrophomonas maltophilia*. *Res. Microbiol.* **2007**, *158*, 702–711.
106. Boon, C.; Deng, Y.; Wang, L.-H.; He, Y.; Xu, J.-L.; Fan, Y.; Pan, S.Q.; Zhang, L.-H. A novel DSF-like signal from *Burkholderia cenocepacia* interferes with *Candida albicans* morphological transition. *ISME J.* **2008**, *2*, 27–36.
107. Ryan, R.P.; McCarthy, Y.; Watt, S.A.; Niehaus, K.; Dow, J.M. Intraspecies signaling involving the diffusible signal factor BDSF (*cis*-2-dodecenoic acid) influences virulence in *Burkholderia cenocepacia*. *J. Bacteriol.* **2009**, *191*, 5013–5019.
108. Sepehr, S.; Rahmani-Badi, A.; Babaie-Naiej, H.; Soudi, M.R. Unsaturated fatty acid, *cis*-2-decenoic acid, in combination with disinfectants or antibiotics removes pre-established biofilms formed by food-related bacteria. *PLoS ONE* **2014**, *9*, e101677.
109. Jennings, J.; Courtney, H.; Haggard, W. *Cis*-2-decenoic acid inhibits *S. aureus* growth and biofilm *in vitro*: A pilot study. *Clin. Orthop. Relat. Res.* **2012**, *470*, 2663–2670.
110. Rahmani-Badi, A.; Sepehr, S.; Mohammadi, P.; Soudi, M.R.; Babaie-Naiej, H. A combination of *cis*-2-decenoic acid and antibiotics eradicates pre- established catheter-associated biofilms. *J. Med. Microbiol.* **2014**, *63*, 1509–1516.
111. Vílchez, R.; Lemme, A.; Ballhausen, B.; Thiel, V.; Schulz, S.; Jansen, R.; Wagner-Döbler, I.; Sztajer, H. *Streptococcus mutans* inhibits *Candida albicans* hyphal formation by the fatty acid signaling molecule *trans*-2-decenoic acid (SDSF). *Chembiochem* **2010**, *11*, 1552–1162.
112. Colnaghi Simionato, A.V.; da Silva, D.S.; Lambais, M.R.; Carrilho, E. Characterization of a putative *Xylella fastidiosa* diffusible signal factor by HRGC-EI-MS. *J. Mass Spectrom.* **2007**, *42*, 490–496.
113. Chatterjee, S.; Wistrom, C.; Lindow, S.E. A cell-cell signaling sensor is required for virulence and insect transmission of *Xylella fastidiosa*. *Proc. Natl. Acad. Sci. USA* **2008**, *105*, 2670–2675.

114. Flavier, A.B.; Clough, S.J.; Schell, M.A.; Denny, T.P. Identification of 3-hydroxypalmitic acid methyl ester as a novel autoregulator controlling virulence in *Ralstonia solanacearum*. *Mol. Microbiol.* **1997**, *26*, 251–259.

115. Oh, K.B.; Miyazawa, H.; Naito, T.; Matsuoka, H. Purification and characterization of an autoregulatory substance capable of regulating the morphological transition in *Candida albicans*. *Proc. Natl. Acad. Sci. USA* **2001**, *98*, 4664–4668.

116. Fouhy, Y.; Scanlon, K.; Schouest, K.; Spillane, C.; Crossman, L.; Avison, M.B.; Ryan, R.P.; Dow, J.M. Diffusible signal factor-dependent cell-cell signaling and virulence in the nosocomial pathogen *Stenotrophomonas maltophilia*. *J. Bacteriol.* **2007**, *189*, 4964–4968.

117. Inoue, T.; Shingaki, R.; Fukui, K. Inhibition of swarming motility of *Pseudomonas aeruginosa* by branched-chain fatty acids. *FEMS Microbiol. Lett.* **2008**, *281*, 81–86.

118. Ryan, R.P.; Fouhy, Y.; Garcia, B.F.; Watt, S.A.; Niehaus, K.; Yang, L.; Tolker-Nielsen, T.; Dow, J.M. Interspecies signalling via the *Stenotrophomonas maltophilia* diffusible signal factor influences biofilm formation and polymyxin tolerance in *Pseudomonas aeruginosa*. *Mol. Microbiol.* **2008**, *68*, 75–86.

119. Rahmani-Badi, A.; Sepehr, S.; Fallahi, H.; Heidari-Keshel, S. Dissection of the *cis*-2-decenoic acid signaling network in *Pseudomonas aeruginosa* using microarray technique. *Front. Microbiol.* **2015**, *6*, 383, doi: 10.3389/fmicb.2015.00383.

120. Amari, D.T.; Marques, C.N.H.; Davies, D.G. The putative enoyl-coenzyme A hydratase DspI is required for production of the *Pseudomonas aeruginosa* biofilm dispersion autoinducer *cis*-2-decenoic acid. *J. Bacteriol.* **2013**, *195*, 4600–4610.

121. Feinbaum, R.L.; Urbach, J.M.; Liberati, N.T.; Djonovic, S.; Adonizio, A.; Carvunis, A.-R.; Ausubel, F.M. Genome-wide identification of *Pseudomonas aeruginosa* virulence-related genes using a Caenorhabditis elegans infection model. *PLoS Pathog.* **2012**, *8*, e1002813.

122. Marques, C.N.H.; Amari, D.T.; Varer, M.; Richards, J.S.; Cavender, D.; Davies, D.G. Influence of *cis*-2-decenoic acid on *Propionibacterium acnes* biofilm development and antimicrobial susceptibility. In Proceedings of the 5th ASM Conference on Biofilms, Cancun, Mexico, 15–19 November, 2009; Volume Abstract A88, p. 67.

123. Marques, C.N.H.; Guttenplan, S.B.; Okkotsu, Y.; Davies, D.G. Biofilm dispersion in oral bacterial species. In Proceedings of the 109th General Meeting of the American Society for Microbiology; Philadelphia, PA, USA, 17–21 May, 2009; Volume Abstrract Q-28, p. 158.

124. Barraud, N.; Schleheck, D.; Klebensberger, J.; Webb, J.S.; Hassett, D.J.; Rice, S.A.; Kjelleberg, S. Nitric oxide signaling in *Pseudomonas aeruginosa* biofilms mediates phosphodiesterase activity, decreased cyclic di-GMP levels, and enhanced dispersal. *J. Bacteriol.* **2009**, *191*, 7333–7342.

125. Marques, C.N.H.; Dolacky, S.D.; Guttenplan, S.B.; Payabyab, E.C.; Davies, D.G. *Pseudomonas aeruginosa* antimicrobial challenge in combination with a biofilm dispersal agent. In Proceedings of the 4th ASM conference on Biofilms, Quebec, QC, Canada, 25–29 March 2007; Volume abstract B-30, p. 166.

126. Amari, D.T.; Marques, C.N.H.; Ferreira, A.; Silverman, I.; Davies, D.G. Resuscitation of bacteria by exposure to the intercellular communication molecule, *cis*-2-decenoic acid. In Proceedings of the 110th General Meeting of the American Society for Microbiology; San Diego, CA, USA, 23–27 May 2010; Volume abstract D-2096, p. 162.

127. Fauvart, M.; de Groote, V.N.; Michiels, J. Role of persister cells in chronic infections: Clinical relevance and perspectives on anti-persister therapies. *J. Med. Microbiol.* **2011**, *60*, 699–709.
128. Mulcahy, L.R.; Burns, J.L.; Lory, S.; Lewis, K. Emergence of *Pseudomonas aeruginosa* strains producing high levels of persister cells in patients with cystic fibrosis. *J. Bacteriol.* **2010**, *192*, 6191–6199.
129. Balaban, N.Q.; Merrin, J.; Chait, R.; Kowalik, L.; Leibler, S. Bacterial persistence as a phenotypic switch. *Science* **2004**, *305*, 1622–1625.
130. Lewis, K. Persister cells and the riddle of biofilm survival. *Biochemistry* **2005**, *70*, 267–274.
131. Kwan, B.W.; Valenta, J.A.; Benedik, M.J.; Wood, T.K. Arrested protein synthesis increases persister-like cell formation. *Antimicrob. Agents Chemother.* **2013**, *57*, 1468–1473.
132. Korch, S.B.; Henderson, T.A.; Hill, T.M. Characterization of the *hipA7* allele of *Escherichia coli* and evidence that high persistence is governed by (p)ppGpp synthesis. *Mol. Microbiol.* **2003**, *50*, 1199–1213.
133. Scherrer, R.; Moyed, H.S. Conditional impairment of cell division and altered lethality in *hipA* mutants of *Escherichia coli* K-12. *J. Bacteriol.* **1988**, *170*, 3321–3326.
134. Pascoe, B.; Dams, L.; Wilkinson, T.S.; Harris, L.G.; Bodger, O.; Mack, D.; Davies, A.P. Dormant cells of *Staphylococcus aureus* are resuscitated by spent culture supernatant. *PLoS ONE* **2014**, *9*, e85998.
135. Allison, K.R.; Brynildsen, M.P.; Collins, J.J. Metabolite-enabled eradication of bacterial persisters by aminoglycosides. *Nature* **2011**, *473*, 216–220.
136. Pan, J.; Bahar, A.A.; Syed, H.; Ren, D. Reverting antibiotic tolerance of *Pseudomonas aeruginosa* PAO1 persister cells by (*Z*)-4-bromo-5-(bromomethylene)-3-methylfuran-2(5*H*)-one. *PLoS ONE* **2012**, *7*, e45778.
137. Mina, E.; Marques, C.N.H. Effect of *cis*-2-decenoic acid on virulence of *Staphylococcus aureus* persister cells. In Proceedings of the 113th General Meeting of the American Society for Microbiology; Denver, CO, USA, 18–21 May 2013; Volume abstract 512, p. 121.

3

Novel Strategies for the Prevention and Treatment of Biofilm Related Infections

Meng Chen [1], **Qingsong Yu** [2] **and Hongmin Sun** [3,*]

[1] Nanova, Inc. Columbia, MO 65211, USA; E-Mail: chenmeng@nanovamed.com or mengchen2002slc@yahoo.com

[2] Department of Mechanical and Aerospace Engineering, University of Missouri, Columbia, MO 65211, USA; E-Mail: yuq@missouri.edu

[3] Division of Cardiovascular Medicine, Department of Internal Medicine, University of Missouri, Columbia, MO 65212, USA

* Author to whom correspondence should be addressed; E-Mail: sunh@health.missouri.edu;

Abstract: Biofilm formation by human bacterial pathogens on implanted medical devices causes major morbidity and mortality among patients, and leads to billions of dollars in healthcare cost. Biofilm is a complex bacterial community that is highly resistant to antibiotics and human immunity. As a result, novel therapeutic solutions other than the conventional antibiotic therapies are in urgent need. In this review, we will discuss the recent research in discovery of alternative approaches to prevent or treat biofilms. Current anti-biofilm technologies could be divided into two groups. The first group focuses on targeting the biofilm forming process of bacteria based on our understanding of the molecular mechanism of biofilm formation. Small molecules and enzymes have been developed to inhibit or disrupt biofilm formation. Another group of anti-biofilm technologies focuses on modifying the biomaterials used in medical devices to make them resistant to biofilm formation. While these novel anti-biofilm approaches are still in nascent phases of development, efforts devoted to these technologies could eventually lead to anti-biofilm therapies that are superior to the current antibiotic treatment.

Keywords: biofilm; antimicrobial; small molecule; biomaterial

---

1. Introduction

Biofilm is a community of bacteria that are attached to a substratum or surface. Bacteria in biofilm are embedded in extracellular polymeric matrix produced by the bacteria. Bacteria develop biofilm on submerged surfaces such as natural aquatic systems, water pipes, living tissues, tooth surfaces, indwelling medical devices and implants [1]. Biofilm formation on indwelling medical devices and implants such catheters, mechanical heart valves, pacemakers, prosthetic joints, and contact lenses pose a critical medical problem. Both Gram-negative and Gram-positive bacteria can form biofilms on indwelling medical devices. The most common biofilm-forming bacteria include *Enterococcus faecalis*, *Staphylococcus aureus*, *Staphylococcus epidermidis*, *Streptococcus viridans*, *Escherichia coli*, *Klebsiella pneumoniae*, *Proteus mirabilis*, and *Pseudomonas aeruginosa* [2].

Among these biofilm-forming bacteria, *S. aureus* and *S. epidermidis* are most commonly found on cardiovascular devices [3,4]. It was estimated that *S. aureus* and *S. epidermidis* caused about 40%–50% of prosthetic heart valve infections, and 50%–70% catheter biofilm infections [5]. Each year about 250,000–500,000 primary blood stream infections occur among the 150 million intravascular devices implanted in the US. Health care cost could be increased from $4000 to $56,000 for each infection [6,7]. Approximately 87% of blood stream infections were caused by staphylococci [5]. Taken together, the burden on healthcare system by *S. aureus* and *S. epidermidis* in biofilm is enormous.

Biofilm formation is initiated when bacterial cells attach and adhere to the surfaces of implants or host tissues. *S. aureus* generates multiple adhesive factors that could bind to host factors [8]. The host factors could mediate bacterial attachment to implant surfaces, which is covered by host plasma and other extracellular components. For example, *S. aureus* produces fibronectin-binding proteins (FnBPA and FnBPB) [9], collagen-binding protein Cna [10] and fibrinogen-binding proteins, clumping factor A and B (ClfA and ClfB) [11,12] to bind host plasma and extracellular matrix (ECM) components. Bacteria attached to the surfaces will proliferate, aggregate and recruit cells from the surrounding to form and differentiate into biofilm structures [13]. Bacterial attachment to the surface will change from reversible to irreversible accompanied by profound physiological, gene expression and protein profile changes. The mature biofilm structures consist of complex architecture and channels. Bacterial cells can detach from mature biofilms and spread to other organ systems [13,14]. As a result, biofilms become sources of persistent and chronic infections.

Bacteria in biofilm behave differently from planktonic bacteria, especially in terms of their response to antibiotic treatment [2]. Biofilm-associated bacteria are highly resistant to antibiotics. The complicated structure of biofilm with extracellular polymeric matrix could prevent antibiotics from reaching the bacteria. Bacteria in biofilm could also adopt a slow growing or starved state due to the altered microenvironment such as depletion of nutrition and accumulation of waste. The changed physiological state of bacteria could make them more resistant to antibiotics, which target more active cell processes [3,15–17].

In addition to the difficulty of treating biofilm with conventional antibiotic therapy, treating biofilm is further hindered by the rising antibiotic resistance among pathogens. Antibiotics targets are essential for bacterial survival. As a result, antibiotic resistant strains have been favored by selective pressure [18]. Antibiotic resistance in *S. aureus* such as the methicillin resistance is one of the most urgent medical problems [19,20]. It was estimated that 94,360 invasive methicillin-resistant *Staphylococcus aureus* (MRSA) infections occurred in the US in 2005, and these infections were associated with death in 18,650 cases [21]. Although *S. epidermidis* is part of the normal human epithelial bacterial flora, it can cause infection when skin or mucous membrane is injured. Biofilm formation on implanted indwelling medical devices is the major manifestation of *S. epidermidis* pathogenesis [3]. Antibiotic resistance is also widespread in *S. epidermidis.* For example, more than 70% of all hospital isolates of *S. epidermidis* are resistant to methicillin [22]. In summary, alternative approaches other than conventional antibiotic therapy are in urgent need to treat biofilm related infections. In this review, we will discuss alternative approaches to prevent or treat biofilms focusing on *S. aureus* and *S. epidermidis*, two of the most important biofilm forming pathogens.

2. Anti-Biofilm Agents

2.1. Small Molecules

High throughput screening of small molecule libraries has been one of the major approaches to search for drug leads. In recent years, high throughput screening has been increasingly adopted in academics to screen for low molecular weight compounds with desired biological properties. A chemical series of small compounds was identified by our group that inhibited the virulence gene expression of Gram positive pathogens such as *Streptococcus pyogenes* and *S. aureus* [23,24].

We performed a high throughput screening of 55,000 chemical compounds to search for inhibitors of gene expression of a key *S. pyogenes* virulence factor streptokinase [24]. A lead compound and its analogs were identified to be able to inhibit streptokinase gene expression. Detailed analysis of the global effect of the inhibitor on *S. pyogenes* gene expression demonstrated that the inhibitor changed gene expression of many key virulence factors. Furthermore, the lead compound also protected mice against *S. pyogenes* infection [24]. Analogs of the lead compounds were subsequently tested in *S. aureus.* Two analogs from the same chemical series inhibited biofilm formation by *S. aureus* [23]. The anti-biofilm compound also inhibited gene expression of a number of important *S. aureus* virulence factors [23]. Among the inhibited genes are genes known to be involved in biofilm formation. Inhibition of these genes could lead to inhibition of biofilm formation. The broad spectrum anti-virulence effect of the compounds on both *S. pyogenes* and *S. aureus* suggested that this class of compounds could target a conserved gene regulatory mechanism. As a result, this class of compounds could potentially be developed into novel anti-microbial agents against multiple pathogens.

Panmanee *et al.* screened 42,865 compounds to identify compounds that inhibited formation of or kill *S. epidermidis* biofilms. Sixteen compounds were confirmed to be able to either kill or inhibit *S. epidermidis* biofilm [25]. The mechanism of action of these anti-biofilm compounds remained to be characterized. Sambanthamoorthy *et al.* performed high throughput screening on 66,000 compounds and natural products to identify small molecules that inhibited induction of *Vibrio cholerae* cyclic di-GMP-inducible transcription [26]. Cyclic di-GMP is a second-messenger signal that is a key regulator of switch between planktonic and attached lifestyle of the majority of bacteria [27,28]. A benzimidazole compound demonstrated broad spectrum inhibition of biofilm formation by several Gram-negative and Gram-positive bacterial pathogens, including *P. aeruginosa* and *S. aureus* [26].

Opperman *et al.* screened 87,250 compounds for inhibitors of *S. epidermidis* biofilm [29]. Twenty three aryl rhodanines were identified to inhibit early phase biofilm formation by multiple strains of *S. aureus*, *S. epidermidis*, and *E. faecalis* [29]. However, the mechanism underlying the function of the aryl rhodanines was unclear.

P. aeruginosa produces an organic compound cis-2-decenoic acid capable of dispersing established biofilms and inhibiting biofilm development by a number of bacteria [30]. It was observed that *P. aeruginosa* dispersed from a continuous culture biofilm after medium flow stopped for several hours. The extracellular message that induced the release of cells from biofilm was purified from the organic fraction of spent medium and identified as *cis*-2-decenoic acid which was able to disperse biofilms by *E. coli*, *K. pneumoniae*, *P. mirabilis*, *S. pyogenes*, *B. subtilis*, *S. aureus*, and *C. albicans* [30]. Similarly, bacteria produce d-amino acids, which inhibited biofilm formation by *S. aureus* and *P. aeruginosa* [31].

N-acetylcysteine is a mucolytic agent that could interfere with exopolysaccharide formation in biofilms and inhibit *S. epidermidis* biofilm formation [32]. Metallic cations such as Ca^{2+} and Mg^{2+} play roles in microbial adherence and biofilm formation. As a result, chelators that can remove these cations could also inhibit biofilm formation [33].

The number of small molecules that can interfere with biofilm formation and thus serve as lead for development of anti-biofilm agents is growing rapidly (Table 1). However, mechanisms of action of many of these small molecules are still unclear which hinders the further development. More pharmacokinetic and *in vivo* studies are needed to optimize these leads to meet the necessary criteria for medical application.

Table 1. Small molecules that can inhibit biofilm formation.

Agent	Mechanism	Effect	Reference
Anti-virulence compounds	Inhibition of gene expression of virulence factors	Inhibition of biofilm formation by *S. aureus*	[23]
Anti-biofilm compounds	Unknown	Inhibition of biofilm formation by *S. epidermidis*	[25]
ABC-1	Inhibition of c-di-GMP-inducible transcription	Inhibition of biofilm formation by multiple Gram-negative and Gram-positive bacterial pathogens	[26]
Aryl rhodanines	Unknown	Inhibition of biofilm formation by *S. aureus* and *S. epidermidis*	[29]
Cis-2-decenoic acid	Unknown	Dispersion of biofilms by *E. coli*, *K. pneumoniae*, *P. mirabilis*, *S. pyogenes*, *B. subtilis*, *S. aureus*, and *C. albicans*	[30]
D-amino acids	Unknown	Inhibition of biofilm formation by *S. aureus* and *P. aeruginosa*	[31]
N-acetylcysteine	Interference with exopolysaccharide formation in biofilms	Inhibition of biofilm formation by *S. epidermidis*	[32]
Chelators	Interference with metal ion's function in biofilm formation	Inhibition of biofilm formation by *S. aureus*	[33]

2.2. Matrix-Targeting Enzymes

Disrupting or degrading the extracellular polymeric matrix of biofilms can weaken and disperse biofilms. There have been a number of studies done to degrade matrix components such as polysaccharide, eDNA and proteins [34]. The Gram-negative, oral bacterium *Actinobacillus actinomycetemcomitans* produces dispersin B that could disperse biofilms by other bacteria. Kaplan *et al.* found that dispersin B could disrupt extracellular matrix of *S. epidermidis* biofilm and disperse the biofilm [35]. Extracellular genomic DNA (eDNA) is released by bacteria as an important component of extracellular matrix of biofilm [36]. As a result, DNase I was shown to be able to disperse *S. aureus* biofilms [37]. Proteinase K and trypsin effectively disrupted *S. aureus* biofilms [38]. There are still a lot of limitations with these approaches. The *in vivo* efficacy of such approaches isn't well established and treating host with proteins could cause inflammatory and allergic reaction, which could affect the therapeutic potential [38].

3. Bioengineering Approaches

3.1. Bactericidal/Bacteriostatic Coating

Altering the surface properties of indwelling medical devices is one of the main focuses to prevent or decrease biofilm infections [3,39]. One of the approaches to make biomaterial surfaces resistant to biofilm formation is to coat the surface with bactericidal/bacteriostatic substances. Antibiotics are commonly used. For example, vancomycin was covalently bonded to the surface of titanium metal implant. As a result, *S. epidermidis* biofilm formation was significantly inhibited on a vancomycin coated titanium alloy [40]. Antibiotics have been used to impregnate catheters to prevent biofilm formation in clinics [41–44]. However, using antibiotics could lead to selection of antibiotic resistance and even induce biofilm formation [45].

Heavy metal silver was also used as an anti-biofilm agent by depositing silver on the surfaces of biomaterials using coating technology [46,47]. Silver is one of the strongest bactericidal agents. The mechanism of the bactericidal function of silver is still unclear. It was observed that when silver ion penetrated into cells, DNA was condensed and lost ability to replicate, which led to cell death. Silver ion could also inactivate proteins by reacting with the thiol groups in cysteine residues [48–50]. Silver nanoparticles have been studied for their antimicrobial property. Because silver nanoparticles have extremely large surface area, they can interact with microorganisms better. The nanoparticles could penetrate inside the bacteria and react with proteins and DNA, and interrupt the respiratory chain and cell division, leading to cell death [48].

Coating medical devices with silver ions or metallic silver has disappointing clinic results, probably due to inactivation of metallic silver when the devices contacting blood and coating wearing off [48]. On the other hand, biofilm formation by a number of pathogens such *E. coli*, *Enterococcus*, *S. aureus*, *coagulase-negative Staphylococci* on silver nanoparticle coated catheters was almost completely prevented [51]. However, silver nanoparticle could have genotoxic and cytotoxic effects on human cells at high dose [49]. Accelerated thrombin formation and platelet activation were also observed on surfaces of catheters coated with the silver nanoparticles, which could increase the thrombosis risk of patients in clinics [52]. As a result, much effort is still needed to improve the silver nanoparticle coating technology to diminish these side-effects.

Red alga *Delisea pulchra* produces halogenated furanones that can inhibit fouling of their surface. Furanones have been studied as a new class of anti-microbial agents [53,54]. Furanone was coated on biomaterial surfaces by physical adsorption and biofilm formation by *S. epidermidis* was significantly inhibited by furanone coating [55]. Furanone was also covalently bonded to Silastic Tenckhoff catheters and rendered inhibitory effect on biofilm formation [56]. Furthermore, in a sheep catheter infection model, furanone coated catheters tended to cause less severe infection than control catheters [56].

Covalently coupled 3-(trimethoxysilyl)-propyldimethyloctadecylammonium chloride (QAS) to silicone rubber will generate quaternary ammonium groups on the surface with antimicrobial activity. Viability of *S. aureus* adhered to QAS-coated silicone rubber was decreased, both *in vitro* and *in vivo* [57]. Quaternary ammonium functionalized silica nanoparticles was used to coat glass surfaces and exhibited inhibition of growth and accumulation of Gram-negative and Gram-positive bacteria on the surface [58].

One of the shortcomings of the bactericidal surfaces is that they could be covered by macromolecules and dead microorganisms, and then lose their antimicrobial function [59].

3.2. Anti-Adhesion Coating

The infection-resistant surface of indwelling medical devices could also be achieved by depositing a thin layer of anti-adhesion coating on the surface to reduce attachment of pathogenic bacteria. The number of bacteria that may adhere and their ability to grow and spread on biomaterial surfaces is greatly influenced by not only the bacteria but also the physicochemical properties of the biomaterial. The surface properties of biomaterials or medical devices can be changed by coating application or surface modification to create the desired anti-adhesion characteristics without altering the bulk properties of materials. These surface properties include chemical composition and reactivity, hydrophilicity and hydrophobicity [60], surface roughness [61,62] or texture [63], and surface charge. Following this approach, our research team has developed trimethylsilane (TMS) plasma nanocoatings using low temperature plasma coating technology to coat surfaces of stainless steel and titanium for reduced bacterial adhesion and biofilm formation [64]. Significant inhibition of *S. epidermidis* biofilm was observed on TMS plasma coated stainless steel and titanium. The biofilm inhibition could be attributed to the coating chemical inertness, low surface free energy, coating smoothness, and surface-bound CH_3 groups. The changed surface properties could result in less protein adsorbed to the coated surfaces than that adsorbed to the uncoated stainless steel and titanium controls, leading to significantly decreased bacterial adhesion.

Harris *et al.* coated titanium surface with Poly(l-lysine)-grafted-poly(ethylene glycol) (PLL-g-PEG) to decrease non-specific adsorption of blood. The PEG coating also decreased *S. aureus* adhesion [65]. Zwitterionic poly(carboxybetaine methacrylate) (pCBMA) film grafted to glass surface was shown to be highly resistant to fibrinogen adhesion and *S. epidermidis* and *P. aeruginosa* attachment and accumulation [66]. It is believed that the surface hydration layer generated by these hydrophilic coatings could serve as a physical and energetic barrier to protein adsorption and thus bacteria adhesion [67].

A superhydrophobic coating on glass surface was synthesized from a mixture of nanostructured silica colloids and a low surface energy fluorinated silane xerogel. The adhesion of *S. aureus* and *P. aeruginosa* to the silica-colloid-doped fluorinated surfaces was decreased by two orders of magnitude *versus* the control [68]. It was found that fibrinogen adsorption on the superhydrophobic surface was very low, leading to low attachment of *S. aureus* [69]. Low surface energy chemistry and nano-textured morphology of the superhydrophobic coating could result in reduced protein adsorption and bacterial attachment. A barrier to wetting could be created by trapping pockets of air in the nano-scale morphology, which in effect presents a reduced surface area onto which protein molecules can diffuse from the solution [69].

The surface roughness of biomaterials has been recognized as one of many important factors for surface-bacterium interactions. Many studies have shown that the surface roughness of biomaterials strongly influences the degree of bacterial attachment to surfaces [62,70,71]. For instance, streptococcal adhesion was sensitive to surface roughness and enhanced as the roughness of composite surfaces increased from 20 nm to 150 and 350 nm [72]. *S. epidermidis* adhesion and growth were markedly higher on rough titanium surfaces than on smooth surfaces [73]. In contrast, there was greater attachment of *S. aureus* cells to mechanochemically polished titanium than the as-received titanium, even though the polished surfaces were much smoother. It was thus speculated that mechanochemical polishing generated nanoscale surface features on the titanium surfaces with a characteristic pattern more suitable for anchoring of spherical *S. aureus* cells [62].

Xu *et al.* reported that submicron (staphylococcal bacterial dimension) surface textures (400–500 nm) on poly(urethane urea) films reduced the material's surface area accessible to bacteria of *S. epidermidis* and *S. aureus*, resulting in a decreased probability of interaction with the material surface or adhesive plasma proteins (e.g., fibrinogen and fibronectin) adsorbed onto the material. Thus, the flow of fluid over the material surface removes bacteria from a textured surface more efficiently than it would from a smooth surface, and subsequently resists bacterial adhesion and biofilm formation [63].

Organoselenium can catalyze the formation of superoxide radicals to prevent bacterial colonization on biomaterial surfaces [74]. Organoselenium antimicrobial agent selenocyanatodiacetic acid (SCAA) was coated on hemodialysis catheters by covalent bonding and demonstrated both *in vitro* and *in vivo* efficacy at preventing *S. aureus* biofilm formation [75].

Polymer brush coatings are another type of promising anti-adhesion coatings for inhibition of biofilms. Polymer brush coatings are formed when hydrophilic polymer long-chains are attached to a surface and stretch out into the surrounding medium [76,77]. Polymer brush coatings that have been mostly studied for preventing biofilm formation are made from poly(ethylene oxide) (PEO) [78–81]. As the PEO long-chains are highly mobile and attain extremely large exclusion volume, compression of the PEO long-chain brushes upon approach by incoming proteins or bacteria would give rise to an increase in the local concentration of PEO, which would lead to a repulsive osmotic pressure to repel the approaching proteins or bacteria and keep them away at a distance [77]. Excellent *in vitro* results have shown significant reduction in protein adsorption and bacterial adhesion, and thus the high effectiveness of polymer brush coatings in preventing bacterial adhesion [82,83]. In contrast, *in vivo* results [84,85] using PEO brush coatings have been discouraging mainly due to the weak surface attachment of polymer chains and the susceptibility of PEO to oxidation damage that prevent successful applications of such coatings for *in vivo* conditions [77].

Anti-adhesion coatings prevent biofilm formation at early stages, which should be more desirable in clinical settings. However, *in vivo* efficacy success is still elusive with many of the coatings. Due to the complexity of interaction between coating surfaces with bacteria and host proteins, the mechanism of anti-adhesion coatings is also difficult to pinpoint. As a result, more effort is needed to further exploit this promising strategy for prevention of biofilm related infections.

In summary, the bioengineering approaches (Table 2) could prevent biofilm formation which is more desirable than treating biofilm related infection. In spite of the shortcomings of many of the approaches, improving biomaterial anti-biofilm properties remains the most effective and promising strategy to prevent the morbidity and mortality associated with biofilm infections.

Table 2. Surface modification approaches that can inhibit biofilm formation.

Coating agent	Coating method	Mechanism	Reference
Antibiotics	Non-covalent, covalent bonding	Bactericidal/Bacteriostatic	[40–44]
Silver	Plasma deposition, sol-gel coating, wet-chemical coating	Bactericidal	[46,47,51]
Furanones	Physical adsorption, covalent bonding	Bactericidal/Bacteriostatic	[55,56]
QAS	Covalent bonding	Inhibition of bacterial adhesion and viability	[57]
Silica nanoparticles with QAS	Covalent bonding	Bactericidal/Bacteriostatic	[58]
TMS	Plasma coating deposition with covalent bonding	Anti-adhesion	[64]
PLL-g-PEG	Physical adsorption & covalent coupling	Anti-adhesion	[65]
pCBMA	Zwitterionic surfaces grafted via radical polymerization	Anti-adhesion	[66]
Silica colloids/Silane xerogel	Synthesis of superhydrophobic coating	Anti-adhesion	[68]
Submicron surface textures	Physical surface roughness modification	Anti-adhesion	[63]
Selenocyanatodiacetic acid	Covalent bonding	Anti-adhesion	[75]
Polymer brush coatings	Surface grafting	Anti-adhesion	[82,83]

4. Conclusions

The approaches under development to prevent and treat biofilm caused infections include small molecules and matrix-targeting enzymes, bactericidal and anti-adhesion coatings. Small molecules and enzymes have been investigated to inhibit or disrupt biofilm formation. Anti-biofilm coatings have been targeting on modifying the surface of medical devices for enhanced inhibition of bacterial adhesion and/or growth leading to high resistance to biofilm formation. These novel anti-biofilm technologies could eventually lead to anti-biofilm therapies that are superior to the current antibiotic treatment.

Acknowledgments

The works of the authors are supported by NIH Grant P01HL573461 to HS, and NIH Grant 1R44HL097485-01A2 to MC. We would also like to thank all our colleagues on the works discussed in the review. We apologize to all colleagues whose works could not be cited due to space limitations.

References

1. Donlan, R.M. Biofilms: Microbial life on surfaces. *Emerg. Infect. Dis.* **2002**, *8*, 881–890.
2. Donlan, R.M. Biofilms and device-associated infections. *Emerg. Infect. Dis.* **2001**, *7*, 277–281.
3. Otto, M. Staphylococcal biofilms. *Curr. Top. Microbiol. Immunol.* **2008**, *322*, 207–228.
4. Otto, M. *Staphylococcus epidermidis*—The "accidental" pathogen. *Nat. Rev. Microbiol.* **2009**, *7*, 555–567.
5. Agarwal, A.; Singh, K.P.; Jain, A. Medical significance and management of staphylococcal biofilm. *FEMS Immunol. Med. Microbiol.* **2010**, *58*, 147–160.
6. Maki, D.G.; Kluger, D.M.; Crnich, C.J. The risk of bloodstream infection in adults with different intravascular devices: A systematic review of 200 published prospective studies. *Mayo Clin. Proc.* **2006**, *81*, 1159–1171.
7. Uckay, I.; Pittet, D.; Vaudaux, P.; Sax, H.; Lew, D.; Waldvogel, F. Foreign body infections due to *Staphylococcus epidermidis*. *Ann. Med.* **2009**, *41*, 109–119.
8. Gotz, F. Staphylococcus and biofilms. *Mol. Microbiol.* **2002**, *43*, 1367–1378.
9. Menzies, B.E. The role of fibronectin binding proteins in the pathogenesis of *Staphylococcus aureus* infections. *Curr. Opin. Infect. Dis.* **2003**, *16*, 225–229.
10. Switalski, L.M.; Patti, J.M.; Butcher, W.; Gristina, A.G.; Speziale, P.; Hook, M. A collagen receptor on *Staphylococcus aureus* strains isolated from patients with septic arthritis mediates adhesion to cartilage. *Mol. Microbiol.* **1993**, *7*, 99–107.
11. Ni Eidhin, D.; Perkins, S.; Francois, P.; Vaudaux, P.; Hook, M.; Foster, T.J. Clumping factor B (ClfB), a new surface-located fibrinogen-binding adhesin of *Staphylococcus aureus*. *Mol. Microbiol.* **1998**, 30, 245–257.
12. McDevitt, D.; Nanavaty, T.; House-Pompeo, K.; Bell, E.; Turner, N.; McIntire, L.; Foster, T.; Hook, M. Characterization of the interaction between the *Staphylococcus* auireus clumping factor (ClfA) and fibrinogen. *Eur. J. Biochem. FEBS* **1997**, *247*, 416–424.
13. Stoodley, P.; Sauer, K.; Davies, D.G.; Costerton, J.W. Biofilms as complex differentiated communities. *Annu. Rev. Microbiol.* **2002**, *56*, 187–209.
14. Fey, P.D.; Olson, M.E. Current concepts in biofilm formation of *Staphylococcus epidermidis*. *Future Microbiol.* **2010**, *5*, 917–933.
15. Costerton, J.W.; Stewart, P.S.; Greenberg, E.P. Bacterial biofilms: A common cause of persistent infections. *Science* **1999**, *284*, 1318–1322.

16. Fux, C.A.; Costerton, J.W.; Stewart, P.S.; Stoodley, P. Survival strategies of infectious biofilms. *Trends Microbiol.* **2005**, *13*, 34–40.
17. Stewart, P.S.; Costerton, J.W. Antibiotic resistance of bacteria in biofilms. *Lancet* **2001**, *358*, 135–138.
18. Martinez, J.L.; Baquero, F. Interactions among strategies associated with bacterial infection: Pathogenicity, epidemicity, and antibiotic resistance. *Clin. Microbiol. Rev.* **2002**, *15*, 647–679.
19. Chambers, H.F.; DeLeo, F.R. Waves of resistance: *Staphylococcus aureus* in the antibiotic era. *Nat. Rev. Microbiol.* **2009**, *7*, 629–641.
20. Grundmann, H.; ires-de-Sousa, M.; Boyce, J.; Tiemersma, E. Emergence and resurgence of meticillin-resistant *Staphylococcus aureus* as a public-health threat. *Lancet* **2006**, *368*, 874–885.
21. Klevens, R.M.; Morrison, M.A.; Nadle, J.; Petit, S.; Gershman, K.; Ray, S.; Harrison, L.H.; Lynfield, R.; Dumyati, G.; Townes, J.M.; *et al.* Invasive methicillin-resistant *Staphylococcus aureus* infections in the United States. *JAMA* **2007**, *298*, 1763–1771.
22. Diekema, D.J.; Pfaller, M.A.; Schmitz, F.J.; Smayevsky, J.; Bell, J.; Jones, R.N.; Beach, M.; Group, S.P. Survey of infections due to *Staphylococcus* species: Frequency of occurrence and antimicrobial susceptibility of isolates collected in the United States, Canada, Latin America, Europe, and the Western Pacific region for the SENTRY Antimicrobial Surveillance Program, 1997–1999. *Clin. Infect. Dis.* **2001**, *32*, S114–S132.
23. Ma, Y.; Xu, Y.; Yestrepsky, B.D.; Sorenson, R.J.; Chen, M.; Larsen, S.D.; Sun, H. Novel inhibitors of *Staphylococcus aureus* virulence gene expression and biofilm formation. *PLoS One* **2012**, 7, e47255.
24. Sun, H.; Xu, Y.; Sitkiewicz, I.; Ma, Y.; Wang, X.; Yestrepsky, B.D.; Huang, Y.; Lapadatescu, M.C.; Larsen, M.J.; Larsen, S.D.; *et al.* Inhibitor of streptokinase gene expression improves survival after group A streptococcus infection in mice. *Proc. Natl. Acad. Sci. USA* **2012**, *109*, 3469–3474.
25. Panmanee, W.; Taylor, D.; Shea, C.J.; Tang, H.; Nelson, S.; Seibel, W.; Papoian, R.; Kramer, R.; Hassett, D.J.; Lamkin, T.J. High-throughput screening for small-molecule inhibitors of *Staphylococcus epidermidis* RP62a biofilms. *J. Biomol. Screening* **2013**, *18*, 820–829.
26. Sambanthamoorthy, K.; Gokhale, A.A.; Lao, W.; Parashar, V.; Neiditch, M.B.; Semmelhack, M.F.; Lee, I.; Waters, C.M. Identification of a novel benzimidazole that inhibits bacterial biofilm formation in a broad-spectrum manner. *Antimicrob. Agents Chemother.* **2011**, *55*, 4369–4378.
27. Jenal, U.; Dorman, C.J. Small molecule signaling. *Curr. Opin. Microbiol.* **2009**, *12*, 125–128.
28. Romling, U.; Balsalobre, C. Biofilm infections, their resilience to therapy and innovative treatment strategies. *J. Inter.Med.* **2012**, *272*, 541–561.
29. Opperman, T.J.; Kwasny, S.M.; Williams, J.D.; Khan, A.R.; Peet, N.P.; Moir, D.T.; Bowlin, T.L. Aryl rhodanines specifically inhibit staphylococcal and enterococcal biofilm formation. *Antimicrob. Agents Chemother.* **2009**, *53*, 4357–4367.
30. Davies, D.G.; Marques, C.N. A fatty acid messenger is responsible for inducing dispersion in microbial biofilms. *J. Bacteriol.* **2009**, *191*, 1393–1403.
31. Kolodkin-Gal, I.; Romero, D.; Cao, S.; Clardy, J.; Kolter, R.; Losick, R. D-amino acids trigger biofilm disassembly. *Science* **2010**, *328*, 627–629.

32. Perez-Giraldo, C.; Rodriguez-Benito, A.; Moran, F.J.; Hurtado, C.; Blanco, M.T.; Gomez-Garcia, A.C. Influence of *N*-acetylcysteine on the formation of biofilm by *Staphylococcus epidermidis*. *J. Antimicrob. Chemother.* **1997**, *39*, 643–646.
33. Abraham, N.M.; Lamlertthon, S.; Fowler, V.G.; Jefferson, K.K. Chelating agents exert distinct effects on biofilm formation in *Staphylococcus aureus* depending on strain background: Role for clumping factor B. *J. Med. Microbiol.* **2012**, *61*, 1062–1070.
34. Kiedrowski, M.R.; Horswill, A.R. New approaches for treating staphylococcal biofilm infections. *Ann. N. Y. Acad. Sci.* **2011**, *1241*, 104–121.
35. Kaplan, J.B.; Ragunath, C.; Velliyagounder, K.; Fine, D.H.; Ramasubbu, N. Enzymatic detachment of *Staphylococcus epidermidis* biofilms. *Antimicrob. Agents Chemother.* **2004**, *48*, 2633–2636.
36. Mann, E.E.; Rice, K.C.; Boles, B.R.; Endres, J.L.; Ranjit, D.; Chandramohan, L.; Tsang, L.H.; Smeltzer, M.S.; Horswill, A.R.; Bayles, K.W. Modulation of eDNA release and degradation affects *Staphylococcus aureus* biofilm maturation. *PLoS One* **2009**, *4*, e5822.
37. Izano, E.A.; Amarante, M.A.; Kher, W.B.; Kaplan, J.B. Differential roles of poly-*N*-acetylglucosamine surface polysaccharide and extracellular DNA in *Staphylococcus aureus* and *Staphylococcus epidermidis* biofilms. *Appl. Environ. Microbiol.* **2008**, *74*, 470–476.
38. Chaignon, P.; Sadovskaya, I.; Ragunah, C.; Ramasubbu, N.; Kaplan, J.B.; Jabbouri, S. Susceptibility of staphylococcal biofilms to enzymatic treatments depends on their chemical composition. *Appl. Microbiol. Biotechnol.* **2007**, *75*, 125–132.
39. Pavithra, D.; Doble, M. Biofilm formation, bacterial adhesion and host response on polymeric implants—Issues and prevention. *Biomed. Mater.* **2008**, *3*, 034003.
40. Antoci, V., Jr.; Adams, C.S.; Parvizi, J.; Davidson, H.M.; Composto, R.J.; Freeman, T.A.; Wickstrom, E.; Ducheyne, P.; Jungkind, D.; Shapiro, I.M.; *et al.* The inhibition of *Staphylococcus epidermidis* biofilm formation by vancomycin-modified titanium alloy and implications for the treatment of periprosthetic infection. *Biomaterials* **2008**, *29*, 4684–4690.
41. Casey, A.L.; Mermel, L.A.; Nightingale, P.; Elliott, T.S. Antimicrobial central venous catheters in adults: A systematic review and meta-analysis. *Lancet Infect. Dis.* **2008**, *8*, 763–776.
42. McConnell, S.A.; Gubbins, P.O.; Anaissie, E.J. Do antimicrobial-impregnated central venous catheters prevent catheter-related bloodstream infection? *Clin. Infect. Dis.* **2003**, *37*, 65–72.
43. Hockenhull, J.C.; Dwan, K.M.; Smith, G.W.; Gamble, C.L.; Boland, A.; Walley, T.J.; Dickson, R.C. The clinical effectiveness of central venous catheters treated with anti-infective agents in preventing catheter-related bloodstream infections: A systematic review. *Crit. Care Med.* **2009**, *37*, 702–712.
44. Niel-Weise, B.S.; Stijnen, T.; van den Broek, P.J. Anti-infective-treated central venous catheters: A systematic review of randomized controlled trials. *Intensive Care Med.* **2007**, *33*, 2058–2068.
45. Hoffman, L.R.; D'Argenio, D.A.; MacCoss, M.J.; Zhang, Z.; Jones, R.A.; Miller, S.I. Aminoglycoside antibiotics induce bacterial biofilm formation. *Nature* **2005**, *436*, 1171–1175.
46. Jiang, H.; Manolache, S.; Wong, A.C.L.; Denes, F.S. Plasma-enhanced deposition of silver nanoparticles onto polymer and metal surfaces for the generation of antimicrobial characteristics. *J. Appl. Polym. Sci.* **2004**, *93*, 1411–1422.

47. Stobie, N.; Duffy, B.; McCormack, D.E.; Colreavy, J.; Hidalgo, M.; McHale, P.; Hinder, S.J. Prevention of Staphylococcus epidermidis biofilm formation using a low-temperature processed silver-doped phenyltriethoxysilane sol-gel coating. *Biomaterials* **2008**, *29*, 963–969.
48. Rai, M.; Yadav, A.; Gade, A. Silver nanoparticles as a new generation of antimicrobials. *Biotechnol. Adv.* **2009**, *27*, 76–83.
49. Jena, P.; Mohanty, S.; Mallick, R.; Jacob, B.; Sonawane, A. Toxicity and antibacterial assessment of chitosan-coated silver nanoparticles on human pathogens and macrophage cells. *Int. J. Nanomed.* **2012**, *7*, 1805–1818.
50. Feng, Q.L.; Wu, J.; Chen, G.Q.; Cui, F.Z.; Kim, T.N.; Kim, J.O. A mechanistic study of the antibacterial effect of silver ions on *Escherichia coli* and *Staphylococcus aureus*. *J. Biomed. Mater. Res.* **2000**, *52*, 662–668.
51. Roe, D.; Karandikar, B.; Bonn-Savage, N.; Gibbins, B.; Roullet, J.B. Antimicrobial surface functionalization of plastic catheters by silver nanoparticles. *J. Antimicrob. Chemother.* **2008**, *61*, 869–876.
52. Stevens, K.N.; Crespo-Biel, O.; van den Bosch, E.E.; Dias, A.A.; Knetsch, M.L.; Aldenhoff, Y.B.; van der Veen, F.H.; Maessen, J.G.; Stobberingh, E.E.; Koole, L.H. The relationship between the antimicrobial effect of catheter coatings containing silver nanoparticles and the coagulation of contacting blood. *Biomaterials* **2009**, *30*, 3682–3690.
53. Khan, M.S.; Husain, A. Syntheses and reactions of some new 2-arylidene-4-(biphenyl-4-yl)-but-3-*en*-4-olides with a study of their biological activity. *Pharmazie* **2002**, *57*, 448–452.
54. Kozminykh, V.O.; Igidov, N.M.; Kozminykh, E.N.; Aliev, Z.G. Reactions of 5-aryl-furan-2,3-diones with acylmethylenetriphenylphosphoranes: Synthesis and biological activity of 3(2H)-furanone derivatives. *Pharmazie* **1993**, *48*, 99–106.
55. Baveja, J.K.; Willcox, M.D.; Hume, E.B.; Kumar, N.; Odell, R.; Poole-Warren, L.A. Furanones as potential anti-bacterial coatings on biomaterials. *Biomaterials* **2004**, *25*, 5003–5012.
56. Hume, E.B.; Baveja, J.; Muir, B.; Schubert, T.L.; Kumar, N.; Kjelleberg, S.; Griesser, H.J.; Thissen, H.; Read, R.; Poole-Warren, L.A.; *et al.* The control of *Staphylococcus epidermidis* biofilm formation and *in vivo* infection rates by covalently bound furanones. *Biomaterials* **2004**, *25*, 5023–5030.
57. Gottenbos, B.; van der Mei, H.C.; Klatter, F.; Nieuwenhuis, P.; Busscher, H.J. *In vitro* and *in vivo* antimicrobial activity of covalently coupled quaternary ammonium silane coatings on silicone rubber. *Biomaterials* **2002**, *23*, 1417–1423.
58. Song, J.; Kong, H.; Jang, J. Bacterial adhesion inhibition of the quaternary ammonium functionalized silica nanoparticles. *Colloids Surf., B* **2011**, *82*, 651–656.
59. Klibanov, A.M. Permanently microbicidal materials coatings. *J. Mater. Chem.* **2007**, *17*, 2479–2482.
60. Boks, N.P.; Kaper, H.J.; Norde, W.; van der Mei, H.C.; Busscher, H.J. Mobile and immobile adhesion of staphylococcal strains to hydrophilic and hydrophobic surfaces. *J. Colloid Interface Sci.* **2009**, *331*, 60–64.
61. Tang, H.; Cao, T.; Liang, X.; Wang, A.; Salley, S.O.; McAllister, J., II; Ng, K.Y. Influence of silicone surface roughness and hydrophobicity on adhesion and colonization of *Staphylococcus epidermidis*. *J. Biomed. Mater. Res., Part A* **2009**, *88*, 454–463.

62. Truong, V.K.; Lapovok, R.; Estrin, Y.S.; Rundell, S.; Wang, J.Y.; Fluke, C.J.; Crawford, R.J.; Ivanova, E.P. The influence of nano-scale surface roughness on bacterial adhesion to ultrafine-grained titanium. *Biomaterials* **2010**, *31*, 3674–3683.
63. Xu, L.C.; Siedlecki, C.A. Submicron-textured biomaterial surface reduces staphylococcal bacterial adhesion and biofilm formation. *Acta Biomater.* **2012**, *8*, 72–81.
64. Ma, Y.; Chen, M.; Jones, J.E.; Ritts, A.C.; Yu, Q.; Sun, H. Inhibition of *Staphylococcus epidermidis* biofilm by trimethylsilane plasma coating. *Antimicrob. Agents Chemother.* **2012**, *56*, 5923–5937.
65. Harris, L.G.; Tosatti, S.; Wieland, M.; Textor, M.; Richards, R.G. *Staphylococcus aureus* adhesion to titanium oxide surfaces coated with non-functionalized and peptide-functionalized poly(L-lysine)-grafted-poly(ethylene glycol) copolymers. *Biomaterials* **2004**, *25*, 4135–4148.
66. Cheng, G.; Zhang, Z.; Chen, S.; Bryers, J.D.; Jiang, S. Inhibition of bacterial adhesion and biofilm formation on zwitterionic surfaces. *Biomaterials* **2007**, *28*, 4192–4199.
67. Chen, S.F.; Li, L.Y.; Zhao, C.; Zheng, J. Surface hydration: Principles and applications toward low-fouling/nonfouling biomaterials. *Polymer* **2010**, *51*, 5283–5293.
68. Privett, B.J.; Youn, J.; Hong, S.A.; Lee, J.; Han, J.; Shin, J.H.; Schoenfisch, M.H. Antibacterial fluorinated silica colloid superhydrophobic surfaces. *Langmuir* **2011**, *27*, 9597–9601.
69. Stallard, C.P.; McDonnell, K.A.; Onayemi, O.D.; O'Gara, J.P.; Dowling, D.P. Evaluation of protein adsorption on atmospheric plasma deposited coatings exhibiting superhydrophilic to superhydrophobic properties. *Biointerphases* **2012**, *7*, 31.
70. Ivanova, E.P.; Mitik-Dineva, N.; Wang, J.; Pham, D.K.; Wright, J.P.; Nicolau, D.V.; Mocanasu, R.C.; Crawford, R.J. Staleya guttiformis attachment on poly(tert-butylmethacrylate) polymeric surfaces. *Micron* **2008**, *39*, 1197–2104.
71. Mitik-Dineva, N.; Wang, J.; Truong, V.K.; Stoddart, P.; Malherbe, F.; Crawford, R.J.; Ivanova, E.P. *Escherichia coli*, *Pseudomonas aeruginosa*, and *Staphylococcus aureus* attachment patterns on glass surfaces with nanoscale roughness. *Curr. Microbiol.* **2009**, *58*, 268–273.
72. Mei, L.; Busscher, H.J.; van der Mei, H.C.; Ren, Y. Influence of surface roughness on streptococcal adhesion forces to composite resins. *Dent. Mater.* **2011**, *27*, 770–778.
73. Wu, Y.; Zitelli, J.P.; TenHuisen, K.S.; Yu, X.; Libera, M.R. Differential response of *Staphylococci* and osteoblasts to varying titanium surface roughness. *Biomaterials* **2011**, *32*, 951–960.
74. Mathews, S.M.; Spallholz, J.E.; Grimson, M.J.; Dubielzig, R.R.; Gray, T.; Reid, T.W. Prevention of bacterial colonization of contact lenses with covalently attached selenium and effects on the rabbit cornea. *Cornea* **2006**, *25*, 806–814.
75. Tran, P.L.; Lowry, N.; Campbell, T.; Reid, T.W.; Webster, D.R.; Tobin, E.; Aslani, A.; Mosley, T.; Dertien, J.; Colmer-Hamood, J.A.; *et al.* An organoselenium compound inhibits *Staphylococcus aureus* biofilms on hemodialysis catheters *in vivo*. *Antimicrob. Agents Chemother.* **2012**, *56*, 972–978.
76. Busscher, H.J.; Rinastiti, M.; Siswomihardjo, W.; van der Mei, H.C. Biofilm formation on dental restorative and implant materials. *J. Dent. Res.* **2010**, *89*, 657–665.
77. Roosjen, A.; Norde, W.; van der Mei, H.C.; Busscher, H.J. The use of positively charged or low surface free energy coatings *versus* polymer brushes in controlling biofilm formation. *Prog. Colloid Polym. Sci.* **2006**, *132*, 138–144.

78. Bridgett, M.J.; Davies, M.C.; Denyer, S.P. Control of staphylococcal adhesion to polystyrene surfaces by polymer surface modification with surfactants. *Biomaterials* **1992**, *13*, 411–416.
79. Huang, N.P.; Michel, R.; Voros, J.; Textor, M.; Hofer, M.; Rossi, A.; Elbert, D.L.; Hubbell, J.A.; Spencer, N.D. Poly(L-lysine)-g-poly(ethylene glycol) layers on metal oxide surfaces: Surface-analytical characterization and resistance to serum and fibrinogen adsorption. *Langmuir* **2001**, *17*, 489–498.
80. Zhu, B.; Eurell, T.; Gunawan, R.; Leckband, D. Chain-length dependence of the protein and cell resistance of oligo(ethylene glycol)-terminated self-assembled monolayers on gold. *J. Biom. Mater. Res.* **2001**, *56*, 406–416.
81. Currie, E.P.; Norde, W.; Cohen Stuart, M.A. Tethered polymer chains: surface chemistry and their impact on colloidal and surface properties. *Adv. Colloid Interface Sci.* **2003**, *100–102*, 205–265.
82. Roosjen, A.; van der Mei, H.C.; Busscher, H.J.; Norde, W. Microbial adhesion to poly(ethylene oxide) brushes: Influence of polymer chain lenghth and temperature. *Langmuir* **2004**, *20*, 10949–10955.
83. Roosjen, A.; Kaper, H.J.; van der Mei, H.C.; Norde, W.; Busscher, H.J. Inhibition of adhesion of yeasts and bacteria by poly(ethylene oxide)-brushes on glass in a parallel plate flow chamber. *Microbiology* **2003**, *149*, 3239–3246.
84. Olsson, J.; van der Heijde, Y.; Holmberg, K. Plaque formation *in vivo* and bacterial attachment *in vitro* on permanently hydrophobic and hydrophilic surfaces. *Caries Res.* **1992**, *26*, 428–433.
85. Park, K.D.; Kim, S.W. PEO-Modified Surfaces—*In Vitro*, *Ex Vivo*, and *In Vivo* Blood Compatibility. In *Poly(ethyleneglycol) Chemistry: Biotechnical and Biomedical Applications*, Harris, J.M., Ed.; Plenum Press: New York, NY, USA, 1992; pp. 283–301.

4

Marine Polysaccharide Networks and Diatoms at the Nanometric Scale

Vesna Svetličić *, Vera Žutić, Galja Pletikapić and Tea Mišić Radić

Division for Marine and Environmental Research, Ruđer Bošković Institute, Bijenička 54, 10000 Zagreb, Croatia; E-Mails: zutic@irb.hr (V.Ž.); gpletik@irb.hr (G.P.); tmisic@irb.hr (T.M.R.)

* Author to whom correspondence should be addressed; E-Mail: svetlicic@irb.hr;

Abstract: Despite many advances in research on photosynthetic carbon fixation in marine diatoms, the biophysical and biochemical mechanisms of extracellular polysaccharide production remain significant challenges to be resolved at the molecular scale in order to proceed toward an understanding of their functions at the cellular level, as well as their interactions and fate in the ocean. This review covers studies of diatom extracellular polysaccharides using atomic force microscopy (AFM) imaging and the quantification of physical forces. Following a brief summary of the basic principle of the AFM experiment and the first AFM studies of diatom extracellular polymeric substance (EPS), we focus on the detection of supramolecular structures in polysaccharide systems produced by marine diatoms. Extracellular polysaccharide fibrils, attached to the diatom cell wall or released into the surrounding seawater, form distinct supramolecular assemblies best described as gel networks. AFM makes characterization of the diatom polysaccharide networks at the micro and nanometric scales and a clear distinction between the self-assembly and self-organization of these complex systems in marine environments possible.

Keywords: atomic force microscopy; marine diatoms; extracellular polymeric substance; extracellular polysaccharides; marine gel network; self assembly; self-organization; *Cylindrotheca closterium*; *Bacteriastrum jadranum*; northern Adriatic Sea

---

1. Introduction

Diatoms, important marine photoautotrophic protists that account for up to 25% of the primary production on Earth [1], produce large quantities of extracellular polymeric substances (EPS), consisting predominantly of polysaccharides [2]. Diatom extracellular polymers participate in various processes, both at the cellular level and in the environment. At the cellular level, extracellular polymers have several important functions, of which some of the most frequently recognized are sessile adhesion, gliding, protection against drying, stabilization of habitats through prevention of sediment erosion, and the formation of biofilms and colonies [2,3]. The total amount of exopolysaccharides produced by diatoms is far in excess of that required for movement or adhesion [3–5] and is referred to as the photosynthetic overflow. In the marine environment, extracellular polysaccharide production by diatoms is a significant route by which photosynthetically produced organic carbon enters the trophic web and may influence the physical environment in the sea. Specifically, species of diatoms in the northern Adriatic Sea can produce large amounts, up to 50 g/m^3 of extracellular polysaccharides in a month [6], resulting in the episodic formation of a macroscopic gel phase [7,8].

Marine polysaccharide gels represent a form of molecular organization in which biopolymer molecules form solvated three-dimensional networks imbedded in seawater. Intense metabolic activities of microorganisms within gel aggregates (e.g., [9]) result in the formation of sharp microbiogeochemical gradients. The diffusion-slowing properties of the gel phase aid in maintaining these gradients over a very small (μm) spatial scale. The continued presence of sharp microbiogeochemical gradients results in the formation of microbial consortia, structured arrangements of microorganisms exhibiting different but highly specific physiological activities over small spatial scales.

Unlike chemical gels that are formed by chemical reaction using a cross-linking agent, it is characteristic of anionic polysaccharide macromolecules to form gels by physical bonds through the intermolecular forces among polymer chains [10,11]. Among the new methodologies (experimental and theoretical) developed for and applied to polysaccharide conformation and dynamics, solution properties, chain aggregation and gelation [12] using atomic force microscopy (AFM) have yielded the most striking results [13].

AFM has been extensively used in nanoscale studies of EPS intimately associated with frustules as coating and adhesive structures (strands, tethers, pads and stalks), as recently reviewed by Higgins and Wetherbee, 2012 [14]. Here, we intend to review recent AFM studies of diatom EPS focusing on the detection of the supramolecular structures of polysaccharide fibrils produced by marine diatoms, either attached to the diatom cell wall or released into the surrounding seawater.

2. Basic Principles of AFM and Its Application to Polysaccharides

AFM connects the nanometer- and micrometer-length scales, utilizing a sharp probe that senses interatomic forces acting between the surface of a sample and the atoms at the apex of the tip. The physical basis behind AFM and its ability to "feel" the surface make AFM a versatile tool in biophysics, allowing high resolution imaging, nanomechanical characterization and measurements of inter- and intramolecular forces in living and non-living structures [15–19]. Thanks to the simple principle on which it is based, the AFM is a surprisingly small and compact instrument. Its use includes an electronic control unit, computer and usually two monitors for the simultaneous checking of the image and imaging parameters. The probe, which scans the sample surface, consists of a cantilever and a tip located at its free end. The deflection of the cantilever is measured by an optical detection system. Registered values of cantilever deflection are electronically converted into a pseudo 3D image of a sample. As a result AFM produces real 3D images of a sample with a vertical resolution of 0.1 nm and lateral resolution of 1 nm. The measured forces range from 10^{-6} N to 10^{-11} N. In AFM force spectroscopy, a single molecule or fiber is stretched between the AFM flexible cantilever tip and a flat substrate. A polysaccharide molecule, protein or other biopolymer is either adsorbed to the substrate or linked to it through the formation of covalent bonds. When the tip and substrate are brought together and then withdrawn, one or more molecules can attach to the tip. The deflection of the cantilever measures the force on the polymer with an accuracy of ~5 pN, while the piezoelectric positioner records the changes in the end-to-end length of the molecule with an accuracy of 0.1 nm. AFM force spectroscopy is a widely used method in polymer biophysics, allowing the measurement of the mechanical properties of single molecules with the possibility of quantifying the forces involved directly in both intra- and inter-molecular polymer interactions [18,20–23]. It has also been adopted in advancing diatom research into the nanotechnology [14].

In the AFM imaging mode that makes visualization at the molecular scale of polysaccharide samples possible [24–32], the molecules are usually spread on freshly cleaved mica (a hydrophilic aluminosilicate mineral). The imaging of hydrated samples is preferably conducted in air to inhibit the unfavorable motion of polysaccharides in liquid medium. Protocols for marine sample AFM imaging have been recently developed for single diatom cells and released polymers, isolated polysaccharides from diatom cultures and polymer networks of the marine gel phase [33–39].

3. AFM Studies of Diatom EPS

Higgins and coworkers [40] were the first to use AFM for the characterization of diatom extracellular polymers. Intra- and intermolecular forces were measured between polymers on the cell wall and between those extruded from the diatom raphe on diatom species *Craspedostauros australis* E.J. Cox and *Pinnularia viridis* (Nitzsch) Ehrenberg [41,42]. The resulting force curves obtained from the pores of cell walls were attributed to the soft and compressible material. However, due to their complexity, it was difficult to assign specific interactions between biopolymers to patterns on the force curves. AFM force spectroscopy was also applied to study extracellular adhesive pads released by the diatom *Toxarium undulatum* Bailey [43–45]. These pads are very sticky and cells use them to form colonies or attach themselves to the surface. The resulting force profiles with numerous sawtooth patterns (e.g., [46]) were attributed to the extensible modular proteins that are associated into the nanofibers. In a differently designed experiment, Arce and coworkers [47] compared the adhesion of whole diatom *Navicula* spp. cells on different surfaces. In that study, individual cells were glued to tipless cantilevers. With such "diatom probes," standard force curves were recorded on surfaces such as Intersleek (a hydrophobic agent inhibiting biofilm formation) and mica. The resulting force curves showed comparable adhesion forces, and it was concluded that the extracellular polymeric material on the surface of *Navicula* spp. has both hydrophobic and hydrophilic properties. Direct imaging of EPS molecules was, however, hampered by the weak interaction of strands and tethers with the substrates used in the liquid [41]. These difficulties were overcome by using mica as a substrate and by imaging in air under ambient conditions (experimental details in [35]).

3.1. EPS of Cylindrotheca *Species*

The ubiquitous marine diatoms of the *Cylindrotheca* spp. were used in AFM studies of exopolysaccharide production [35–38]. The *Cylindrotheca closterium* (*Ceratonea Closterium* Ehrenberg) strain CCNA1 [35] was isolated from northern Adriatic seawater, while the CCMP1544 and *Cylindrotheca fusiformis* Reimann & J.C. Lewin, CCMP343 strains were obtained from the Bigelow Laboratory for Ocean Sciences.

The molecular organization of the EPS biopolymers released by *C. closterium* was approached at different levels: (i) EPS released by a single cell; (ii) EPS released in the culture medium and (iii) as biofilms grown on mica slides inserted in the culture [35]. The release of extracellular polymers by single cells was investigated by AFM in exponential and stationary growth phases. While the release of extracellular polymers in the exponential phase of growth was negligible, in the stationary phase extracellular polymers were visualized on more than 25% of the cells. The AFM data are in line with literature data on increased production of extracellular polymers in the stationary growth phase [48–52]. Parallel experiments with Alcian Blue staining and light microscopy performed in the cell culture have shown that the fibrils extending from the cell rostrum were mainly polysaccharides that existed in the liquid phase before the cell deposition to the mica surface [53].

Figure 1. Extracellular polymers released by *C. closterium* (CCNA1) obtained by AFM imaging in contact mode after deposition on mica surface. (**a**) AFM image of the whole cell presented as deflection data. The arrow indicates the position of the polymer excretion site; (**b**) The released polymers still attached to the apex of the cell rostrum, deflection data, scan size 5 μm × 5 μm; (**c**) Released polymers presented as height data, scan size 4 μm × 4 μm and vertical scale shown as the color bar (reproduced from [33]).

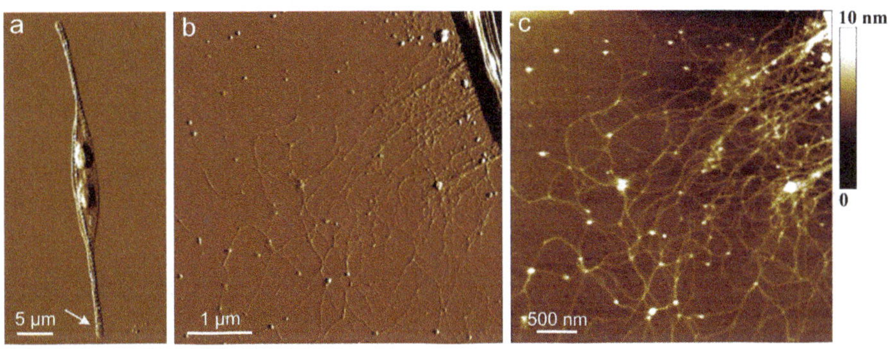

AFM images of extracellular polymers released by *C. closterium* are shown in Figure 1. The image of the whole cell (Figure 1a) presents the general features of the *C. closterium* cell with two chloroplasts and its drawn-out flexible rostra. The arrow indicates the position of polymer release shown in b and c of Figure 1. Bundles of polymer fibrils appear at the position close to the site of excretion. Their heights were 5–7 nm. These bundles unfold into a fibrillar network with gradually decreasing fibril heights reaching a distance of up to 10 μm from the cell wall. At the distance of 1 μm, a dense network is observed with fibril heights of 2–3 nm. At larger distances, *C. closterium* EPS appeared as a relaxed network of fibrils with incorporated spherical nanoparticles/globules. The fibrils appeared flexible, usually with a curved shape, some of them even forming loops. The globules were found to interconnect two or more fibrils but were also imbedded along a single fibril. The fibril heights were in the range of 0.6 to 1.2 nm, while the average height of the incorporated globules was 5 nm. It was hypothesized that the globules are positively charged proteins whose function is the intracellular packing of negatively charged polysaccharide fibrils [35].

3.2. Biofilm of Cylindrotheca Closterium

Although several AFM studies exist that explored the nature of diatom biofilms and adhesives [40,45,47,54], visualization of a biofilm has only been achieved by AFM imaging in air [35]. The mica slides were withdrawn from the flasks when the cultures of diatom *C. closterium* entered the stationary growth phase (after 18 days), washed with ultrapure water and imaged in air after drying.

The observed biofilm appeared as a continuous fibrillar network between predominantly individual cells (Figure 2). Some cells were associated (Figure 2a) and 2–6 cells were captured on a 35 μm × 35 μm scanned biofilm surface. At a higher resolution, the biofilm was visualized as a dense fibrillar network with pore sizes ranging from 50 nm to 300 nm (Figure 2b,c). A gradual increase in pore sizes was detected going from the cell wall to the distance of 10 μm (Figure 2b). The globules (3–12 nm high) found exclusively on the fibrillar network appear as silica nanoparticles nucleated from the culture medium. Shchipunov and coworkers applied AFM to demonstrate that gel-forming polysaccharides (e.g., carrageenans) promote silica precipitation, serving as a template [55,56]. Comparison of the fibril and globule height analyses in the EPS of the two *C. closterium* strains and the biofilm grown in a *C. closterium* (CCNA1) culture is given in Table 1.

Figure 2. AFM image of biofilm grown in a CCNA1 culture formed on a mica slide. (**a**) Association of cells in the biofilm (height data, vertical scale 1.5 μm); (**b**) Cell rostrum surrounded by the biofilm network (deflection data); (**c**) The biofilm network (height data, vertical scale 10 nm). Images were acquired in contact mode (reproduced from [35]).

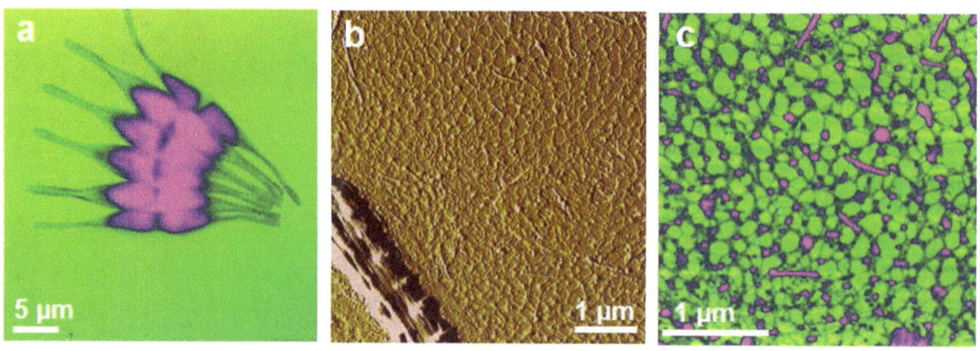

Table 1. Height ranges of the fibrils and globules in the EPS and biofilm of *C. closterium* strains CCNA1 and CCMP1544 (data from [35]).

Cell culture	Fibril height/nm			Globule height/nm		
	EPS bound to cell	EPS in bulk culture	Biofilm	EPS bound to cell	EPS in bulk culture	Biofilm
CCNA1	0.4–1.9	0.4–2.2	1.7–4.0	3–12	2–12	3–12
CCMP1544	0.6–1.6	0.7–2.6	–	3–9	2–13	–

3.3. Polysaccharide Network in Diatom Colony Formation

Diatoms have evolved a variety of colonial life forms in which cells are connected by organic threads, mucilage pads or silicate structures. In general, these connecting structures are clearly visible by light or electron microscopy. The colonial planktonic diatom *Bacteriastrum jadranum* Godrijan, Marić & Phannkuchen [57] was found to be unusual, since the chain formation does not involve fusion of the setae of adjacent cells [39], unlike all other known colonial *Bacteriastrum* species [58]. Moreover, no thread or any other organic or inorganic substance clearly visible by light or electron microscopy connects the cells.

Figure 3. Structural details of the cell jacket network obtained by AFM. (**A**) branching fibril with height analysis along the indicated line; (**B**) 2D network of interconnect patches; (**C**) Spatial arrangement of interconnected patches in the 3D collapsed network (3D view). Images were acquired in tapping mode (reproduced from [39]).

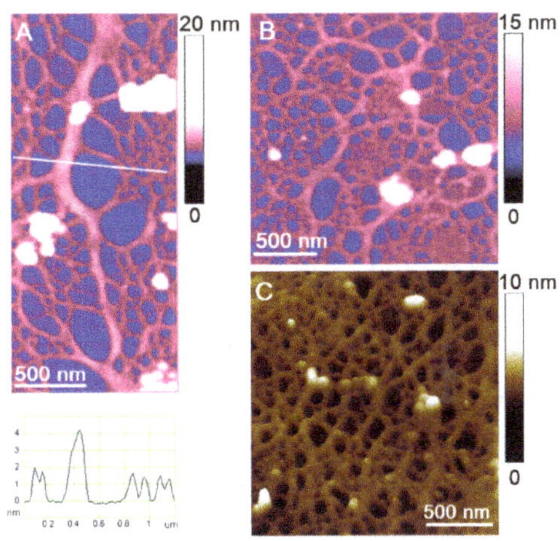

AFM imaging [39] provided the first evidence that *B. jadranum* cells in colonies are enclosed in a fibrillar polysaccharide network, termed a cell jacket. At nanoscale resolution, the cell jacket appeared as a cross-linked fibrillar network organized into recognizable patterns (Figure 3). Circular high-density domains (patches, Figure 3B), were surrounded and interconnected by thicker fibrils in a continuous network and appear as the basic structural motive. Pores inside a patch were of the same hexagonal shape, 8–100 nm in size. Their size was continuously smaller from the patch edge toward the center. The branching fibrils (Figure 3A) can be considered as the backbone of the network, critical for assuring its integrity. The pore-forming fibrils within the patches were only 0.6–1.6 nm high, the surrounding fibrils connecting the patches were 2.0–2.8 nm high, while the branching fibrils were considerably wider but not higher than 4.0 nm. Quantitative analysis of the cell jacket network is given in Table 2.

Table 2. Cell jacket network: pores and pore-forming fibrils analyzed over a surface area of 4×4 μm^2 (data from [39]).

Pores		Pore-forming fibrils
size/nm	number	height/nm
8–100	900	0.6–1.6
100–160	200	2.0–2.8
500–1000	10	2.5–4.0

It was concluded that the *Bacteriastrum* polysaccharide jacket represents an essential part of the cell, as the conjunction of the polymer network with the frustule appears to be extremely tight and such specific and unique patterns were not found in the polysaccharide networks of marine gel imaged by AFM.

4. AFM of Marine Gel

The broad polydispersity of marine gels ranges from microscopic to macroscopic dimensions [59]. Santschi and coworkers [60] were the first to use AFM to image individual fibrillar polysaccharides in marine macromolecular organic matter. The massive appearance of gelatinous macroaggregates known as mucilage events [7,8,61] offered a possibility for systematic studies of the marine gel phase using AFM [53,62]. The macromolecular characterization of the isolated polysaccharide fraction using physico-chemical techniques [63] showed that they are polydisperse high-molecular weight heteropolysaccharides, in which at least some of the hydroxyl groups of sugar residues are substituted by ester sulfate groups and to a minor extent by carboxylic groups-uronic acid.

These features, conferring a marked polyelectrolytic behavior of the polysaccharides in salt solutions, were also found for the polysaccharide fraction from *C. closterium* EPS [36]. The fact that the EPS isolated polysaccharide fraction has the capacity to self-assemble into a gel network (Figure 4) is an important finding, with implications on the mechanism of gel phase formation in marine systems [36].

Figure 4. AFM images of polysaccharides isolated from the CCNA1 culture and dissolved in ultrapure water: (**a**) Single fibrils (concentration 5 mg/L) vertical scale 2.5 nm; (**b,c**) Fibril networks (concentration 10 mg/L), vertical scales: 5 nm (**b**) and 10 nm (**c**). The height analyses are shown along the indicated lines. The images were acquired in tapping mode and are presented as height data (reproduced from [35]).

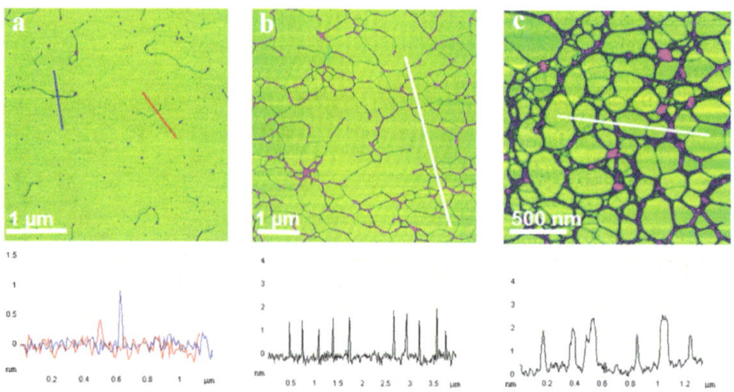

4.1. Imaging a Polysaccharide Network of Marine Gel

Development of the protocol for the AFM imaging of the marine gel phase was a critical step. The sampling procedure and specimen preparation protocol for AFM imaging is described in detail by Mišić, Radić *et al.* [34]. The main organizational features of the polymer gel network were preserved during the transfer of marine gel from seawater to a mica substrate in air. The appearance of a marine gel network at micro and nanometric scales is shown in Figure 5. The gel network imaged by AFM is to a certain extent distorted from the 3D structure in the aqueous phase due to attachment and stretching on the mica surface. Nevertheless, such constraints make studies of fibril associations at the molecular level that would not be accessible by other techniques possible. AFM imaging of marine gel provided insight into the molecular organization of the gel network and associations between polysaccharide fibrils forming the network. Among the complex network structures, associations of fibrils forming junction zones were identified. Modes of fibril associations into junction zones are exemplified in Figure 6.

Figure 5. Marine gel fibrils at micro and nanometric scales: (**a**) A polysaccharide network of marine gel aggregate imaged in seawater by confocal microscopy after FITC-Concavalin A staining; (**b,c**) High resolution AFM images of marine gel fibrils with different degrees of cross-linking. AFM images were acquired in tapping mode in air using mica as a substrate (reproduced from [53]).

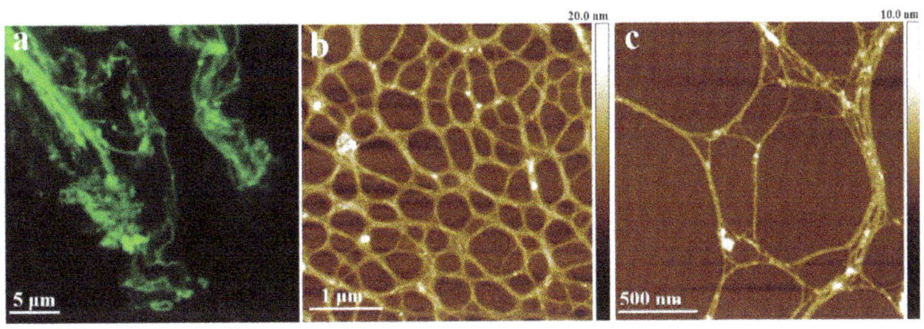

Figure 6. Mode of fibril associations into junction zones obtained by high resolution AFM imaging: (**a**) Two 0.7 nm high fibrils forming a 260 nm long junction zone; (**b**) Two 1.3 nm high fibrils forming 70 and 120 nm long junction zones; (**c**) Side-by-side association of two 1.6 nm high fibrils. The images were acquired in contact mode (adapted from [34]).

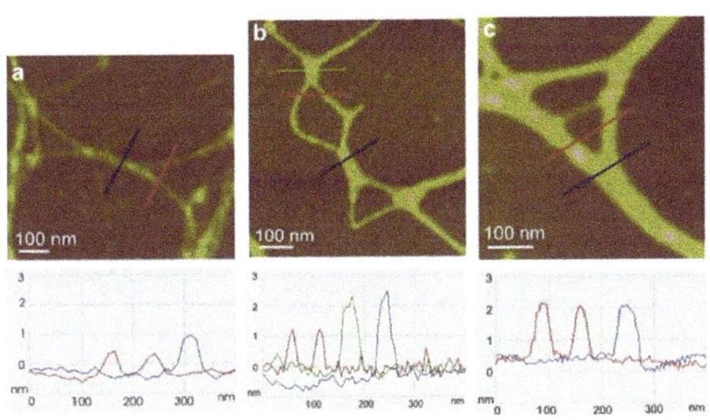

The evolution of polysaccharide fibrils into marine gel during the mucilage event in the northern Adriatic Sea was captured by AFM [36]. The samples were prepared from macroaggregates collected after different residence times in the water column, from the early stage of gel phase formation to the condensed (mature) gel network of an older macroaggregate. The long polymer strands with small patches of initial fibril associations (Figure 7a) coexisted with the continuous gel network shown in Figure 7b. With a prolonged residence time (one month), a more condensed network was formed, as presented in Figure 7c.

Figure 7. Evolution of polymer networks in the macroscopic gel phase from the early stage of gel phase formation (**a**,**b**) to the condensed gel network of an older macroaggregate (**c**). The AFM images were acquired in contact mode and presented as height data, scan size 4 μm × 4 μm (reproduced from [36]).

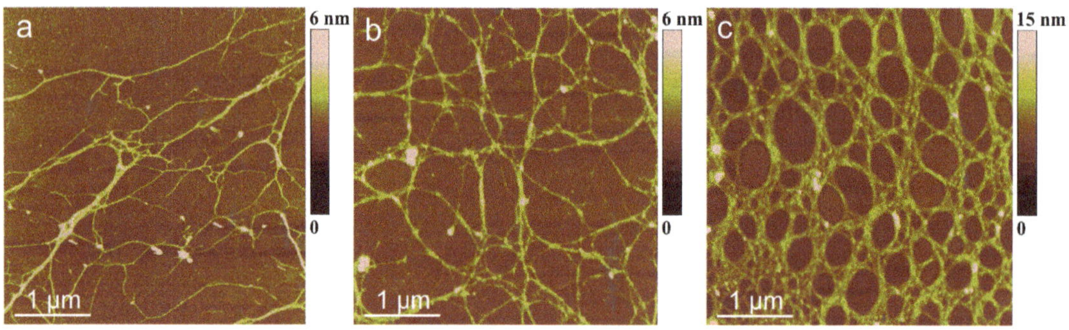

4.2. Force Spectroscopy of a Marine Gel Network

The knowledge of the mechanical strength of individual molecular assemblies within a marine gel network contributes to the understanding of the gel phase formation and its persistence in the marine environment. Due to the inherent complexity and heterogeneity of the marine gel phase, it is difficult to isolate the physical forces in the biopolymer network assemblies. However, based on AFM imaging and differential scanning calorimetry, the marine gel was characterized as a thermoreversible physical gel and the dominant mode of gelation was proposed to be the crosslinking of polysaccharide fibrils by hydrogen bonding, which results in helical structures and their associations [34].

Force spectroscopy and high resolution AFM imaging were applied to quantify the intramolecular, interdomain and intermolecular forces within the marine gel network [63,64]. The ability to control the degree of gel network entanglements by dilution and stirring was used to probe marine polysaccharides at different levels of association. The typical events that lead to specific patterns upon stretching include the entropic behavior of individual fibrils (Figure 8) and more complex events, such as the unfolding of polysaccharide entanglements (Figure 9).

Figure 8. Force curves of a single polysaccharide fibril from a disentangled gel network (AFM image shown as the insert) acquired in filtered seawater: approach curve in red and extension curve in blue (adapted from [63]).

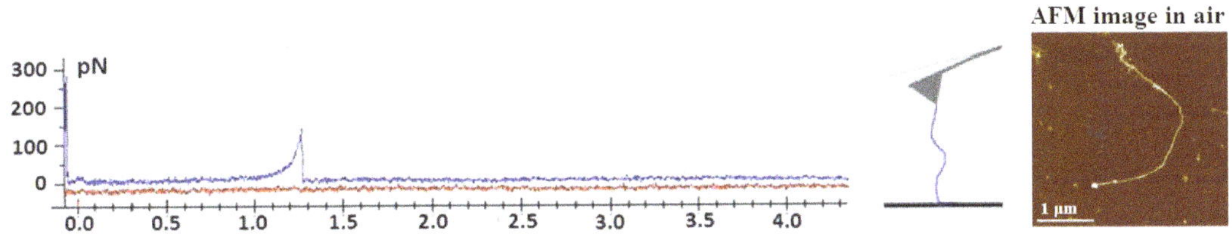

Figure 9. Force curves of polysaccharide fibrils in a marine gel network (AFM image shown as the insert) acquired in filtered seawater: approach curve in red and extension curve in blue (adapted from [63]).

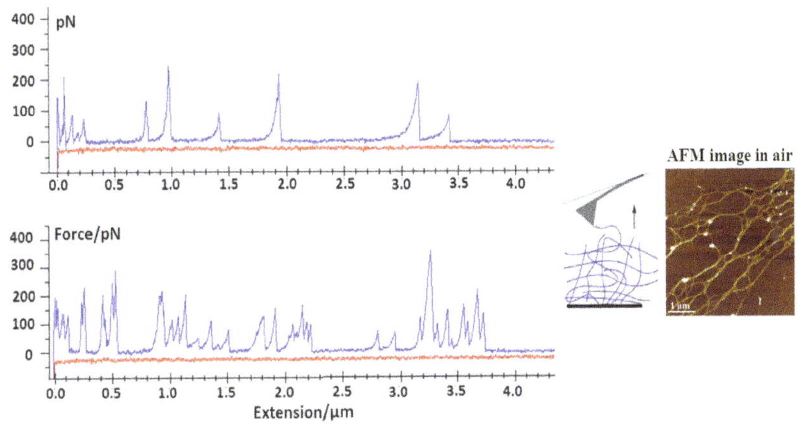

Marine gels are highly extensible, as they may be stretched with very little force to distances of up to several micrometers. Fairly large forces are required to unzip the fibrils, suggesting that lateral stability may be important in maintaining the structural integrity of the marine gel.

5. Conclusions

AFM made characterization of diatom polysaccharide networks at the nanometric scale and a clear distinction between the self-assembly and self-organization of these complex systems possible. Marine polysaccharides produced by diatoms are shown to form distinct supramolecular assemblies that are best described as gel networks. These photosynthetically produced macromolecules may stay attached to the cell wall or be released into the environment.

This review encompasses three specific examples of polysaccharide polymer networks attached to the diatom cell wall: (i) *Cylindrotheca closterium* polymers at the moment of their release; (ii) a *C. closterium* biofilm grown on an atomically smooth substrate and (iii) a 3D polysaccharide network enclosing a cell colony of planktonic diatom *Bacteriastrum jadranum*.

Extracellular polymers produced by *C. closterium* exhibit a certain level of self-organization, while the highest level of self-organization was found in a 3D polysaccharide network produced by *B. jadranum*. The polysaccharide fibrillar network of *B. jadranum* represents an essential part of the cell and not a random extracellular polysaccharide structure. With a more generalized use of AFM, it would probably be discovered that there are many similar types of coverings among planktonic diatoms, as well as diatoms adapted to other habitats (e.g., benthos, tychoplankton *etc.*). This may be additionally extended to a wider range of microalgae and protists.

In contrast to the networks attached to the diatom cells, free-floating marine gel aggregates are formed by a self-assembly of diatom-released polymers. Structural details of the gel network visualized down to the molecular level revealed monomolecular, helical and superhelical associations.

The physical forces in marine gel network assemblies have been quantified using force spectroscopy together with high resolution AFM imaging. The marine gels appeared highly extensible, as they could be stretched with very little force to distances of up to several micrometers. Such self-assembled networks with randomly distributed microscopic features and high extensibility are capable of responding to environmental conditions, such as change in salinity, temperature, pH and shear stress, maintaining favorable physiological conditions for microbial communities.

Acknowledgments

The research was supported by the Croatian Ministry of Science, Education and Sport, the project Surface Forces on an Atomic Scale Applied in Marine Science and Nanotechnology (No. 0982934-2744).

References

1. Falkowski, P.; Raven, J.A. *Aquatic Photosynthesis*; Blackwell: Malden, MA, USA, 1997.
2. Hoagland, K.D.; Rosovski, J.R.; Gretz, M.R.; Roemer, S.C. Diatom extracellular polymeric substances: Function, fine structure, chemistry and physiology. *J. Phycol.* **1993**, *29*, 537–566.
3. Underwood, G.J.C.; Paterson, D.M. The importance of extracellular carbohydrate production by marine epipelic diatoms. *Adv. Bot. Res.* **2003**, *40*, 184–240.
4. Decho, A.W. Microbial exopolymer secretions in ocean environments: Their role(s) in food webs and marine processes. *Oceanogr. Mar. Biol. Annu. Rev.* **1990**, *28*, 73–153.
5. Staats, N.; Stal, L.I.; de Winder, B.; Mur, L.R. Oxygenic photosynthesis as driving process in exopolysaccharide production of benthic diatoms. *Mar. Ecol. Prog. Ser.* **2000**, *193*, 261–269.
6. Myklestad, S.M. Release of extracellular products by phytoplankton with special emphasis on polysaccharides. *Sci. Total Environ.* **1995**, *165*, 155–164.
7. Vollenweider, R.; Rinaldi, A. Marine mucilages. *Sci. Total Environ.* **1995**, *165*, 1–235.
8. Giani, M.; Degobbis, D.; Rinaldi, A. Mucilages in the Adriatic and Tyrrhenian Seas. *Sci. Total Environ.* **2005**, *353*, 1–379.
9. Müller-Niklas, G.; Schuster, S.; Kaltenbök, E.; Herndl, G.J. Organic content and bacterial metabolism in amorphous aggregations of the northern Adriatic Sea. *Limnol. Oceanogr.* **1994**, *39*, 58–68.

10. Rinaudo, M. Advances in Characterization of Polysaccharides in Aqueous Solution and Gel State. In *Polysaccharides-Structural Diversity and Functional Versatiliy*, Dumitriu, S., Ed.; Marcel Dekker: New York, NY, USA, 2005; pp. 237–252.
11. Israelachvili, J.N. *Intermolecular and Surface Forces*, 3rd ed.; Academic Press: London, UK, 2010.
12. Dimitriu, S. *Polysaccharides: Structural Diversity and Functional Versatility*, 2nd ed.; Marcel Dekker: New York, NY, USA, 2005.
13. Cesàro, A.; Bellich, B.; Borgogna, M. Biophysical functionality in polysaccharides: From Lego-blocks to nano-particles. *Eur. Biophys. J.* **2012**, *41*, 379–395.
14. Higgins, M.J.; Wetherbee, R. The Role of Atomic Force Microscopy in Advancing Diatom Research into the Nanotechnology Era. In *Life at the Nanoscale: Atomic Force Microscopy of Live Cells*, Dufrene, Y., Ed.; Pan Stanford Publishing: Singapore, 2011; pp. 405–420.
15. Eaton, P.; West, P. *Atomic Force Microscopy*; Oxford University Press: New York, NY, USA, 2010.
16. Jena, B.P.; Hörber, J.K.H. *Force Microscopy: Applications in Biology and Medicine*; John Wiley & Sons: New York, NY, USA, 2006.
17. Dufrene, Y. *Life at the Nanoscale: Atomic Force Microscopy of Live Cells*; Pan Stanford Publishing: Singapore, 2011.
18. Zhang, W.; Zhang, X. Single molecule mechanochemistry of macromolecules. *Prog. Polym. Sci.* **2003**, *28*, 1271–1295.
19. Akai, A. Review on: Atomic force microscopy applied to nano-mechanics of the cell. *Adv. Biochem. Eng. Biotechnol.* **2010**, *119*, 47–61.
20. Bustamante, C.; Macosko, J.C.; Wuite, G.J. Grabbing the cat by the tail: Manipulating molecules one by one. *Nat. Rev. Mol. Cell Biol.* **2000**, *1*, 130–136.
21. Noy, A. *Handbook of Molecular Force Spectroscopy*; Springer: Livermore, CA, USA, 2007.
22. Giannotti, M.I.; Vancso, G.J. Interrogation of single synthetic polymer chains and polysaccharides by AFM-based force spectroscopy. *Chem. Phys. Chem.* **2007**, *8*, 2290–2307.
23. Zhang, X.; Liu, C.; Wang, Z. Force spectroscopy of polymers: Studying on intramolecular and intermolecular interactions in single molecular level. *Polymer* **2008**, *49*, 3353–3361.
24. Kirby, A.R.; Gunning, A.P.; Morris, V.J. Imaging polysaccharides by atomic force microscopy. *Biopolymers* **1996**, *38*, 355–366.
25. Morris, V.J.; Kirby, A.R.; Gunning, A.P. *Atomic Force Microscopy for Biologists*; Imperial College Press: London, UK, 1999.
26. Abu-Lail, N.I.; Camesano, T.A. Polysaccharide properties probed with atomic force microscopy. *J. Microsc.* **2003**, *212*, 217–238.
27. Abu-Lail, N.I.; Camesano, T.A. Atomic Force Microscope and Single-Molecule Force Microscopy Studies of Biopolymers. In *Dekker Encyclopedia of Nanoscience and Nanotechnology*, Schwarz, J., Contescu, C., Putyera, K., Eds.; Taylor & Francis: London, UK, 2004; pp. 119–131.
28. McIntire, T.M.; Brant, D.A. Imaging of individual biopolymers and supramolecular assemblies using noncontact atomic force microscopy. *Biopolymers* **1997**, *42*, 133–146.
29. Ikeda, S.; Morris, V.J.; Nishinari, K. Microstructure of aggregated and nonaggregated kappa-carrageenan helices visualized by atomic force microscopy. *Biomacromolecules* **2001**, *2*, 1331–1337.

30. Keresztes, Z.; Rigó, T.; Telegdi, J.; Kálmán, E. Investigation of biopolymer networks by means of AFM. *Appl. Phys. A* **2001**, *72*, 113–116.
31. Funami, T.; Hiroe, M.; Noda, S.; Asai, I.; Ikeda, S.; Nishimari, K. Influence of molecular structure imaged with atomic force microscopy on the rheological behavior of carrageenan aqueous systems in the presence or absence of cations. *Food Hydrocolloids* **2007**, *21*, 617–629.
32. Noda, S.; Funami, T.; Nakauma, M.; Asai, I.; Takahashi, R.; Al-Assaf, S.; Ikeda S. Molecular structures of gellan gum imaged with atomic force microscopy in relation to the rheological behavior in aqueous systems. 1. Gellan gum with various acyl contents in the presence and absence of potassium. *Food Hydrocolloids* **2008**, *22*, 1148–1159.
33. Svetličić, V.; Balnois, E.; Žutić, V.; Chevalet, J.; Hozić Zimmermann, A.; Kovač, S.; Vdović, N. Electrochemical detection of gel microparticles in seawater. *Croat. Chem. Acta* **2006**, *79*, 107–113.
34. Mišić Radić, T.; Svetličić, V.; Žutić, V.; Boulgaropoulos, B. Seawater at the nanoscale: Marine gel imaged by atomic force microscopy. *J. Mol. Recognit.* **2011**, *24*, 397–405.
35. Pletikapić, G.; Mišić Radić, T.; Hozić Zimmermann, A.; Svetličić, V.; Pfannkuchen, M.; Marić, D.; Godrijan, J.; Žutić, V. Extracellular polymer release AFM imaging of extracellular polymer release by marine diatom *Cylindrotheca closterium* (Ehrenberg) Reiman & J.C. Lewin. *J. Mol. Recognit.* **2011**, *24*, 436–445.
36. Svetličić, V.; Žutić, V.; Mišić Radić, T.; Pletikapić, G.; Hozić Zimmerman, A.; Urbani, R. Polymer networks produced by marine diatoms in the northern Adriatic Sea. *Mar. Drugs* **2011**, *9*, 666–679.
37. Pletikapić, G.; Berquand, A.; Mišić Radić, T.; Svetličić, V. Quantitative nanomechanical mapping of marine diatom. *J. Phycol.* **2012**, *48*, 174–185.
38. Pletikapić, G.; Vinković Vrček, I.; Žutić, V.; Svetličić, V. Atomic force microscopy characterization of silver nanoparticles interactions with marine diatom cells and extracellular polymeric substance. *J. Mol. Recognit.* **2012**, *25*, 309–317.
39. Bosak, S.; Pletikapić, G.; Hozić, A.; Svetličić, V.; Sarno, D.; Viličić, D. A novel type of colony formation in marine planktonic diatoms revealed by atomic force microscopy. *PLoS One* **2012**, *7*, e44851.
40. Higgins, M.J.; Crawford, S.A.; Mulvaney, P.; Wetherbee, R. Characterization of the adhesive mucilages secreted by live diatom cells using atomic force microscopy. *Protist* **2002**, *153*, 25–38.
41. Higgins, M.J.; Molino, P.; Mulvaney, P.; Wetherbee, R. The structure and nanomechanical properties of the adhesive mucilage that mediates diatom-substratum adhesion and motility. *J. Phycol.* **2003**, *39*, 1181–1193.
42. Higgins, M.J.; Sader, J.E.; Mulvaney, P.; Wetherbee, R. Probing the surface of living diatoms with atomic force microscopy: The nanostructure and nanomechanical properties of the mucilage layer. *J. Phycol.* **2003**, *39*, 722–734.
43. Dugdale, T.M.; Dagastine, R.; Chiovitti, A.; Mulvaney, P.; Wetherbee, R. Single adhesive nanofibres from a live diatom have the signature fingerprint of modular proteins. *Biophys. J.* **2005**, *89*, 4252–4260.
44. Dugdale, T.M.; Dagastine, R.; Chiovitti, A.; Wetherbee, R. Diatom adhesive mucilage contains distinct supramolecular assemblies of a single modular protein. *Biophys. J.* **2006**, *90*, 2987–2993.

45. Dugdale, T.M.; Willis, A.; Wetherbee, R. Adhesive modular proteins occur in the extracellular mucilage of the motile, pennate diatom phaeodactylum tricornutum. *Biophys. J.* **2006**, *90*, L58–L60.
46. Rief, M.; Gautel, M.; Oesterhelt, F.; Fernandez, J.M.; Gaub, H.E. Reversible unfolding of individual titin immunoglobulin domains by AFM. *Science* **1997**, *16*, 1109–1112.
47. Arce, F.T.; Avci, R.; Beech, I.B.; Cooksey, K.E.; Wigglesworth-Cooksey, B. A live bioprobe for studying diatom-surface interactions. *Biophys. J.* **2004**, *87*, 4284–4297.
48. Staats, N.; de Winder, B.; Stal, L.J.; Mur, L.R. Isolation and characterization of extracellular polysaccharides from the epipelic diatoms *Cylindrotheca closterium* and *Navicula salinarum*. *Eur. J. Phycol.* **1999**, *34*, 161–169.
49. Staats, N.; Stal, L.J.; Mur, L.R. Exopolysaccharide production by the epipelic diatom *Cylindrotheca closterium*: Effects of nutrient conditions. *J. Exp. Mar. Biol. Ecol.* **2000**, *249*, 13–27.
50. Alcoverro, T.; Conte, E.; Mazzella, L. Production of mucilage by the Adriatic epipelic diatom *Cylindrotheca closterium* (Bacillariophyceae) under nutrient limitation. *J. Phycol.* **2000**, *36*, 1087–1095.
51. Smith, D.J.; Underwood, G.J.C. Production of extracellular carbohydrates by estuarine benthic diatoms: The effects of growth phase and light and dark treatment. *J. Phycol.* **2000**, *36*, 321–333.
52. De Brouwer, J.F.C.; Stal, L.J. Daily fluctuations of exopolymers in cultures of the benthic diatoms *Cylindrotheca closterium* and *Nitzschia* sp. (Bacillariophyceae). *J. Phycol.* **2002**, *38*, 464–472.
53. Pletikapić, G. Nanomechanical Properties of Diatom Cell Surfaces and Extracellular Polymers. Ph.D. Thesis, University of Zagreb, Zagreb, Croatia, March 2013.
54. Stal, L.J.; de Brouwer, J.F.C. Biofilm formation by benthic diatoms and their influence on the stabilization of intertidal mudflats. *Berichte-Forschungszentrum Terramare* **2003**, *12*, 109–111.
55. Shchipunov, Y.A. Sol-gel-derived biomaterials of silica and carrageenans. *J. Colloid Interface Sci.* **2003**, *268*, 68–76.
56. Shchipunov, Y.A.; Kojima, A.; Imae, T. Polysaccharides as a template for silicate generated by sol-gel processes. *J. Colloid Interface Sci.* **2005**, *285*, 574–580.
57. Godrijan, J.; Marić, D.; Imešek, M.; Janeković, I.; Schweikert, M.; Pfannkuchen , M. Diversity, occurrence, and habitats of the diatom genus *Bacteriastrum* (Bacillariophyta) in the northern Adriatic Sea, with the description of *B. jadranum* sp. nov. *Bot. Mar.* **2012**, *55*, 1–12.
58. Sarno, D.; Zingone, A.; Marino, D. *Bacteriastrum parallelum* sp. nov., a new diatom from the Gulf of Naples, and new observations on *B. furcatum* (Chaetocerotaceae, Bacillariophyta). *Phycologia* **1997**, *36*, 257–266.
59. Verdugo, P.; Alldredge, A.L.; Azam, F.; Kirchman, D.L.; Passow, U.; Santschi, P.H. The oceanic gel phase: A bridge in the DOM-POM continuum. *Mar. Chem.* **2004**, *92*, 67–85.
60. Santschi, P.H.; Balnois, E.; Wilkinson, K.J.; Zhang. J.; Buffle, J. Fibrillar polysaccharides in marine macromolecular organic matter as imaged by atomic force microscopy and transmission electron microscopy. *Limnol. Oceanogr.* **1998**, *43*, 896–908.
61. Svetličić, V.; Žutić, V.; Hozić Zimmermann, A. Biophysical scenario of giant gel formation in the northern Adriatic Sea. *Ann. N. Y. Acad.* **2005**, *1048*, 524–527.
62. Mišić Radić, T. Supramolecular organization of marine gel biopolymers studied by atomic force microscopy. Ph.D. Thesis, University of Zagreb, Zagreb, Croatia, October 2010.

63. Urbani, R.; Sist, P.; Pletikapić, G.; Mišić Radić, T.; Svetličić, V.; Žutić, V. Diatom Polysaccharides: Extracellular Production, Isolation and Molecular Characterization. In *The Complex World of Polysaccharide*, Karunaratn, D.N., Ed.; Intech: Rijeka, Croatia, 2012; pp. 346–356.
64. Svetličić, V.; Pletikapić, G.; Mišić Radić, T.; Kellermayer, M.S.Z.; Brujić, J. Force spectroscopy of marine polysaccharides: From gel networks to individual fibrils. Technical Report for AFMBiomed Conference: Shanghai, China, May 2013.

5

Biofilms: The Stronghold of *Legionella pneumophila*

Mena Abdel-Nour [1,2,3], Carla Duncan [1], Donald E. Low [2,3,†] and Cyril Guyard [1,2,3,*]

[1] Ontario Agency for Health Protection and Promotion (OAHPP), Toronto, ON M9P 3T1, Canada; E-Mails: mena_abdelnour@hotmail.com (M.A.-N.); carla.duncan@oahpp.ca (C.D.)

[2] Department of Laboratory Medicine and Pathobiology, University of Toronto, Toronto, ON M5S 1A8, Canada

[3] Mount Sinai Hospital, Toronto, ON M5G 1X5, Canada

† Deceased.

* Author to whom correspondence should be addressed; E-Mail: cyril.guyard@gmail.com;

Abstract: Legionellosis is mostly caused by *Legionella pneumophila* and is defined as a severe respiratory illness with a case fatality rate ranging from 5% to 80%. *L. pneumophila* is ubiquitous in natural and anthropogenic water systems. *L. pneumophila* is transmitted by inhalation of contaminated aerosols produced by a variety of devices. While *L. pneumophila* replicates within environmental protozoa, colonization and persistence in its natural environment are also mediated by biofilm formation and colonization within multispecies microbial communities. There is now evidence that some legionellosis outbreaks are correlated with the presence of biofilms. Thus, preventing biofilm formation appears as one of the strategies to reduce water system contamination. However, we lack information about the chemical and biophysical conditions, as well as the molecular mechanisms that allow the production of biofilms by *L. pneumophila*. Here, we discuss the molecular basis of biofilm formation by *L. pneumophila* and the roles of other microbial species in *L. pneumophila* biofilm colonization. In addition, we discuss the protective roles of biofilms against current *L. pneumophila* sanitation strategies along with the initial data available on the regulation of *L. pneumophila* biofilm formation.

Keywords: *Legionella pneumophila*; legionellosis; amoeba; prozoa; multispecies biofilms

---

1. Introduction

The Gram-negative bacterium *Legionella pneumophila* is responsible for the majority of legionellosis cases and is a significant contributor of community acquired, and hospital acquired pneumonia with a case fatality rate ranging from 5% to 80% [1,2]. *L. pneumophila* is an aquatic pathogen that is ubiquitously found in nature, in both anthropogenic structures and in environmental waters [3–7]. *In vitro*, *L. pneumophila* is able to produce monospecies biofilms (Figure 1) that is believed to contain an extracellular matrix [8,9]. In the environment, *L. pneumophila* can be found in several different habitats, including multispecies biofilms. In naturally occurring multispecies biofilms, the colonization with *L. pneumophila* can be influenced by several other species of microorganisms [10,11]. Of these microorganisms, protozoa are arguably one of the most important in determining *L. pneumophila* persistence, as the pathogen uses protozoa to replicate intracellularly [12]. Co-evolution with multiple species of protozoa has resulted in the development of mechanisms that allow *L. pneumophila* to occupy a very broad host range, and to infect human cells [13–15].

Figure 1. Monospecies biofilm of *L. pneumophila* labelled with the DNA stain Syto62. Scale bar represents 100 μm.

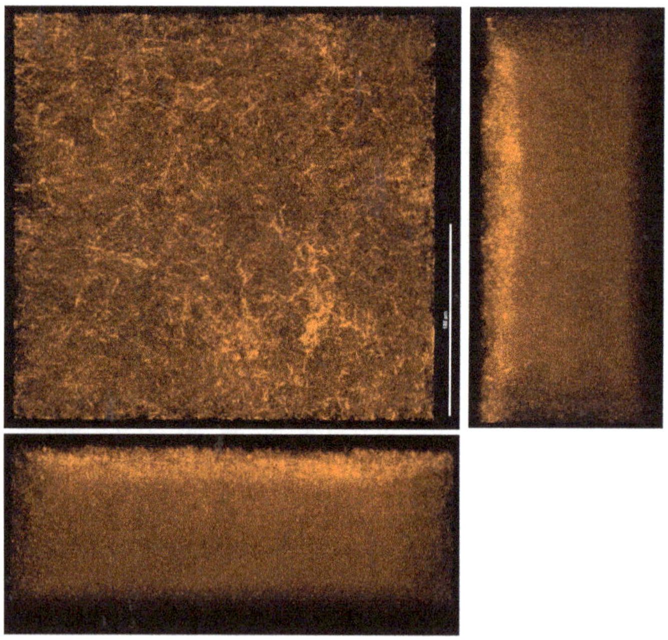

Recent reports suggest that the growth of *Legionella* in biofilms may lead to enhanced virulence. *L. pneumophila* isolates from serogroups 1, 10, and 12 that were collected from biofilms were more cytotoxic towards amoeba than reference outbreak and worldwide epidemic strains [16]. Moreover, initial data suggest that biofilm-derived *Legionella pneumophila* evades the innate immune response in macrophages [17]. As legionellosis is not transmitted from person to person, insights into the ecology of *L. pneumophila* may yield information that can be used to prevent the colonization of anthropogenic systems by *L. pneumophila*. In light of recent discoveries, this review intends to provide an overview of the findings on biofilm production and colonization by *L. pneumophila*.

2. Protozoa and *L. pneumophila* Biofilm Formation

Protozoa play a crucial role in the lifecycle of *Legionella* species as they provide a habitat for their environmental survival and replication [12,18–21]. In biofilm communities, several amoeba species have been found associated with *L. pneumophila* [22]. To feed, protozoan species often graze on bacteria present in multispecies biofilms, a phenomenon that *L. pneumophila* exploits in order to replicate (Figure 2) [23,24]. As a consequence, the presence of protozoa in anthropogenic water sources has been deemed a risk factor for *L. pneumophila* outbreaks [23]. In fact, the amount of *L. pneumophila* in biofilms is directly correlated with the biomass of protozoa [25]. This is in accordance with *in vitro* models showing that the presence of amoeba species promotes the biofilm formation of *L. pneumophila* on pins of "inverse" microtiter plates [26]. *L. pneumophila* is also capable of growing off the debris from dead amoebae, thus, amoeba may also encourage the replication of *L. pneumophila* indirectly [27]. Floating biofilms, also contain protozoa in association with *L. pneumophila* suggesting that *L. pneumophila* may attach to protozoa in floating biofilms in the absence of available abiotic surfaces [28,29]. In addition to the role of protozoa as a means of replication, the intracellular stage of *L. pneumophila* provides protection from environmental stressors [30,31] including biocides used to disinfect water systems [32,33]. Indeed, biofilms produced with *L. pneumophila* in the presence of thermotolerant amoebae allow *L. pneumophila* to persist after heat treatment [34], demonstrating that amoebae can provide a protective niche for *L. pneumophila* [30].

Figure 2. *A. castellanii* infected with *L. pneumophila* expressing green fluorescent protein. Scale bar represents 10 μm.

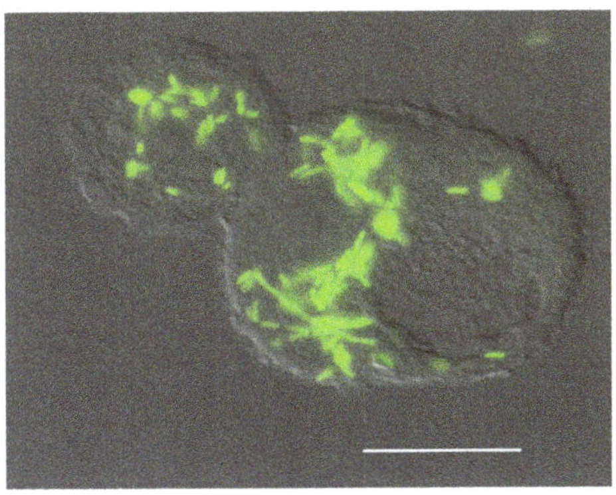

3. Physiochemical Determinants in *L. pneumophila* Biofilm Formation and Colonization

Attachment is the process of one object adhering to another, and when producing surface associated biofilms, attachment of *L. pneumophila* serves as an initial crucial step, whether it is on biotic or abiotic surfaces. Once bacteria are attached to a substratum and a biofilm is formed, the process of spreading and persisting within a new area is defined as colonization. Although *L. pneumophila* can often be found attached to various surfaces in the environment, colonization of existing biofilms in addition to attachment to abiotic substrates is determined by a wide variety of parameters (Figure 3) [35]. One important factor that governs the adherence of *L. pneumophila* in anthropogenic water systems is the composition of the surface material to which the bacteria are adhering [36]. *L. pneumophila* can adhere well to several different plastics that are commonly used in plumbing, whereas copper inhibits its attachment [36–38]. It remains unclear however whether this is due to differences in surface and *L. pneumophila* interactions or because different plumbing materials select for different pioneering species, which establishes the initial biofilm and *L. pneumophila* colonizes afterwards.

Figure 3. *L. pneumophila* (denoted by Lpn and shown in orange) replicate within environmental protozoa. Uptake within protozoan hosts is promoted by the presence of other amoebae and bacterial species such as *P. aeruginosa* (**top left**) and by environmental cues such as quorum sensing (green hexagons, **top right**). Environmental cues can also influence changes in *L. pneumophila* cell metabolism that favour biofilm production and colonization, which may occur following replication within protozoa or independently of protozoa infection (**middle**). Other microbial species such as *P. aeruginosa* (Lpn antagonistic microbial species, brown) can inhibit *L. pneumophila* colonization (**bottom left**). The presence of other microorganisms such as *K. pneumoniae* alleviates the inhibitory effect of *P. aeruginosa* (Lpn permissive microbial species, blue) and allows *L. pneumophila* to be incorporated within biofilms. *L. pneumophila* produces a surfactant (brown circles), which is toxic to other *Legionella* species (red), and may therefore prevent incorporation of these bacteria within biofilms. Physio-chemical parameters such as divalent cations (grey circles) can favour *L. pneumophila* biofilm colonization while other factors such as the presence of nanoparticles and copper (blue circles) can hinder *L. pneumophila* colonization.

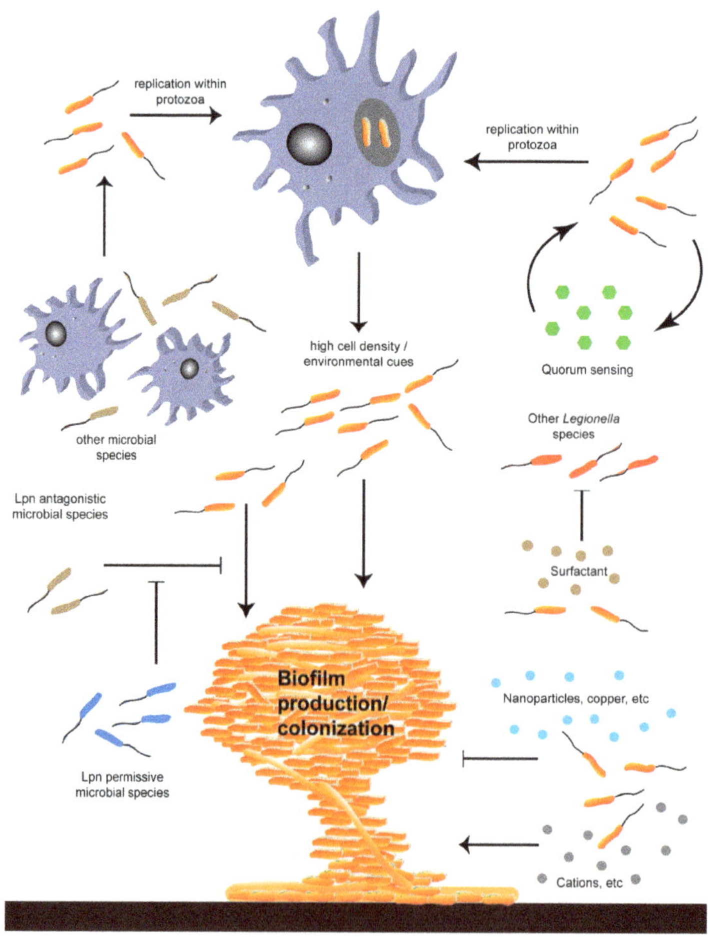

Cations are implicated in the attachment of bacteria to different substrata, and can contribute to biofouling [39]. Similarly, both calcium and magnesium were demonstrated to facilitate the attachment of *L. pneumophila* to abiotic surfaces [40]. Elevated zinc, magnesium, and manganese levels are correlated with increased *L. pneumophila* contamination and zinc increases the ability of *L. pneumophila* to bind to host cells such as human lung epithelial cells, suggesting that cations may increase the attachment of *L. pneumophila* to biotic surfaces in addition to abiotic substrata [41–43]. Interestingly as it pertains to the cation dependent attachment of *L. pneumophila*, an orthologue of the *Pseudomonas fluorescens* calcium-dependent cyclic-diGMP regulated protease LapG was identified in *L. pneumophila*. LapG regulates biofilm formation of *Pseudomonas fluorescens* by cleaving the surface adhesin LapA required for biofilm formation [44,45].

In addition to the presence of cations, the availability of carbon favours the colonization of biofilms with *L. pneumophila*, presumably because it provides nutrients for the bacteria to replicate [46]. Notably, the increase in biofilm production due to organic-carbon has only been reported at 20 °C, suggesting that carbon may only influence biofilm production at certain temperatures [47]. Temperature is also an important determinant for *L. pneumophila* biofilm colonization. Studies have shown that heating water above 55 °C can reduce the detectable amount of *L. pneumophila* in water systems, even in the presence of organic carbon sources, however this may be due to a decrease in other biofilm species which may serve as a platform for *L. pneumophila* colonization [48,49].

Static and flow conditions of water play important roles in the biofilm formation and biofilm colonization with *L. pneumophila* in water systems. Stagnation of water in distribution systems seems to favour colonization with *L. pneumophila* [50]. Moreover, Legionnaires' disease cases have been linked to stagnant water in hospital settings [51]. In accordance with these data, a constant flow in anthropogenic water can decrease the presence of *L. pneumophila* through the use of Venturi systems by preventing the attachment of the bacteria to surfaces [52]. However, biofilms in aquatic environments can persist under turbulent flow conditions [53,54] and maintain a population of *L. pneumophila* [55]. To explain the persistence of *L. pneumophila* under turbulent flow, it was proposed that the bacteria can localize to the sediment where it is less affected by turbulence [56,57]. The settling of *L. pneumophila* in environmental sediments, which was recently linked to quorum sensing, will be discussed below [58,59].

Although the majority of *L. pneumophila* biofilm formation/colonization research has been focused on determining the physiochemical parameters that allow *L. pneumophila* to colonize and form biofilms, little is known regarding the *L. pneumophila* molecular factors that contribute directly to this process. The *Legionella* collagen-like protein (Lcl) was initially identified as an adhesin required for infection of protozoa and macrophages [60]. Subsequently, Lcl was found to be an important mediator of *L. pneumophila* biofilm formation [61]. Lcl facilitates biofilm production by promoting attachment to abiotic substrates as well as cell-cell/cell-matrix interactions [62]. Type IV pili was also implicated in *L. pneumophila* biofilm colonization based in its role in adherence to protozoan cells [63]. However, a site directed type IV pili mutant was shown to colonize biofilms of other organisms as well as wild-type bacteria [64]. In addition to surface exposed adhesins, the twin arginine transport (Tat) secretion system has also been implicated in biofilm formation. Deletion of the *tatB* and *tatC* genes resulted in a significant reduction in biofilm formation, however the specific role that this secretion system plays is unknown [65].

4. Regulation of *L. pneumophila* Endogenous Factors that May Influence Biofilm Colonization

For *L. pneumophila,* as well as for other microorganisms, biofilm formation is an environmental response that can promote survival. *L. pneumophila,* like many other microorganisms, responds to environmental cues, which can greatly influence biofilm formation and or colonization (Figure 3). One important environmental prompt is iron, which has important roles in the growth of many organisms, and can influence *L. pneumophila* replication [66]. The addition of lactoferrin, an iron chelator, can directly kill *L. pneumophila* demonstrating the importance of iron in *L. pneumophila* viability [67]. Furthermore, bacterial ferrous iron transport promotes the intracellular replication of *L. pneumophila* in protozoa, which may influence multispecies biofilm colonization [68]. Iron is also required for the production of melanin and it is believed that deletion of the *lbtA* and *lbtB* genes, which encode iron siderophores prevent growth within aquatic biofilms [66]. Interestingly, although iron is essential for biofilm formation, high iron concentrations can inhibit biofilm formation, yet to date the reasons for this are unknown [8].

The ability of bacteria to monitor and respond to cell density is known as quorum sensing and it is a crucial process during biofilm production. Among quorum sensing molecules, α-hydroxy ketones (AHKs) have been identified in *L. pneumophila,* and are similar to the AHKs produced by *Vibrio cholera* [69,70]. Although there is currently no evidence that AHKs regulate *L. pneumophila* biofilm production directly, these molecules regulate a wide variety of traits that may influence *L. pneumophila* biofilm production/colonization indirectly, including virulence (infection of protozoan hosts in multispecies biofilms), extracellular filament production, and sedimentation through the *lqs* gene cluster, which encodes for the AHK synthase LqsA, the AHK sensor LqsS and the response regulator LqsR [58,71,72]. In addition to the products of this gene cluster, an orphan sensor kinase named LqsT regulates competence, a process that is correlated to biofilm formation in other species [59].

The second-messenger molecule cyclic di-GMP (c-di-GMP) is also an important signaling system that allows bacteria to respond to environmental changes [73] *L. pneumophila* has 22 predicted genes related to c-di-GMP production, degradation and/or recognition [74]. One of these genes, *lpg1057*, was found to encode an enzyme responsible for the production of cyclic di-GMP which promotes biofilm formation, and is the only c-di-GMP related gene to date found to directly influence monospecies biofilm production of *L. pneumophila* [75]. In response to amino acid starvation, the alarmone guanosine tetraphosphate (ppGpp) can also regulate *L. pneumophila* gene expression [76]. Although the ppGpp system is mainly linked to the regulation of virulence related traits, this system may indirectly affect environmental biofilm colonization by influencing *L. pneumophila*-amoeba interactions. In addition, sensitivity to ppGpp signaling requires the sigma factor RpoS [77]. RpoS in turn influences LqsR expression, suggesting that virulence related traits regulated by AHKs require multiple environmental signals [78]. In parallel with the ppGpp-RpoS regulation of virulence, downstream is the two-component system LetA/LetS [79]. The LetA/LetS system relieves the repression of virulence related genes by the RNA binding protein CsrA [80]. Despite the initially suspected roles of these transcriptional regulators in surface attachment and biofilm formation, none of the mutants lacking *rpoS*, *letA* or *csrA* were affected in biofilm formation in the *L. pneumophila* strain JR32 [26]. Of the known *L. pneumophila* sigma factors, to date only the flagellar sigma factor FliA has been implicated in the regulation of biofilm production and deletion of *fliA* results in a decrease in biofilm formation in JR32, however it is unclear what downstream or upstream factors are involved in this process [26].

Temperature was mentioned above as being an important determinant for biofilm colonization [48,49]. In addition, temperature can regulate the properties of the biofilms produced by *L. pneumophila* [81]. *In vitro*, at 37–42 °C, monospecies biofilms are mycelial mat-like and are composed of filamentous bacteria whereas biofilms produced at 25 °C are thinner and made up of rod shaped cells [81]. These findings coincide with other studies demonstrating that the filamentation of *L. pneumophila* is regulated by temperature [82]. Filamentous growth occurs in other bacterial species to increase fitness against adverse environmental conditions [83]. In turn, intracellular filamentatous *L. pneumophila* can produce progeny more efficiently than short rod forms [84]. Furthermore, the length of *L. pneumophila* cells has been linked to ppGpp signalling [80]. *In vitro*, biofilms produced at 37 °C are much more robust than at 25 °C [26], and interestingly biofilms produced at 25 °C are more adherent [81]. In addition, the production of the *L. pneumophila* type II secretion system, and type IV pili are temperature regulated, and may influence attachment at different temperatures [85].

5. The Role of Non-Protozoa Microbial Species in *L. pneumophila* Biofilm Colonization

Environmental biofilms often contain several different bacterial species [86]. These bacterial species may promote the persistence of *L. pneumophila* in biofilms, while other species inhibit *L. pneumophila*'s colonization (Figure 3). For example, *Flavobacterum breve* and cyanobacterial species can promote *L. pneumophila* growth and colonization in biofilms by providing a source of nutrients [87,88]. *In vitro*, the growth of *L. pneumophila* is necrotrophic when heat killed *Pseudomonas putida* bacteria are given as a nutrient source, however heat killed Gram-positive organisms such *Bacillus subtilis* and *Lactobacillus plantarum* did not alter the growth of *L. pneumophila*, suggesting that *L. pneumophila* is capable of replicating without the presence of protozoan species, and that necrotrophic growth of *L. pneumophila* is restricted to certain microbial species [27].

Summer seasons, which coincide with legionellosis outbreaks, favour the proliferation of *L. pneumophila* in cooling tower microbial populations while other *Legionella* species decrease in number [89]. Based on this shift in abundance it has been hypothesized that *L. pneumophila* may inhibit the growth of other *Legionella* species. In fact, *L. pneumophila* produces a surfactant secreted by the protein TolC, which is toxic to other *Legionella* species, but has no effect on *Pseudomonas aeruginosa*, *Klebsiella pneumoniae* and *Listeria monocytogenes* [90]. Therefore it is tempting to speculate that *L. pneumophila* may influence the growth of other *Legionella* species in their natural environment.

One of the most studied bacteria that can influence *L. pneumophila*'s biofilm colonization ability is *P. aeruginosa*. Although there is a body of evidence suggesting that *L. pneumophila* can coexist in biofilms with *P. aeruginosa*, these studies were performed with inoculums from natural environmental sources, which may contain several different bacterial species [11,38]. In contrast to these studies, monospecies biofilms with *P. aeruginosa* were shown to prevent *L. pneumophila* colonization [26,91]. This phenomenon may be mediated by acylhomoserine lactones (AHLs) produced by *P. aeruginosa* as these AHLs not only inhibit the growth of *L. pneumophila* but also its biofilm production [92]. Furthermore, specific AHLs produced by *P. aeruginosa* can downregulate Lcl production, which is essential for biofilm formation in *L. pneumophila* [62]. Interestingly, the *in vitro* inhibition of

L. pneumophila colonization by *P. aeruginosa* is alleviated if *K. pneumoniae* is present in the produced biofilm [91]. In fact complex multispecies biofilms that contain both *P. aeruginosa* and *K. pneumoniae* are permissive for *L. pneumophila* colonization [28]. The presence of amoeba seems to also effect whether *P. aeruginosa* is antagonistic to *L. pneumophila* colonization, as biofilms which contain both *Acanthamoeba castellanii* and *P. aeruginosa*, increase the uptake of *L. pneumophila* within *A. castellanii*, and the colonization of *L. pneumophila* in biofilms [93].

6. The Resistance of *L. pneumophila* Containing Biofilms to Biocides

There is a great interest in improving methods for disinfecting *L. pneumophila* containing biofilms because of the ongoing threat to human health posed by these organisms in anthropogenic water sources. Due to the intracellular lifestyle of *L. pneumophila* within protozoa, however, it is difficult to tease out whether the resistance of *L. pneumophila* in environmental biofilms is due to the biofilm structure, its association with amoeba or both. It is however evident that environmental *L. pneumophila* found in biofilms are extremely resilient to treatment with biocides [94]. *L. pneumophila* exposed to environmental stresses and/or found within biofilms can enter a viable but non-culturable (VBNC) state, and treatment of water systems with biocides can make *L. pneumophila* enter the VBNC state [95]. This property makes the accurate assessment of the contamination levels with *L. pneumophila* cumbersome since it requires the co-culturing of *L. pneumophila* with amoeba to lift the VBNC state [96].

Recently, nanoparticles have been suggested to be powerful tools to prevent *L. pneumophila* biofilm formation, as nanoparticles are able to disrupt *L. pneumophila*-amoebae interactions and biofilm structure [97,98]. Nanoparticles can also effectively clear *L. pneumophila* from mixed species biofilms and appear to be an attractive treatment option for disinfecting anthropogenic water sources [99]. The most common biocides used to control water-borne pathogens are generally chlorine derivatives, and chlorine derivatives are more efficacious than UV for disinfecting *L. pneumophila* [100–102]. Yet chloramine, one of the most potent chlorine derivative biocides, does not completely eradicate *L. pneumophila* from aquatic biofilms [103,104].

The location of the biofilm can also play a role in resistance to disinfection strategies. This is particularly the case for biofilms formed in sediments, which provide protection to *L. pneumophila* from UV radiation [105]. Furthermore, *L. pneumophila* bacteria grown on a solid surface are more resistant to killing by iodine than bacteria grown in broth, suggesting that there are metabolic differences between surface associated and planktonic phase bacteria [106]. This is consistent with data suggesting that sessile and planktonic *L. pneumophila* in biofilms have different gene expression profiles [8].

7. Conclusions

L. pneumophila is an environmental pathogen, and understanding the ecology of this pathogen can help to determine methods for preventing its environmental dissemination and the transmission of legionellosis. There are a multitude of factors that can influence whether *L. pneumophila* produces biofilms, and likely many more that remain to be uncovered. Although there have been significant advances in the understanding *L. pneumophila* biofilm formation and colonization in the last several years, there is much that remains unknown. The presence of other microbial species, physiochemical parameters, and *L. pneumophila* gene regulation are all factors that could potentially be exploited to prevent colonization of *L. pneumophila* in anthropogenic systems. The physiochemical parameters, which favour biofilm, formation is a topic of great interest. Research has yielded insight into factors, which could potentially limit *L. pneumophila* growth and may be useful for the prevention of legionellosis. One area of increasing interest is the role of other bacterial species in *L. pneumophila* biofilm production, and the mechanism with which certain species promote *L. pneumophila* growth while other species inhibit it. Another question that remains to be answered is to what extent the intracellular lifestyle contributes to *L. pneumophila* biofilm resistance to disinfection *in situ*. Finally, there is still much unknown about the endogenous factors that *L. pneumophila* utilizes to facilitate biofilm formation, such as what *L. pneumophila* quorum sensing systems regulate biofilm formation and what other factors are involved, (for example an extracellular matrix has been found in *L. pneumophila* monospecies biofilms but has yet to be characterized [8]). Ultimately, this research can yield valuable information that can lead to translational research for prevention and protection against *L. pneumophila* infections.

Acknowledgments

This work was supported by the Ontario Agency for Health protection and Promotion, and the Canadian Institutes of Health Research (MOP-102514).

References

1. Carratala, J.; Fernandez-Sabe, N.; Ortega, L.; Castellsague, X.; Roson, B.; Dorca, J.; Fernandez-Aguera, A.; Verdaguer, R.; Martinez, J.; Manresa, F.; *et al.* Outpatient care compared with hospitalization for community-acquired pneumonia: A randomized trial in low-risk patients. *Ann. Intern. Med.* **2005**, *142*, 165–172.
2. Yu, V.L.; Plouffe, J.F.; Pastoris, M.C.; Stout, J.E.; Schousboe, M.; Widmer, A.; Summersgill, J.; File, T.; Heath, C.M.; Paterson, D.L.; *et al.* Distribution of *Legionella* species and serogroups isolated by culture in patients with sporadic community-acquired Legionellosis: An international collaborative survey. *J. Infect. Dis.* **2002**, *186*, 127–128.
3. Bentham, R.H.; Broadbent, C.R. A model for autumn outbreaks of Legionnaires' disease associated with cooling towers, linked to system operation and size. *Epidemiol. Infect.* **1993**, *111*, 287–295.
4. Fields, B.S.; Benson, R.F.; Besser, R.E. *Legionella* and Legionnaires' disease: 25 years of investigation. *Clin. Microbiol. Rev.* **2002**, *15*, 506–526.
5. Feazel, L.M.; Baumgartner, L.K.; Peterson, K.L.; Frank, D.N.; Kirk Harris, J.; Pace, N.R. Opportunistic pathogens enriched in showerhead biofilms. *Proc. Natl. Acad. Sci. USA* **2009**, *106*, 16393–16398.
6. McEvoy, M.; Batchelor, N.; Hamilton, G.; MacDonald, A.; Faiers, M.; Sills, A.; Lee, J.; Harrison, T. A cluster of cases of Legionnaires' disease associated with exposure to a spa pool on display. *Commun. Dis. Public Health* **2000**, *3*, 43–45.

7. Wadowsky, R.M.; Yee, R.B.; Mezmar, L.; Wing, E.J.; Dowling, J.N. Hot water systems as sources of *Legionella pneumophila* in hospital and nonhospital plumbing fixtures. *Appl. Environ. Microbiol.* **1982**, *43*, 1104–1110.

8. Hindre, T.; Bruggemann, H.; Buchrieser, C.; Hechard, Y. Transcriptional profiling of *Legionella Pneumophila* biofilm cells and the influence of iron on biofilm formation. *Microbiology* **2008**, *154*, 30–41.

9. Pecastaings, S.; Berge, M.; Dubourg, K.M.; Roques, C. Sessile *Legionella pneumophila* is able to grow on surfaces and generate structured monospecies biofilms. *Biofouling* **2010**, *26*, 809–819.

10. Kwon, S.; Moon, E.; Kim, T.; Hong, S.; Park, H. Pyrosequencing demonstrated complex microbial communities in a membrane filtration system for a drinking water treatment plant. *Microbes Environ.* **2011**, *26*, 149–155.

11. Murga, R.; Forster, T.S.; Brown, E.; Pruckler, J.M.; Fields, B.S.; Donlan, R.M. Role of biofilms in the survival of *Legionella pneumophila* in a model potable-water system. *Microbiology* **2001**, *147*, 3121–3126.

12. Rowbotham, T.J. Pontiac fever, ameobae and *Legionllae*. *Lancet* **1981**, *317*, 40–41.

13. Molmeret, M.; Horn, M.; Wagner, M.; Santic, M.; Abu Kwaik, Y. Amoebae as training grounds for intracellular bacterial pathogens. *Appl. Environ. Microbiol.* **2005**, *71*, 20–28.

14. Steinert, M.; Hentschel, U.; Hacker, J. *Legionella pneumophila*: An aquatic microbe goes astray. *FEMS Microbiol. Rev.* **2002**, *26*, 149–162.

15. Yan, L.; Cerny, R.L.; Cirillo, J.D. Evidence that hsp90 is involved in the altered interactions of *Acanthamoeba castellanii* variants with bacteria. *Eukaryotic Cell* **2004**, *3*, 567–578.

16. Chaabna, Z.; Forey, F.; Reyrolle, M.; Jarraud, S.; Atlan, D.; Fontvieille, D.; Gilbert, C. Molecular diversity and high virulence of *Legionella pneumophila* strains isolated from biofilms developed within a warm spring of a thermal spa. *BMC Microbiol.* **2013**, *13*, 17.

17. Abu Khweek, A.; Fernandez Davila, N.S.; Caution, K.; Akhter, A.; Abdulrahman, B.A.; Tazi, M.; Hassan, H.; Novotny, L.A.; Bakaletz, L.O.; Amer, A.O. Biofilm-derived *Legionella pneumophila* evades the innate immune response in macrophages. *Front. Cell Infect. Microbiol.* **2013**, *3*, 18.

18. Barker, J.; Lambert, P.A.; Brown, M.R. Influence of intra-amoebic and other growth conditions on the surface properties of *Legionella pneumophila*. *Infect. Immun.* **1993**, *61*, 3503–3510.

19. Newsome, A.L.; Farone, M.B.; Berk, S.G.; Gunderson, J.H. Free living amoebae as opportunistic hosts for intracellular bacterial parasites. *J. Eukaryot. Microbiol.* **2001**, *48*, 13S–14S.

20. Rowbotham, T.J. Isolation of *Legionella pneumophila* from clinical specimens via amoebae, and the interaction of those and other isolates with amoebae. *J. Clin. Pathol.* **1983**, *36*, 978–986.

21. Skinner, A.R.; Anand, C.M.; Malic, A.; Kurtz, J.B. *Acanthamoebae* and environmental spread of *Legionella pneumophila*. *Lancet* **1983**, *2*, 289–290.

22. Taylor, M.; Ross, K.; Bentham, R. *Legionella*, protozoa, and biofilms: Interactions within complex microbial systems. *Microb. Ecol.* **2009**, *58*, 538–547.

23. Valster, R.M.; Wullings, B.A.; van der Kooij, D. Detection of protozoan hosts for *Legionella pneumophila* in engineered water systems by using a biofilm batch test. *Appl. Environ. Microbiol.* **2010**, *76*, 7144–7153.

24. Valster, R.M.; Wullings, B.A.; van den Berg, R.; van der Kooij, D. Relationships between free-living protozoa, cultivable *Legionella* spp., and water quality characteristics in three drinking water supplies in the caribbean. *Appl. Environ. Microbiol.* **2011**, *77*, 7321–7328.
25. Liu, R.; Yu, Z.; Guo, H.; Liu, M.; Zhang, H.; Yang, M. Pyrosequencing analysis of eukaryotic and bacterial communities in faucet biofilms. *Sci. Total Environ.* **2012**, *435–436*, 124–131.
26. Mampel, J.; Spirig, T.; Weber, S.S.; Haagensen, J.A.; Molin, S.; Hilbi, H. Planktonic replication is essential for biofilm formation by *Legionella pneumophila* in a complex medium under static and dynamic flow conditions. *Appl. Environ. Microbiol.* **2006**, *72*, 2885–2895.
27. Temmerman, R.; Vervaeren, H.; Noseda, B.; Boon, N.; Verstraete, W. Necrotrophic growth of *Legionella pneumophila*. *Appl. Environ. Microbiol.* **2006**, *72*, 4323–4328.
28. Declerck, P.; Behets, J.; van Hoef, V.; Ollevier, F. Replication of *Legionella pneumophila* in floating biofilms. *Curr. Microbiol.* **2007**, *55*, 435–440.
29. Hsu, B.; Huang, C.; Chen, J.; Chen, N.; Huang, J. Comparison of potentially pathogenic free-living amoeba hosts by *Legionella* spp. in substrate-associated biofilms and floating biofilms from spring environments. *Water Res.* **2011**, *45*, 5171–5183.
30. Atlas, R.M. *Legionella*: From environmental habitats to disease pathology, detection and control. *Environ. Microbiol.* **1999**, *1*, 283–293.
31. Hoffman, P.S. Invasion of eukaryotic cells by *Legionella pneumophila*: A common strategy for all hosts? *Can. J. Infect. Dis.* **1997**, *8*, 139–146.
32. Donlan, R.M.; Forster, T.; Murga, R.; Brown, E.; Lucas, C.; Carpenter, J.; Fields, B. *Legionella pneumophila* associated with the protozoan *Hartmannella vermiformis* in a model multi-species biofilm has reduced susceptibility to disinfectants. *Biofouling* **2005**, *21*, 1–7.
33. Thomas, V.; Bouchez, T.; Nicolas, V.; Robert, S.; Loret, J.F.; Levi, Y. Amoebae in domestic water systems: Resistance to disinfection treatments and implication in *Legionella* persistence. *J. Appl. Microbiol.* **2004**, *97*, 950–963.
34. Storey, M.V.; Winiecka-Krusnell, J.; Ashbolt, N.J.; Stenstrom, T. The efficacy of heat and chlorine treatment against thermotolerant *Acanthamoebae* and *Legionellae*. *Scand. J. Infect. Dis.* **2004**, *36*, 656–662.
35. Wright, J.B.; Ruseska, I.; Athar, M.A.; Corbett, S.; Costerton, J.W. *Legionella pneumophila* grows adherent to surfaces *in vitro* and *in situ*. *Infect. Control Hosp. Epidemiol.* **1989**, *10*, 408–415.
36. Rogers, J.; Dowsett, A.B.; Dennis, P.J.; Lee, J.V.; Keevil, C.W. Influence of plumbing materials on biofilm formation and growth of *Legionella pneumophila* in potable water systems. *Appl. Environ. Microbiol.* **1994**, *60*, 1842–1851.
37. Rogers, J.; Dowsett, A.B.; Dennis, P.J.; Lee, J.V.; Keevil, C.W. Influence of temperature and plumbing material selection on biofilm formation and growth of *Legionella pneumophila* in a model potable water system containing complex microbial flora. *Appl. Environ. Microbiol.* **1994**, *60*, 1585–1592.
38. Moritz, M.M.; Flemming, H.; Wingender, J. Integration of *Pseudomonas aeruginosa* and *Legionella pneumophila* in drinking water biofilms grown on domestic plumbing materials. *Int. J. Hyg. Environ. Health* **2010**, *213*, 190–197.

39. Geesey, G.G.; Wigglesworth-Cooksey, B.; Cooksey, K.E. Influence of calcium and other cations on surface adhesion of bacteria and diatoms: A review. *Biofouling* **2000**, *15*, 195–205.
40. Koubar, M.; Rodier, M.H.; Frere, J. Involvement of minerals in adherence of *Legionella pneumophila* to surfaces. *Curr. Microbiol.* **2013**, *66*, 437–442.
41. Yaradou, D.F.; Raze, D.; Ginevra, C.; Ader, F.; Doleans-Jordheim, A.; Vandenesch, F.; Menozzi, F.D.; Etienne, J.; Jarraud, S. Zinc-dependent cytoadherence of *Legionella pneumophila* to human alveolar epithelial cells *in vitro*. *Microb. Pathog.* **2007**, *43*, 234–242.
42. Bargellini, A.; Marchesi, I.; Righi, E.; Ferrari, A.; Cencetti, S.; Borella, P.; Rovesti, S. Parameters predictive of *Legionella* contamination in hot water systems: Association with trace elements and heterotrophic plate counts. *Water Res.* **2011**, *45*, 2315–2321.
43. Borella, P.; Montagna, M.T.; Romano-Spica, V.; Stampi, S.; Stancanelli, G.; Triassi, M.; Marchesi, I.; Bargellini, A.; Neglia, R.; Paglionico, N.; *et al.* Relationship between mineral content of domestic hot water and microbial contamination. *J. Trace Elem. Med. Biol.* **2003**, *17*, 37–43.
44. Boyd, C.D.; Chatterjee, D.; Sondermann, H.; O'Toole, G.A. LapG, required for modulating biofilm formation by *Pseudomonas fluorescens* Pf0-1, is a calcium-dependent protease. *J. Bacteriol.* **2012**, *194*, 4406–4414.
45. Chatterjee, D.; Boyd, C.D.; O'Toole, G.A.; Sondermann, H. Structural characterization of a conserved, calcium-dependent periplasmic protease from *Legionella pneumophila*. *J. Bacteriol.* **2012**, *194*, 4415–4425.
46. Giao, M.S.; Wilks, S.A.; Azevedo, N.F.; Vieira, M.J.; Keevil, C.W. Comparison between standard culture and peptide nucleic acid 16S rRNA hybridization quantification to study the influence of physico-chemical parameters on *Legionella pneumophila* survival in drinking water biofilms. *Biofouling* **2009**, *25*, 343–351.
47. Pang, C.M.; Liu, W. Biological filtration limits carbon availability and affects downstream biofilm formation and community structure. *Appl. Environ. Microbiol.* **2006**, *72*, 5702–5712.
48. Van der Kooij, D.K.; Veenendaal, H.R.; Scheffer, W.J.H. Biofilm formation and multiplication of *Legionella* in a model warm water system with pipes of copper, stainless steel and cross-linked polyethylene. *Water Res.* **2005**, *39*, 2789–2798.
49. Martinelli, F.; Caruso, A.; Moschini, L.; Turano, A.; Scarcella, C.; Speziani, F. A comparison of *Legionella pneumophila* occurrence in hot water tanks and instantaneous devices in domestic, nosocomial, and community environments. *Curr. Microbiol.* **2000**, *41*, 374–376.
50. Ciesielski, C.A.; Blaser, M.J.; Wang, W.L. Role of stagnation and obstruction of water flow in isolation of *Legionella pneumophila* from hospital plumbing. *Appl. Environ. Microbiol.* **1984**, *48*, 984–987.
51. Fisher-Hoch, S.; Bartlett, C.L.; Tobin, J.O.; Gillett, M.B.; Nelson, A.M.; Pritchard, J.E.; Smith, M.G.; Swann, R.A.; Talbot, J.M.; Thomas, J.A. Investigation and control of an outbreaks of Legionnaires' disease in a district general hospital. *Lancet* **1981**, *1*, 932–936.
52. Makin, T.; Pride, M. Venturi system could be major breakthrough. *Health Estate* **2010**, *64*, 27–33.
53. Storey, M.V.; Langmark, J.; Ashbolt, N.J.; Stenstrom, T.A. The fate of *Legionella*e within distribution pipe biofilms: Measurement of their persistence, inactivation and detachment. *Water Sci. Technol.* **2004**, *49*, 269–275.

54. Lehtola, M.J.; Torvinen, E.; Kusnetsov, J.; Pitkanen, T.; Maunula, L.; von Bonsdorff, C.; Martikainen, P.J.; Wilks, S.A.; Keevil, C.W.; Miettinen, I.T. Survival of *Mycobacterium avium*, *Legionella pneumophila*, *Escherichia coli*, and caliciviruses in drinking water-associated biofilms grown under high-shear turbulent flow. *Appl. Environ. Microbiol.* **2007**, *73*, 2854–2859.
55. Liu, Z.; Lin, Y.E.; Stout, J.E.; Hwang, C.C.; Vidic, R.D.; Yu, V.L. Effect of flow regimes on the presence of *Legionella* within the biofilm of a model plumbing system. *J. Appl. Microbiol.* **2006**, *101*, 437–442.
56. Fliermans, C.B.; Cherry, W.B.; Orrison, L.H.; Smith, S.J.; Tison, D.L.; Pope, D.H. Ecological distribution of *Legionella pneumophila*. *Appl. Environ. Microbiol.* **1981**, *41*, 9–16.
57. Stout, J.E.; Yu, V.L.; Best, M.G. Ecology of *Legionella pneumophila* within water distribution systems. *Appl. Environ. Microbiol.* **1985**, *49*, 221–228.
58. Tiaden, A.; Spirig, T.; Sahr, T.; Walti, M.A.; Boucke, K.; Buchrieser, C.; Hilbi, H. The autoinducer synthase LqsA and putative sensor kinase lqss regulate phagocyte interactions, extracellular filaments and a genomic island of *Legionella pneumophila*. *Environ. Microbiol.* **2010**, *12*, 1243–1259.
59. Kessler, A.; Schell, U.; Sahr, T.; Tiaden, A.; Harrison, C.; Buchrieser, C.; Hilbi, H. The *Legionella pneumophila* orphan sensor kinase LqsT regulates competence and pathogen-host interactions as a component of the LAI-1 circuit. *Environ. Microbiol.* **2013**, *15*, 646–662.
60. Vandersmissen, L.; de Buck, E.; Saels, V.; Coil, D.A.; Anne, J. A *Legionella pneumophila* collagen-like protein encoded by a gene with a variable number of tandem repeats is involved in the adherence and invasion of host cells. *FEMS Microbiol. Lett.* **2010**, *306*, 168–176.
61. Duncan, C.; Prashar, A.; So, J.; Tang, P.; Low, D.E.; Terebiznik, M.; Guyard, C. Lcl of *Legionella pneumophila* is an immunogenic GAG binding adhesin that promotes interactions with lung epithelial cells and plays a crucial role in biofilm formation. *Infect. Immun.* **2011**, *6*, 2168–2181.
62. Mallegol, J.; Duncan, C.; Prashar, A.; So, J.; Low, D.E.; Terebeznik, M.; Guyard, C. Essential roles and regulation of the *Legionella pneumophila* collagen-like adhesin during biofilm formation. *PLoS One* **2012**, *7*, e46462.
63. Stone, B.J.; Kwaik, Y.A. Expression of multiple pili by *Legionella pneumophila*: Identification and characterization of a type IV pilin gene and its role in adherence to mammalian and protozoan cells. *Infect. Immun.* **1998**, *66*, 1768–1775.
64. Lucas, C.E.; Brown, E.; Fields, B.S. Type IV Pili and Type II secretion play a limited role in *Legionella pneumophila* biofilm colonization and retention. *Microbiology* **2006**, *152*, 3569–3573.
65. De Buck, E.; Maes, L.; Meyen, E.; van Mellaert, L.; Geukens, N.; Anne, J.; Lammertyn, E. *Legionella pneumophila* Philadelphia-1 tatB and tatC affect intracellular replication and biofilm formation. *Biochem. Biophys. Res. Commun.* **2005**, *331*, 1413–1420.
66. Cianciotto, N.P. Iron acquisition by *Legionella pneumophila*. *Biometals* **2007**, *20*, 323–331.
67. Orsi, N. The antimicrobial activity of lactoferrin: Current status and perspectives. *Biometals* **2004**, *17*, 189–196.
68. Robey, M.; Cianciotto, N.P. *Legionella pneumophila feoAB* promotes ferrous iron uptake and intracellular infection. *Infect. Immun.* **2002**, *70*, 5659–5669.

69. Tiaden, A.; Hilbi, H. A-hydroxyketone synthesis and sensing by *Legionella* and *Vibrio*. *Sensors* **2012**, *12*, 2899–2919.
70. Spirig, T.; Tiaden, A.; Kiefer, P.; Buchrieser, C.; Vorholt, J.A.; Hilbi, H. The *Legionella* autoinducer synthase LqsA produces an α-hydroxyketone signaling molecule. *J. Biol. Chem.* **2008**, *283*, 18113–18123.
71. Tiaden, A.; Spirig, T.; Hilbi, H. Bacterial gene regulation by α-hydroxyketone signaling. *Trends Microbiol.* **2010**, *18*, 288–297.
72. Tiaden, A.; Spirig, T.; Carranza, P.; Bruggemann, H.; Riedel, K.; Eberl, L.; Buchrieser, C.; Hilbi, H. Synergistic contribution of the *Legionella pneumophila Lqs* genes to pathogen-host interactions. *J. Bacteriol.* **2008**, *190*, 7532–7547.
73. McDonough, K.A.; Rodriguez, A. The myriad roles of cyclic AMP in microbial pathogens: From signal to sword. *Nat. Rev. Microbiol.* **2011**, *10*, 27–38.
74. Levi, A.; Folcher, M.; Jenal, U.; Shuman, H.A. Cyclic diguanylate signaling proteins control intracellular growth of *Legionella pneumophila*. *mBio* **2011**, *2*, e00316-10.
75. Carlson, H.K.; Vance, R.E.; Marletta, M.A. H-NOX regulation of c-di-GMP metabolism and biofilm formation in *Legionella pneumophila*. *Mol. Microbiol.* **2010**, *77*, 930–942.
76. Hammer, B.K.; Swanson, M.S. Co-ordination of *Legionella pneumophila* virulence with entry into stationary phase by ppGpp. *Mol. Microbiol.* **1999**, *33*, 721–731.
77. Bachman, M.A.; Swanson, M.S. RpoS co-operates with other factors to induce *Legionella pneumophila* virulence in the stationary phase. *Mol. Microbiol.* **2001**, *40*, 1201–1214.
78. Tiaden, A.; Spirig, T.; Weber, S.S.; Bruggemann, H.; Bosshard, R.; Buchrieser, C.; Hilbi, H. The *Legionella pneumophila* response regulator LqsR promotes host cell interactions as an element of the virulence regulatory network controlled by RpoS and LetA. *Cell. Microbiol.* **2007**, *9*, 2903–2920.
79. Hammer, B.K.; Tateda, E.S.; Swanson, M.S. A two-component regulator induces the transmission phenotype of stationary-phase *Legionella pneumophila*. *Mol. Microbiol.* **2002**, *44*, 107–118.
80. Molofsky, A.B.; Swanson, M.S. *Legionella pneumophila* CsrA is a pivotal repressor of transmission traits and activator of replication. *Mol. Microbiol.* **2003**, *50*, 445–461.
81. Piao, Z.; Sze, C.C.; Barysheva, O.; Iida, K.; Yoshida, S. Temperature-regulated formation of mycelial mat-like biofilms by *Legionella pneumophila*. *Appl. Environ. Microbiol.* **2006**, *72*, 1613–1622.
82. Konishi, T.; Yamashiro, T.; Koide, M.; Nishizono, A. Influence of temperature on growth of *Legionella pneumophila* biofilm determined by precise temperature gradient incubator. *J. Biosci. Bioeng.* **2006**, *101*, 478–484.
83. Justice, S.S.; Hunstad, D.A.; Cegelski, L.; Hultgren, S.J. Morphological plasticity as a bacterial survival strategy. *Nat. Rev. Microbiol.* **2008**, *6*, 162–168.
84. Prashar, A.; Bhatia, S.; Tabatabaeiyazdi, Z.; Duncan, C.; Garduno, R.A.; Tang, P.; Low, D.E.; Guyard, C.; Terebiznik, M.R. Mechanism of invasion of lung epithelial cells by filamentous *Legionella pneumophila*. *Cell. Microbiol.* **2012**, *14*, 1632–1655.

85. Liles, M.R.; Viswanathan, V.K.; Cianciotto, N.P. Identification and temperature regulation of *Legionella pneumophila* genes involved in type IV pilus biogenesis and type II protein secretion. *Infect. Immun.* **1998**, *66*, 1776–1782.
86. Messi, P.; Anacarso, I.; Bargellini, A.; Bondi, M.; Marchesi, I.; de Niederhausern, S.; Borella, P. Ecological behaviour of three serogroups of *Legionella pneumophila* within a model plumbing system. *Biofouling* **2011**, *27*, 165.
87. Tison, D.L.; Pope, D.H.; Cherry, W.B.; Fliermans, C.B. Growth of *Legionella pneumophila* in association with blue-green algae (Cyanobacteria). *Appl. Environ. Microbiol.* **1980**, *39*, 456–459.
88. Wadowsky, R.M.; Yee, R.B. Satellite growth of *Legionella pneumophila* with an environmental isolate of *Flavobacterium breve. Appl. Environ. Microbiol.* **1983**, *46*, 1447–1449.
89. Wery, N.; Bru-Adan, V.; Minervini, C.; Delgenes, J.; Garrelly, L.; Godon, J. Dynamics of *Legionella* spp. and bacterial populations during the proliferation of L. *pneumophila* in a cooling tower facility. *Appl. Environ. Microbiol.* **2008**, *74*, 3030–3037.
90. Stewart, C.R.; Burnside, D.M.; Cianciotto, N.P. The surfactant of *Legionella pneumophila* is secreted in a TolC-dependent manner and is antagonistic toward other *Legionella* species. *J. Bacteriol.* **2011**, *193*, 5971–5984.
91. Stewart, C.R.; Muthye, V.; Cianciotto, N.P. *Legionella pneumophila* persists within biofilms formed by *Klebsiella pneumoniae*, *Flavobacterium* sp., and *Pseudomonas fluorescens* under dynamic flow conditions. *PLoS One* **2012**, *7*, e50560.
92. Kimura, S.; Tateda, K.; Ishii, Y.; Horikawa, M.; Miyairi, S.; Gotoh, N.; Ishiguro, M.; Yamaguchi, K. *Pseudomonas aeruginosa* las quorum sensing autoinducer suppresses growth and biofilm production in *Legionella* species. *Microbiology* **2009**, *155*, 1934–1939.
93. Declerck, P.; Behets, J.; Delaedt, Y.; Margineanu, A.; Lammertyn, E.; Ollevier, F. Impact of non-*Legionella* bacteria on the uptake and intracellular replication of *Legionella pneumophila* in *Acanthamoeba castellanii* and *Naegleria lovaniensis*. *Microb. Ecol.* **2005**, *50*, 536–549.
94. Emtiazi, F.; Schwartz, T.; Marten, S.M.; Krolla-Sidenstein, P.; Obst, U. Investigation of natural biofilms formed during the production of drinking water from surface water embankment filtration. *Water Res.* **2004**, *38*, 1197–1206.
95. Giao, M.S.; Wilks, S.; Azevedo, N.F.; Vieira, M.J.; Keevil, C.W. Incorporation of natural uncultivable *Legionella pneumophila* into potable water biofilms provides a protective niche against chlorination stress. *Biofouling* **2009**, *25*, 335–341.
96. Alleron, L.; Merlet, N.; Lacombe, C.; Frere, J. Long-term survival of *Legionella pneumophila* in the viable but nonculturable state after monochloramine treatment. *Curr. Microbiol.* **2008**, *57*, 497–502.
97. Raftery, T.D.; Lindler, H.; McNealy, T.L. Altered host cell–bacteria interaction due to nanoparticle interaction with a bacterial biofilm. *Microb. Ecol.* **2013**, *65*, 496–503.
98. Stojak, A.R.; Raftery, T.; Klaine, S.J.; McNealy, T.L. Morphological responses of *Legionella pneumophila* biofilm to nanoparticle exposure. *Nanotoxicology* **2011**, *5*, 730–742.
99. Rogers, J.; Dowsett, A.B.; Keevil, C.W. A paint incorporating silver to control mixed biofilms containing *Legionella pneumophila*. *J. Ind. Microbiol.* **1995**, *15*, 377–383.

100. Liu, Z.; Stout, J.E.; Boldin, M.; Rugh, J.; Diven, W.F.; Yu, V.L. Intermittent use of copper-silver ionization for *Legionella* control in water distribution systems: A potential option in buildings housing individuals at low risk of infection. *Clin. Infect Dis.* **1998**, *26*, 138–140.
101. Schwartz, T.; Hoffmann, S.; Obst, U. Formation of natural biofilms during chlorine dioxide and u.v. disinfection in a public drinking water distribution system. *J. Appl. Microbiol.* **2003**, *95*, 591–601.
102. Walker, J.T.; Mackerness, C.W.; Mallon, D.; Makin, T.; Williets, T.; Keevil, C.W. Control of *Legionella pneumophila* in a hospital water system by chlorine dioxide. *J. Ind. Microbiol.* **1995**, *15*, 384–390.
103. Pryor, M.; Springthorpe, S.; Riffard, S.; Brooks, T.; Huo, Y.; Davis, G.; Sattar, S.A. Investigation of opportunistic pathogens in municipal drinking water under different supply and treatment regimes. *Water Sci. Technol.* **2004**, *50*, 83–90.
104. Williams, M.M.; Braun-Howland, E. Growth of *Escherichia coli* in model distribution system biofilms exposed to hypochlorous acid or monochloramine. *Appl. Environ. Microbiol.* **2003**, *69*, 5463–5471.
105. Kusnetsov, J.M.; Keskitalo, P.J.; Ahonen, H.E.; Tulkki, A.I.; Miettinen, I.T.; Martikainen, P.J. Growth of *Legionella* and other heterotrophic bacteria in a circulating cooling water system exposed to ultraviolet irradiation. *J. Appl. Microbiol.* **1994**, *77*, 461–466.
106. Cargill, K.L.; Pyle, B.H.; Sauer, R.L.; McFeters, G.A. Effects of culture conditions and biofilm formation on the iodine susceptibility of *Legionella pneumophila*. *Can. J. Microbiol.* **1992**, *38*, 423–429.

6

Environmental Stimuli Shape Biofilm Formation and the Virulence of Periodontal Pathogens

Marja T. Pöllänen [1,†,*], Annamari Paino [2] and Riikka Ihalin [2,†]

[1] Institute of Dentistry, University of Turku, FI-20014 Turku, Finland

[2] Department of Biochemistry and Food Chemistry, University of Turku, FI-20014 Turku, Finland; E-Mails: aepain@utu.fi (A.P.); riikka.ihalin@utu.fi (R.I.)

† These authors contributed equally to this work.

* Author to whom correspondence should be addressed; E-Mail: marpol@utu.fi;

Abstract: Periodontitis is a common inflammatory disease affecting the tooth-supporting structures. It is initiated by bacteria growing as a biofilm at the gingival margin, and communication of the biofilms differs in health and disease. The bacterial composition of periodontitis-associated biofilms has been well documented and is under continual investigation. However, the roles of several host response and inflammation driven environmental stimuli on biofilm formation is not well understood. This review article addresses the effects of environmental factors such as pH, temperature, cytokines, hormones, and oxidative stress on periodontal biofilm formation and bacterial virulence.

Keywords: periodontal pathogens; environmental stimuli; biofilm EPS; virulence factors

---

1. Introduction

Periodontitis is a common disease affecting the tooth-supporting structures of millions of people worldwide, and it is a multifactorial disease initiated by bacteria growing as a biofilm at the gingival margin. Periodontal biofilms are diverse, and the nature and communication of these biofilms differs in health and disease. The bacterial composition of periodontitis-associated biofilms has been well documented and is under constant analysis. Several key pathogens have been identified: *Porphyromonas gingivalis*, *Treponema denticola*, *Tannerella forsythia*, *Aggregatibacter actinomycetemcomitans* and more recently also *Filifactor alocis*, *Staphylococcus aureus* and the genus *Desulfobulbus* [1–4]. In addition to bacteria, viruses are commonly detected from periodontal lesions and subgingival plaque samples of patients with aggressive periodontitis [5,6]. Furthermore, the association of the biofilm-forming yeast *Candida albicans* has been detected in approximately 50% of severe chronic periodontitis patients and in only 15% of subgingival samples isolated from healthy patients [7]. Periodontal biofilm formation is a stepwise and continuous process. In the initial phase, the gram-positive and aerobic bacteria dominate. Later, the gram-negative anaerobic periodontopathogens increase in the biofilm. Clinically, the course of periodontitis leads to increased subgingival inflammation and formation of periodontal pockets. The subgingival environment is ideal for the periodontopathogens, being alkaline, hemin and protein rich with various cytokines and hormones. The bacterial cells in biofilm are surrounded by extracellular polymeric substance (EPS), which is composed of polysaccharides, proteins and extracellular DNA, and may account for as much as 90% of the total mass of the biofilm [8–10]. The EPS protects pathogens from host defence cells, such as macrophages, and humoural immune defence factors, such as antibodies and complement, as well as antibiotics. In addition, the high bacterial cell densities in biofilm enable small molecule mediated inter- and intra-species crosstalk, *i.e.*, quorum sensing, which itself may regulate virulence gene expression via pathogens. Moreover, both the host response and the environmental factors in periodontitis can affect the biofilm formation and virulence gene expression of periodontal pathogens (Figure 1). We discuss this crosstalk network of molecular interactions in this review article.

Figure 1. Interactions between periodontal pathogens and host in the subgingival environment.

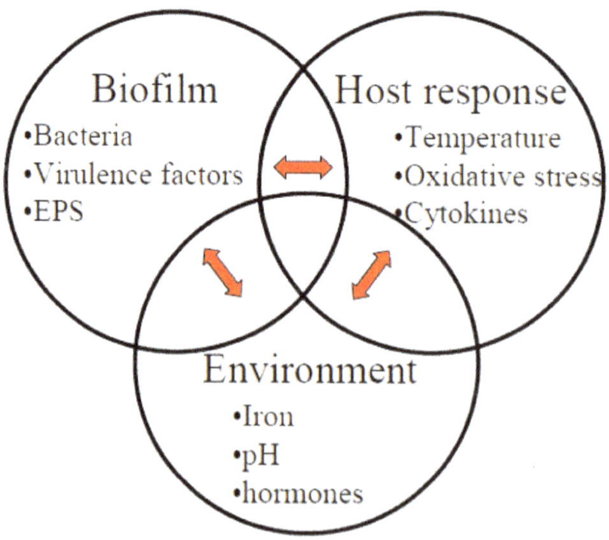

2. Host Inflammatory Reaction-Related Stimuli

The oral environment is an ideal, nutrient-rich, warm and growth-promoting place for bacteria to form communities. This first part of the digestive tract offers the bacteria perfect non-shedding surfaces of the teeth to attach and multiply. The bacterial biofilm on the tooth surface activates both the innate and adaptive host responses, which, in turn, have an effect on the biofilm. The first-line host defence is initiated by the polymorphonuclear (PMN) leukocytes that are recruited to the site by chemotactic factors, e.g., the gradient of IL-8 and ICAM-1 in the junctional epithelium [11]. In the subgingival area, the temperature increases, and the inflammatory reaction causes the release of reactive oxygen species (ROS), such as hydrogen peroxide (H_2O_2) and superoxide ($O_2^{\bullet -}$), and cytokines from the host cells to destroy the bacteria. However, the bacterial biofilm is a community, and not only the bacteria but also their virulence and the EPS surrounding them may be affected and changed by the changes produced during the inflammation process.

2.1. Temperature

Changes in the environmental temperature have an effect on various functions and virulence of diverse microbial species. Bacteria can sense changes in temperature by proteins, e.g., transcriptional regulators, kinases (e.g., histidine kinase in association with a cytoplasmic response regulator) and chaperones, and via their membrane lipids (for a review see [12]). In the periodontal pocket, an approximately 2 °C increase in the local subgingival temperature has been reported in diseased sites compared with healthy sites [13]. The temperature increase is a host defence mechanism that triggers virulence and heat shock gene expression in bacteria in response [12]. However, the temperature increase also affects the attachment of bacteria, coaggregation of bacteria, and protease production [14–17]. *P. gingivalis* has displayed decreased expression of proteases and down-regulation of the genes coding fimbrial proteins in response to temperature elevation [15,16]. Furthermore, the structure of lipid-A in *P. gingivalis* also appears to be affected by temperature. In elevated temperature environments, increasing amounts of monophosphorylated penta-acylated lipid A are expressed [18]. This lipid A form in turn appears to be a more potent activator of the host Toll-like receptor 4 (TLR4) and further renders the bacteria more susceptible to host defensins [18]. However, the net effect of the temperature increase seems to favour the periodontal pathogens in subgingival biofilms because increased proportions of *Prevotella intermedia*, *P. gingivalis* and *A. actinomycetemcomitans* have been reported in sites with elevated temperatures [19].

2.2. Oxidative Stress

ROS generation, as a part of the host defence mechanism or from the initial *Streptococcal* colonisers in the biofilm, induces oxidative stress in other biofilm bacteria and especially in the anaerobic species of the community. The anaerobic bacteria sense the levels of ROS by transcriptional regulators such as OxyR, PerR and OhrR. Superoxide dismutases, alkyl hydroperoxide reductase (AhpC) and catalases are produced by the bacteria for the detoxification of the ROS [20,21]. In the periodontal biofilm community, *F. nucleatum* is important as an intermediate coloniser between the early *Streptococcal* and late anaerobic periodontopathogenic colonisers. *F. nucleatum* appears to facilitate the survival of other anaerobic bacteria in the biofilm [22]. Although an obligate anaerobe, *F. nucleatum* can adapt to oxidative stress in the biofilm and even increase in number in aerobic conditions for up to 2 days [23,24]. The response of *F. nucleatum* to oxidative stress appears to be mediated by the AhpC redox system [25]. Furthermore, the oxidative stress appears to also alter the carbohydrate metabolism of *F. nucleatum* by modifying or increasing the intracellular concentrations of glycolytic enzymes, thereby decreasing ATP production. Increases in the chaperone proteins ClpB and DnaK, the heat shock protein HtpG and the transcription repressor HrcA in *F. nucleatum* in oxidative stress conditions appear to be aimed at diminishing the harmful effects of ROS [25]. In addition, the periodontopathogens have developed strategies to adapt to oxidative stress. For example, *P. gingivalis* produces superoxide dismutase, AhpC, rubrerythrin (rbr), heat shock proteins (HtpG) and chaperons GroEL and DnaK when challenged with oxidative stress [26]. Similarly, elevated inflammatory temperatures upregulate the expression of superoxide dismutase in *P. gingivalis* [27]. Furthermore, the expression of several genes with unknown functions has been observed to be altered in *P. gingivalis* under oxidative stress [28].

2.3. Inflammatory Cytokines

In periodontitis, during the active phases of the tissue destruction, the periodontal tissue and gingival fluid typically contain high levels of proinflammatory mediators such as interleukin (IL)-1β [3,29,30]. This overexpression of IL-1β results in host tissue damage, which is characteristic for periodontal disease and aimed at eliminating pathogens together with the regulation of the immune response [31–33]. In periodontium, cytokines are released by first-line human defence cells after identification of periodontopathogens and bacterial products. For instance, gingival epithelial cells and fibroblasts can secrete IL-1β after *P. gingivalis* and *A. actinomycetemcomitans* infection or after being cultured in the presence of *Treponema denticola* lipooligosaccharides [34,35]. Oral bacteria-derived compounds, such as leucotoxins and lipopolysaccharides of *A. actinomycetemcomitans*, induce IL-1β secretion in macrophages [36,37]. Macrophages and monocytes are considered the main producers of IL-1β during inflammation [38–40]. As a key conductor of the inflammatory response, IL-1β induces the release of other proinflammatory mediators [34,41]. Thus, the release of IL-6 and IL-8 from gingival epithelial cells during periodontal infection can be hindered by blocking induction of IL-1β induction [34]. Tumour necrosis factor-alpha (TNF-α), IL-6 and IL-8 are cytokines commonly released by epithelial cells after bacterial infection [42–45]. Recent studies suggest that the gingival cell response differs when infected with oral biofilm [46] or with multispecies biofilms ([47]; reviewed in [48]) as compared with planktonic bacterial infection.

Human IL-1β, crucial for the host battle against pathogens, might also be sensed by host-colonising microbes. The first evidence of bacterial response against cytokines was observed in studies with virulent *Escherichia coli* strains over 20 years ago. It was shown that *E. coli* cells can bind IL-1β and that the growth of these strains increased after IL-1β exposure, whereas treatment with IL-4 or TNF-α was ineffective. Moreover, the use of decoy receptor IL-Ra for IL-1β reversed the growth-promoting effect in *E. coli.* [49] Later, it was demonstrated that other bacterial species such as *Staphylococcus aureus, Pseudomonas aeruginosa and Acinetobacter* spp. can alter their growth properties as a consequence of exposure to IL-1β, IL-6 or TNF-α [50].

Gram-positive *S. aureus* has been a central model species for studies of the role of IL-1β in pathogen-cytokine crosstalk. Surprisingly, *S. aureus*, which is most often associated with nasal passages and skin, was also identified from subgingival plaques in 60% of aggressive periodontitis patients belonging to a group of non-smokers [4]. According to an *in vitro* experiment, the growth of

S. aureus biofilms increases when cultured in the presence of IL-1β [51,52]. The growth enhancement was also observed with two linear peptide fragments (<5 kDa) of human IL-1β [53]. In addition to growth enhancement, the cytokine modulated the gene expression of *S. aureus* biofilms. The cytokine decreased the gene expression of some toxin-encoding genes and increased the expression of host tissue-attachment responsible genes [54].

The hypothesis concerning the capability of the bacteria to specifically bind IL-1β was strengthened by the characterisation of a specific bacterial outer membrane receptor for IL-1β in gram-negative *Yersinia pestis*. The IL-1β-interacting protein is known as a capsule antigen F1 assembly protein (Caf1A), which contributes to capsule antigen (Caf1) transportation across the outer membrane [55]. Interestingly, Caf1 displays 28% sequence homology with human IL-Ra [56]. The second bacterial cytokine receptor, the outer membrane protein OprF, which binds only human interferon-γ, was initially observed in gram-negative *P. aeruginosa.* As a response to cytokine binding, *P. aeruginosa* increased lectin-encoding gene (*lecA)* expression in a quorum sensing dependent manner [57]. The hydrophobic galactose-binding lectin localised in the EPS of *P. aeruginosa* biofilms contributes to species biofilm formation and endothelial cell adherence [58].

Cytokines, such as IL-1α, IL-1β, IL-2, IL-3, IL-4, IL6, IL-7 [59] and TNF [60], use their carbohydrate-binding domains to recognise specific oligosaccharide ligands (reviewed in [61]). The receptor-binding site of IL-1 associates with its cognate receptor, and the second domain, localised opposite to the receptor binding sites, interacts with carbohydrate. For instance, IL-1α and IL-β have different carbohydrate binding activities. Whereas IL-1α binds N-glycan with two α-2-3-linked sialic acid residues, IL-1β recognises the α2-3-sialylated β-galactosyl-ceramides, which have very long and unusual long-chain bases [62]. Moreover, some lipooligosaccharides of *Haemophilus* species are sialylated, and they encode sialyltransferases [63,64].

A. actinomycetemcomitans is the only major periodontopathogen to our knowledge that has been shown to both sense and bind IL-1β. The clinical strains of the species display a physiological response to cytokines by decreasing their metabolism and by increasing their biofilm mass [65]. *A. actinomycetemcomitans* biofilms co-cultured with an organotypic gingival mucosa model bind IL-1β, but the use of antibiotics during co-culturing inhibits IL-1β binding [66]. Moreover, the species appears to uptake IL-1β as the cytokine has been detected in the intracellular space of the bacterium [66]. Two intracellular proteins, the trimeric form of ATP synthase subunit β and bacterial histone-like protein HU, have displayed interaction with human IL-β [65,66]. The interaction of internalised IL-1β with a key protein in cellular energy production and genomic DNA condensing HU protein might explain the above-described physiological responses of *A. actinomycetemcomitans.* These results suggest that viable *A. actinomycetemcomitans* cells possess a specific uptake mechanism for IL-1β. According to our recent results, *A. actinomycetemcomitans* encodes a *Pasteurellaceae*-specific outer membrane lipoprotein responsible for IL-1β interaction [67]. In addition to bacterial species, herpesviruses and yeasts found in subgingival biofilms can sense and bind the cytokines produced by the host [68,69].

3. Periodontitis-Associated Environmental Factors

Various environmental factors, other than those strictly related to the inflammatory response of the host, may change during the progression of periodontitis. Such factors include, for example, pH, the concentration of iron and hemin, and the presence of various host hormones. All these factors can be sensed by at least some of the periodontal pathogens, although studies at the molecular level are scarce. However, the results obtained thus far suggest that elevated pH and iron limitation may enhance the biofilm formation of some species and influence virulence gene expression, which, in turn, might alter the host inflammatory response.

3.1. Alkaline pH

The environment of the periodontitis-associated gingival pocket, and especially the gingival crevicular fluid, is characterised by alkaline pH, which may rise above 8.5 [70–72]. Some periodontal pathogens, such as *P. gingivalis*, *P. intermedia* and *F. nucleatum*, are able to elevate the ambient pH by fermenting amino acids *in vitro* [73], a feature which may also alkalify the microenvironment in subgingival biofilm locally.

An alkaline pH of 8.2 has been shown to increase cell surface hydrophobicity as well as induce the co-adhesion and biofilm formation of *F. nucleatum*, which is accompanied by decreased intracellular polyglucose content and elongation of individual cells [74]. When grown at a slightly lower pH than 8.2, *i.e.*, at pH 7.8, *F. nucleatum* cells display upregulated expression of the enzyme formiminotetrahydrofolate cyclodeaminase, which could be involved in raising the surrounding pH value [75]. In addition, the production of a non-iron redox acceptor flavodoxin, the expression of which is typically upregulated in iron-limited growth conditions [76], was also observed to be upregulated in alkaline pH conditions [75]. *F. nucleatum* cells appear to change their metabolism in response to high pH, because some enzymes of the glycolytic pathway, as well as glutamic acid and histidine catabolism, are downregulated in planktonic cells grown at an alkaline pH of 7.8 [75]. However, when the pH is further raised to 8.2, which induces biofilm formation, the amounts of glycolytic enzymes do not increase, whereas glucose storage and lactate production increases [77]. In contrast to increased glucose storage, the production of various proteins involved in protein synthesis decrease at high pH values in *F. nucleatum* biofilm [77]. At pH 8.2, another change in the expression of metabolic enzymes is the increased production of glutamate dehydrogenase, which might indicate that the bacterium adjusts its metabolism to the increased concentration of glutamate in the gingival crevicular fluid associated with inflamed periodontal tissue [77]. The cellular stress response in bacteria might also be activated at an alkaline pH because *F. nucleatum* cells upregulate the expression of peptidyl-prolyl *cis*-trans isomerase (PPI) and the heat-shock protein GroEL [75,77]. Of these proteins, GroEL might have a role in the host-bacterium crosstalk because GroEL-like proteins may modulate the host immune response, as shown with *F. nucleatum* [78] and *A. actinomycetemcomitans* [79]. Moreover, GroEL might be a link between periodontitis and systemic diseases, such as atherosclerosis, as *F. nucleatum* GroEL induces various risk factors of atherosclerosis in mice [80].

Whereas some of the pH-regulated intracellular proteins may alter the virulence of the bacterium, one interesting group of pH-regulated proteins are expressed in the cellular envelope of the bacterial cell. In particular, the proteins that are involved in adhesion to other periodontal pathogens as well as to host cells may play important roles in virulence. When *F. nucleatum* was grown at an alkaline pH of 8.2, elevated levels of FomA adhesion isoforms were detected [77]. FomA may function as recruiter of other periodontal pathogens, such as *P. gingivalis*, [81], and has been proven to be a possible vaccine target in a mouse study [82]. Some of the downregulated cellular envelope proteins of *F. nucleatum* are involved in ATP synthesis and maintenance of a neutral cyto- or periplasmic pH, indicating decreased metabolic activity and adjustment to alkaline conditions, respectively [83]. Some of the downregulated proteins likely have dual roles, as in the case of butyrate-acetoacetate CoA transferase, which is involved both in energy metabolism and, as a virulence factor, butyric acid production [83]. The group of cellular envelope proteins that were upregulated at an alkaline pH contained at least two putative surface antigens: outer membrane protein (OMP), belonging to the Omp IP family of porins, and a pathogen-specific membrane antigen that was predicted to have high affinity to Fe^{2+} [83]. In addition to the Omp IP family porins, the expression levels of various transporter proteins was altered when *F. nucleatum* cells were grown in biofilm at a pH of 8.2, which could be an indication of a changed need to uptake various solutes from the microenvironment in the biofilm [77]. Most of the studies that have investigated the effects of alkaline environment on periodontal pathogens have been performed in *F. nucleatum*. However, up to 50% of the alkaline pH-regulated genes coding *F. nucleatum* cell envelope proteins could have been acquired through horizontal gene transfer, and similar proteins are also found in other periodontal pathogens, such as *P. gingivalis* and *Treponema denticola* [83].

3.2. Iron and Hemin

Because free iron catalyses the formation of toxic free radicals from H_2O_2 and is essential for the function of both the host and the pathogenic bacteria, human organs have developed complex ways to limit the availability of free iron in the environment [84]. Thus, the environment that surrounds potential colonisers when they enter a human host is suboptimal in terms of free iron concentration. Most of the host iron is bound to iron-binding proteins such as transferrin, ferritin, lactoferrin and haemoglobin that contain hemin or haem. However, the situation may change during the progression of periodontitis. It has been hypothesised that the concentration of hemin may increase due to the increasing concentration of haemoglobin leaking from vascular ulcers of the gingival pocket. Some periodontal pathogens, such as *P. gingivalis*, *T. denticola* and *A. actinomycetemcomitans*, have been demonstrated to express hemin-binding proteins on their surfaces [85–87] that may facilitate iron acquisition in free iron-limited conditions. Moreover, hemin may also directly regulate the virulence characteristics of *P. gingivalis* [88].

Iron chelation has been shown to increase the expression of the EPS-, fimbrial-, and LPS-related genes, *pgaC*, *tadV*, and *rmlB*, respectively, in *A. actinomycetemcomitans*, which also leads to increased biofilm formation [89]. The effect of limited concentrations of iron most likely is mediated by small regulatory RNAs (sRNA) in *A. actinomycetemcomitans*, though the target genes for these sRNAs, have not been yet identified [90]. The most studied periodontal pathogen, regarding the need and effects of iron on the bacterium, is likely *P. gingivalis*. In agreement with being an essential growth factor and important cause of oxidative stress (for review see [9]), iron limitation upregulates the genes involved in iron uptake and downregulates the genes associated with the storage of iron as well as the oxidative stress response of *P. gingivalis* [9]. Moreover, by limiting iron and hemin availability, the host can also increase biofilm formation, the invasion of single bacterial cells to the host cells [9], and increase vesicle secretion and protease production of *P. gingivalis* [91]. Another potential virulence factor of *P. gingivalis* that is regulated by hemin is LPS, and more specifically, the lipid A form of it [88]. A form of lipid A, which is a Toll-like receptor (TLR) 4 antagonist, is produced at high hemin concentrations, whereas at low hemin concentrations TLR4-agonist lipid A is the major form, suggesting that *P. gingivalis* can alter the host response with the changing hemin microenvironment [88].

3.3. Hormones

Although various bacterial species are known to respond to the stress-related hormones adrenaline and noradrenaline, and the molecular players of their sensory machinery have been clarified in detailed (reviewed in [92]), the periodontal pathogens have been little studied. In the first study that investigated the effects of catecholamines, noradrenaline and adrenaline, on periodontal bacteria, both negative and positive growth effects on planktonic species were reported [93]. However, both catecholamines inhibited the growth of "red complex" [94] periodontal pathogens *P. gingivalis* and *T. forsythia* (formerly *Bacteroides forsythus*) as well as *A. actinomycetemcomitans* serotypes a and b and *F. nucleatum* [93]. Thus, the authors stressed the importance of negative growth effects and hypothesised that these species might also use catecholamines to enhance virulence gene expression [93]. A recent study by Saito *et al.* [95] demonstrated that growth inhibition of *P. gingivalis* by noradrenaline is accompanied with enhanced production of the virulence-associated protease arg-gingipain B and downregulation of the genes coding polysaccharide biosynthesis-related proteins.

The most significant rise in the levels of the female sex hormones, oestrogen and progesterone, occurs during pregnancy [96]. Although pregnancy gingivitis is currently categorised under the class of "dental plaque-induced gingival diseases modified by the endocrine system" [97], the effects of hormone levels on the progression of gingivitis during pregnancy is still under debate [98–100]. It appears that even though gingival inflammation may intensify during pregnancy, the hormones themselves may cause controversial changes in the periodontium, including decreased inflammatory reactions [101–104] and changes in the composition of subgingival biofilm [105–107]. Moreover, some gram-negative periodontal pathogens, such as *Prevotella melaninogenica*, *P. intermedia* and *P. gingivalis* are able to take up estradiol and progesterone, which the *Prevotella* species may use as a growth factor instead of vitamin K [108]. However, whether these hormones affect the biofilm formation and the virulence of periodontal pathogens is not known.

4. Conclusions

In periodontitis, the crosstalk between the host and the bacterial biofilm is diverse and bidirectional. The host response and environmental changes induce stress in the biofilm bacteria (Table 1). Elevated temperature of the subgingival environment, though aimed to eliminate the pathogens appears to only decrease virulence factors in pathogens (e.g., proteases in *P. gingivalis*) and does not sufficiently eliminate the pathogens [19]. The change in the local pH towards an alkaline environment appears to play an important role in the shift towards periodontopathogenic biofilm composition. The biofilm mass is increased in alkaline conditions and, in particular, the intermediate coloniser *F. nucleatum* displays increased adhesion and coaggregation with other bacteria. Oxidative stress and the inflammatory cytokine IL-1β result in decreased metabolism in periodontal biofilm but still increase various virulence factors as well as biofilm formation. The limited amount of free iron appears to enhance biofilm EPS formation [89]. During periodontal inflammation, the increased amount of hemin might downregulate the expression of bacterial virulence factors and upregulate the expression of immune-suppressing molecules. In summary, the environmental changes generated in inflammation favour biofilm formation and appear to drive the bacteria into the shelter provided by the EPS and the lower metabolic activity. The inflammatory environment with active immune cells and a hostile humoural response is not ideal for planktonic bacteria, which are released by mature biofilm when expanding to new habitats. Biofilm formation might explain the onset of less progressive phases in periodontal inflammation and tissue destruction and allow the periodontal pathogens to persist in subgingival spaces.

Table 1. Environmental stimuli affecting periodontal biofilm and bacterial virulence factors.

Stimuli	Effect	Species	References
Elevated temperature	Proteases ↓	*Porphyromonas gingivalis*	[16]
	Fimbrial proteins ↓		[15]
	TLR4 activating lipid-A ↑		[18]
Oxidative stress	ATP production ↓	*Fusobacterium nucleatum*	[25]
	Chaperones ClpB, DnaK ↑		[25]
	Heat shock protein HtpG ↑		[25]
	Transcription repressor HrcA ↑		[25]
Oxidative stress	Chaperones ClpB, DnaK ↑	*Porphyromonas gingivalis*	[26]
	Heat shock protein HtpG ↑		[26]
	superoxide dismutase ↑		[26]
Inflammatory cytokine IL-1β	Biofilm formation ↑	*Aggregatibacter actinomycetemcomitans*	[65]
	Metabolism ↓		[65]
Alkaline pH	Co-adhesion ↑	*Fusobacterium nucleatum*	[74]
	Biofilm formation ↑		[74]
	Flavodoxin ↑		[75]
	Glucose storage ↑		[77]
	Lactate production ↑		[77]
	Protein synthesis enzymes ↓		[77]
	Glutamate dehydrogenase ↑		[77]
	PPI and GroEL ↑		[75,77]
	FomA adhesion isoforms ↑		[77]
	ATP synthesis proteins ↓		[83]
	Butyrate-acetoacetate CoA transferase ↓		[83]
	Surface antigens Omp IP ↑		[83]

Table 1. *Cont.*

Stimuli	Effect	Species	References
Iron-limitation	EPS (*pgaC*) ↑	*Aggregatibacter actinomycetemcomitans*	[89]
	Fimbrial (*tadV*) ↑		[89]
	LPS (*rmlB*) ↑		[89]
	Biofilm formation ↑		[89]
Iron limitation	Iron uptake ↑	*Porphyromonas gingivalis*	[9]
	Iron storage ↓		[9]
	Oxidative stress response ↓		[9]
	Biofilm formation ↑		[9]
	Host cell invasion ↑		[9]
High hemin concentration	Proteases ↓	*Porphyromonas gingivalis*	[91]
	Vesicles ↓		[91]
	TLR4 inactivating lipid A ↑		[88]
Noradrenaline	Growth ↓	*Porphyromonas gingivalis*	[93]
	Arg-gingipain B ↑		[95]

Acknowledgments

Biofilm research in the Ihalin laboratory has been funded by the Academy of Finland, the Paulo Foundation, the Ella and Georg Ehrnrooth Foundation, the Turku University Foundation and the Finnish Dental Society Apollonia.

References

1. Bartold, P.M.; van Dyke, T.E. Periodontitis: A host-mediated disruption of microbial homeostasis. Unlearning learned concepts. *Periodontol. 2000* **2013**, *62*, 203–217.
2. Griffen, A.L.; Beall, C.J.; Campbell, J.H.; Firestone, N.D.; Kumar, P.S.; Yang, Z.K.; Podar, M.; Leys, E.J. Distinct and complex bacterial profiles in human periodontitis and health revealed by 16S pyrosequencing. *ISME J.* **2012**, *6*, 1176–1185.
3. Rescala, B.; Rosalem, W., Jr.; Teles, R.P.; Fischer, R.G.; Haffajee, A.D.; Socransky, S.S.; Gustafsson, A.; Figueredo, C.M. Immunologic and microbiologic profiles of chronic and aggressive periodontitis subjects. *J. Periodontol.* **2010**, *81*, 1308–1316.
4. Fritschi, B.Z.; Albert-Kiszely, A.; Persson, G.R. *Staphylococcus aureus* and other bacteria in untreated periodontitis. *J. Dent. Res.* **2008**, *87*, 589–593.
5. Saygun, I.; Kubar, A.; Sahin, S.; Sener, K.; Slots, J. Quantitative analysis of association between herpesviruses and bacterial pathogens in periodontitis. *J. Periodontal. Res.* **2008**, *43*, 352–359.
6. Imbronito, A.V.; Okuda, O.S.; Maria de Freitas, N.; Moreira Lotufo, R.F.; Nunes, F.D. Detection of herpesviruses and periodontal pathogens in subgingival plaque of patients with chronic periodontitis, generalized aggressive periodontitis, or gingivitis. *J. Periodontol.* **2008**, *79*, 2313–2321.

7. Canabarro, A.; Valle, C.; Farias, M.R.; Santos, F.B.; Lazera, M.; Wanke, B. Association of subgingival colonization of *Candida albicans* and other yeasts with severity of chronic periodontitis. *J. Periodontal. Res.* **2013**, *48*, 428–432.
8. Flemming, H.C.; Wingender, J. The biofilm matrix. *Nat. Rev. Microbiol.* **2010**, *8*, 623–633.
9. Lewis, J.P. Metal Uptake in host-pathogen interactions: Role of iron in *Porphyromonas gingivalis* interactions with host organisms. *Periodontol. 2000* **2010**, *52*, 94–116.
10. McDougald, D.; Rice, S.A.; Barraud, N.; Steinberg, P.D.; Kjelleberg, S. Should we stay or should we go: Mechanisms and ecological consequences for biofilm dispersal. *Nat. Rev. Microbiol.* **2012**, *10*, 39–50.
11. Tonetti, M.S.; Imboden, M.A.; Lang, N.P. Neutrophil Migration into the gingival sulcus is associated with transepithelial gradients of interleukin-8 and ICAM-1. *J. Periodontol.* **1998**, *69*, 1139–1147.
12. Shapiro, R.S.; Cowen, L.E. Thermal control of microbial development and virulence: Molecular mechanisms of microbial temperature sensing. *MBio* **2012**, *3*, doi:10.1128/mBio.00238-12.
13. Fedi, P.F., Jr.; Killoy, W.J. Temperature differences at periodontal sites in health and disease. *J. Periodontol.* **1992**, *63*, 24–27.
14. Percival, R.S.; Marsh, P.D.; Devine, D.A.; Rangarajan, M.; Aduse-Opoku, J.; Shepherd, P.; Curtis, M.A. Effect of temperature on growth, hemagglutination, and protease activity of *Porphyromonas gingivalis*. *Infect. Immun.* **1999**, *67*, 1917–1921.
15. Amano, A.; Fujiwara, T.; Nagata, H.; Kuboniwa, M.; Sharma, A.; Sojar, H.T.; Genco, R.J.; Hamada, S.; Shizukuishi, S. *Prophyromonas gingivalis* fimbriae mediate coaggregation with *Streptococcus oralis* through specific domains. *J. Dent. Res.* **1997**, *76*, 852–857.
16. Murakami, Y.; Nagata, H.; Amano, A.; Takagaki, M.; Shizukuishi, S.; Tsunemitsu, A.; Aimoto, S. Inhibitory effects of human salivary histatins and lysozyme on coaggregation between *Porphyromonas gingivalis* and *Streptococcus mitis*. *Infect. Immun.* **1991**, *59*, 3284–3286.
17. Sato, T.; Nakazawa, F. Coaggregation between *Prevotella oris* and *Porphyromonas gingivalis*. *J. Microbiol. Immunol. Infect.* **2012**, doi:10.1016/j.jmii.2012.09.005.
18. Curtis, M.A.; Percival, R.S.; Devine, D.; Darveau, R.P.; Coats, S.R.; Rangarajan, M.; Tarelli, E.; Marsh, P.D. Temperature-dependent modulation of *Porphyromonas gingivalis* lipid A structure and interaction with the innate host defenses. *Infect. Immun.* **2011**, *79*, 1187–1193.
19. Haffajee, A.D.; Socransky, S.S.; Smith, C.; Dibart, S.; Goodson, J.M. Subgingival temperature (III). Relation to microbial counts. *J. Clin. Periodontol.* **1992**, *19*, 417–422.
20. Imlay, J.A. Cellular defenses against superoxide and hydrogen peroxide. *Annu. Rev. Biochem.* **2008**, *77*, 755–776.
21. Cabiscol, E.; Tamarit, J.; Ros, J. Oxidative stress in bacteria and protein damage by reactive oxygen species. *Int. Microbiol.* **2000**, *3*, 3–8.
22. Bradshaw, D.J.; Marsh, P.D.; Watson, G.K.; Allison, C. Role of *Fusobacterium nucleatum* and coaggregation in anaerobe survival in planktonic and biofilm oral microbial communities during aeration. *Infect. Immun.* **1998**, *66*, 4729–4732.
23. Silva, V.L.; Diniz, C.G.; Cara, D.C.; Santos, S.G.; Nicoli, J.R.; Carvalho, M.A.; Farias, L.M. Enhanced pathogenicity of *Fusobacterium nucleatum* adapted to oxidative stress. *Microb. Pathog.* **2005**, *39*, 131–138.

24. Gursoy, U.K.; Pöllänen, M.; Könönen, E.; Uitto, V.J. Biofilm formation enhances the oxygen tolerance and invasiveness of *Fusobacterium nucleatum* in an oral mucosa culture model. *J. Periodontol.* **2010**, *81*, 1084–1091.
25. Steeves, C.H.; Potrykus, J.; Barnett, D.A.; Bearne, S.L. Oxidative stress response in the opportunistic oral pathogen *Fusobacterium nucleatum*. *Proteomics* **2011**, *11*, 2027–2037.
26. Okano, S.; Shibata, Y.; Shiroza, T.; Abiko, Y. Proteomics-based analysis of a counter-oxidative stress system in *Porphyromonas gingivalis*. *Proteomics* **2006**, *6*, 251–258.
27. Amano, A.; Sharma, A.; Sojar, H.T.; Kuramitsu, H.K.; Genco, R.J. Effects of temperature stress on expression of fimbriae and superoxide dismutase by *Porphyromonas gingivalis*. *Infect. Immun.* **1994**, *62*, 4682–4685.
28. McKenzie, R.M.; Johnson, N.A.; Aruni, W.; Dou, Y.; Masinde, G.; Fletcher, H.M. Differential response of *Porphyromonas gingivalis* to varying levels and duration of hydrogen peroxide-induced oxidative stress. *Microbiology* **2012**, *158*, 2465–2479.
29. Gamonal, J.; Acevedo, A.; Bascones, A.; Jorge, O.; Silva, A. Levels of interleukin-1 beta, -8, and -10 and RANTES in gingival crevicular fluid and cell populations in adult periodontitis patients and the effect of periodontal treatment. *J. Periodontol.* **2000**, *71*, 1535–1545.
30. Silva, N.; Dutzan, N.; Hernandez, M.; Dezerega, A.; Rivera, O.; Aguillon, J.C.; Aravena, O.; Lastres, P.; Pozo, P.; Vernal, R.; *et al.* Characterization of progressive periodontal lesions in chronic periodontitis patients: Levels of chemokines, cytokines, matrix metalloproteinase-13, periodontal pathogens and inflammatory cells. *J. Clin. Periodontol.* **2008**, *35*, 206–214.
31. Graves, D.T.; Cochran, D. The contribution of interleukin-1 and tumor necrosis factor to periodontal tissue destruction. *J. Periodontol.* **2003**, *74*, 391–401.
32. Ishihara, Y.; Nishihara, T.; Maki, E.; Noguchi, T.; Koga, T. Role of interleukin-1 and prostaglandin in *in vitro* bone resorption induced by *Actinobacillus actinomycetemcomitans* lipopolysaccharide. *J. Periodontal. Res.* **1991**, *26*, 155–160.
33. Cochran, D.L. Inflammation and bone loss in periodontal disease. *J. Periodontol.* **2008**, *79*, 1569–1576.
34. Eskan, M.A.; Benakanakere, M.R.; Rose, B.G.; Zhang, P.; Zhao, J.; Stathopoulou, P.; Fujioka, D.; Kinane, D.F. Interleukin-1beta modulates proinflammatory cytokine production in human epithelial cells. *Infect. Immun.* **2008**, *76*, 2080–2089.
35. Tanabe, S.; Bodet, C.; Grenier, D. *Treponema denticola* lipooligosaccharide activates gingival fibroblasts and upregulates inflammatory mediator production. *J. Cell. Physiol.* **2008**, *216*, 727–731.
36. Kelk, P.; Claesson, R.; Chen, C.; Sjöstedt, A.; Johansson, A. IL-1beta secretion induced by *Aggregatibacter* (*Actinobacillus*) *actinomycetemcomitans* is mainly caused by the leukotoxin. *Int. J. Med. Microbiol.* **2008**, *298*, 529–541.
37. Tanabe, S.I.; Grenier, D. Macrophage tolerance response to *Aggregatibacter actinomycetemcomitans* lipopolysaccharide induces differential regulation of tumor necrosis factor-alpha, interleukin-1 beta and matrix metalloproteinase 9 secretion. *J. Periodontal. Res.* **2008**, *43*, 372–377.
38. Matsuki, Y.; Yamamoto, T.; Hara, K. Interleukin-1 mRNA-expressing macrophages in human chronically inflamed gingival tissues. *Am. J. Pathol.* **1991**, *138*, 1299–1305.

39. Hsi, E.D.; Remick, D.G. Monocytes are the major producers of interleukin-1 beta in an *ex vivo* model of local cytokine production. *J. Interferon Cytokine Res.* **1995**, *15*, 89–94.
40. Takeichi, O.; Saito, I.; Tsurumachi, T.; Saito, T.; Moro, I. Human polymorphonuclear leukocytes derived from chronically inflamed tissue express inflammatory cytokines *in vivo*. *Cell. Immunol.* **1994**, *156*, 296–309.
41. Vardar-Sengul, S.; Arora, S.; Baylas, H.; Mercola, D. Expression profile of human gingival fibroblasts induced by interleukin-1beta reveals central role of nuclear factor-kappa b in stabilizing human gingival fibroblasts during inflammation. *J. Periodontol.* **2009**, *80*, 833–849.
42. Uchida, Y.; Shiba, H.; Komatsuzawa, H.; Takemoto, T.; Sakata, M.; Fujita, T.; Kawaguchi, H.; Sugai, M.; Kurihara, H. Expression of IL-1 beta and IL-8 by human gingival epithelial cells in response to *Actinobacillus actinomycetemcomitans*. *Cytokine* **2001**, *14*, 152–161.
43. Dickinson, B.C.; Moffatt, C.E.; Hagerty, D.; Whitmore, S.E.; Brown, T.A.; Graves, D.T.; Lamont, R.J. Interaction of oral bacteria with gingival epithelial cell multilayers. *Mol. Oral Microbiol.* **2011**, *26*, 210–220.
44. Stathopoulou, P.G.; Benakanakere, M.R.; Galicia, J.C.; Kinane, D.F. Epithelial cell pro-inflammatory cytokine response differs across dental plaque bacterial species. *J. Clin. Periodontol.* **2010**, *37*, 24–29.
45. Umeda, J.E.; Demuth, D.R.; Ando, E.S.; Faveri, M.; Mayer, M.P. Signaling transduction analysis in gingival epithelial cells after infection with *Aggregatibacter actinomycetemcomitans*. *Mol. Oral Microbiol.* **2012**, *27*, 23–33.
46. Peyyala, R.; Kirakodu, S.S.; Novak, K.F.; Ebersole, J.L. Oral microbial biofilm stimulation of epithelial cell responses. *Cytokine* **2012**, *58*, 65–72.
47. Guggenheim, B.; Gmur, R.; Galicia, J.C.; Stathopoulou, P.G.; Benakanakere, M.R.; Meier, A.; Thurnheer, T.; Kinane, D.F. *In vitro* modeling of host-parasite interactions: The 'subgingival' biofilm challenge of primary human epithelial cells. *BMC Microbiol.* **2009**, *9*, doi:10.1186/1471-2180-9-280.
48. Peyyala, R.; Ebersole, J.L. Multispecies biofilms and host responses: "Discriminating the trees from the forest". *Cytokine* **2013**, *61*, 15–25.
49. Porat, R.; Clark, B.D.; Wolff, S.M.; Dinarello, C.A. Enhancement of growth of virulent strains of *Escherichia coli* by interleukin-1. *Science* **1991**, *254*, 430–432.
50. Meduri, G.U.; Kanangat, S.; Stefan, J.; Tolley, E.; Schaberg, D. Cytokines IL-1beta, IL-6, and TNF-alpha enhance *in vitro* growth of bacteria. *Am. J. Respir. Crit. Care Med.* **1999**, *160*, 961–967.
51. Stashenko, P.; Fujiyoshi, P.; Obernesser, M.S.; Prostak, L.; Haffajee, A.D.; Socransky, S.S. Levels of interleukin 1 beta in tissue from sites of active periodontal disease. *J. Clin. Periodontol.* **1991**, *18*, 548–554.
52. McLaughlin, R.A.; Hoogewerf, A.J. Interleukin-1beta-induced growth enhancement of *Staphylococcus aureus* occurs in biofilm but not planktonic cultures. *Microb. Pathog.* **2006**, *41*, 67–79.
53. Kanangat, S.; Bronze, M.S.; Meduri, G.U.; Postlethwaite, A.; Stentz, F.; Tolley, E.; Schaberg, D. Enhanced extracellular growth of *Staphylococcus aureus* in the presence of selected linear

peptide fragments of human interleukin (IL)-1beta and IL-1 receptor antagonist. *J. Infect. Dis.* **2001**, *183*, 65–69.

54. Kanangat, S.; Postlethwaite, A.; Cholera, S.; Williams, L.; Schaberg, D. Modulation of virulence gene expression in *Staphylococcus aureus* by interleukin-1beta: Novel implications in bacterial pathogenesis. *Microbes Infect.* **2007**, *9*, 408–415.
55. Zav'yalov, V.P.; Chernovskaya, T.V.; Navolotskaya, E.V.; Karlyshev, A.V.; MacIntyre, S.; Vasiliev, A.M.; Abramov, V.M. Specific high affinity binding of human interleukin 1 beta by Caf1A usher protein of *Yersinia pestis*. *FEBS Lett.* **1995**, *371*, 65–68.
56. Zav'yalov, V.; Denesyuk, A.; Zav'yalova, G.; Korpela, T. Molecular modeling of the steric structure of the envelope F1 antigen of *Yersinia pestis*. *Immunol. Lett.* **1995**, *45*, 19–22.
57. Wu, L.; Estrada, O.; Zaborina, O.; Bains, M.; Shen, L.; Kohler, J.E.; Patel, N.; Musch, M.W.; Chang, E.B.; Fu, Y.X.; *et al.* Recognition of host immune activation by *Pseudomonas aeruginosa*. *Science* **2005**, *309*, 774–777.
58. Diggle, S.P.; Stacey, R.E.; Dodd, C.; Camara, M.; Williams, P.; Winzer, K. The galactophilic lectin, LecA, contributes to biofilm development in *Pseudomonas aeruginosa*. *Environ. Microbiol.* **2006**, *8*, 1095–1104.
59. Cebo, C.; Dambrouck, T.; Maes, E.; Laden, C.; Strecker, G.; Michalski, J.C.; Zanetta, J.P. Recombinant human interleukins IL-1alpha, IL-1beta, IL-4, IL-6, and IL-7 show different and specific calcium-independent carbohydrate-binding properties. *J. Biol. Chem.* **2001**, *276*, 5685–5691.
60. Sherblom, A.P.; Decker, J.M.; Muchmore, A.V. The lectin-like interaction between recombinant tumor necrosis factor and uromodulin. *J. Biol. Chem.* **1988**, *263*, 5418–5424.
61. Cebo, C.; Vergoten, G.; Zanetta, J.P. Lectin activities of cytokines: Functions and putative carbohydrate-recognition domains. *Biochim. Biophys. Acta* **2002**, *1572*, 422–434.
62. Vergoten, G.; Zanetta, J.P. Structural differences between the putative carbohydrate-recognition domains of human IL-1 alpha, IL-1 beta and IL-1 receptor antagonist obtained by *in silico* modeling. *Glycoconj. J.* **2007**, *24*, 183–193.
63. Mandrell, R.E.; McLaughlin, R.; Aba Kwaik, Y.; Lesse, A.; Yamasaki, R.; Gibson, B.; Spinola, S.M.; Apicella, M.A. Lipooligosaccharides (LOS) of some *Haemophilus* species mimic human glycosphingolipids, and some LOS are sialylated. *Infect. Immun.* **1992**, *60*, 1322–1328.
64. Li, Y.; Sun, M.; Huang, S.; Yu, H.; Chokhawala, H.A.; Thon, V.; Chen, X. The *hd0053* gene of *Haemophilus ducreyi* encodes an α2,3-sialyltransferase. *Biochem. Biophys. Res. Commun.* **2007**, *361*, 555–560.
65. Paino, A.; Tuominen, H.; Jääskeläinen, M.; Alanko, J.; Nuutila, J.; Asikainen, S.E.; Pelliniemi, L.J.; Pöllänen, M.T.; Chen, C.; Ihalin, R. Trimeric form of intracellular ATP synthase subunit beta of *Aggregatibacter actinomycetemcomitans* binds human interleukin-1beta. *PLoS One* **2011**, *6*, e18929.
66. Paino, A.; Lohermaa, E.; Sormunen, R.; Tuominen, H.; Korhonen, J.; Pöllänen, M.T.; Ihalin, R. Interleukin-1β is internalised by viable *Aggregatibacter actinomycetemcomitans* biofilm and locates to the outer edges of nucleoids. *Cytokine* **2012**, *60*, 565–574.

67. Paino, A.; Ahlstrand, T.; Nuutila, J.; Navickaite, I.; Lahti, M.; Tuominen, H.; Välimaa, H.; Lamminmäki, U.; Pöllänen, M.T.; Ihalin, R. Identification of a novel bacterial outer membrane interleukin-1beta-binding protein from *Aggregatibacter actinomycetemcomitans*. *PLoS One* **2013**, *8*, e70509.
68. Alcami, A. Viral mimicry of cytokines, chemokines and their receptors. *Nat. Rev. Immunol.* **2003**, *3*, 36–50.
69. Treseler, C.B.; Maziarz, R.T.; Levitz, S.M. Biological activity of interleukin-2 bound to *Candida albicans*. *Infect. Immun.* **1992**, *60*, 183–188.
70. Bickel, M.; Munoz, J.L.; Giovannini, P. Acid-base properties of human gingival crevicular fluid. *J. Dent. Res.* **1985**, *64*, 1218–1220.
71. Bickel, M.; Cimasoni, G. The pH of human crevicular fluid measured by a new microanalytical technique. *J. Periodontal. Res.* **1985**, *20*, 35–40.
72. Eggert, F.M.; Drewell, L.; Bigelow, J.A.; Speck, J.E.; Goldner, M. The pH of gingival crevices and periodontal pockets in children, teenagers and adults. *Arch. Oral Biol.* **1991**, *36*, 233–238.
73. Takahashi, N. Acid-neutralizing activity during amino acid fermentation by *Porphyromonas gingivalis*, *Prevotella intermedia* and *Fusobacterium nucleatum*. *Oral Microbiol. Immunol.* **2003**, *18*, 109–113.
74. Zilm, P.S.; Rogers, A.H. Co-adhesion and biofilm formation by *Fusobacterium nucleatum* in response to growth pH. *Anaerobe* **2007**, *13*, 146–152.
75. Zilm, P.S.; Bagley, C.J.; Rogers, A.H.; Milne, I.R.; Gully, N.J. The proteomic profile of *Fusobacterium nucleatum* is regulated by growth pH. *Microbiology* **2007**, *153*, 148–159.
76. Kapatral, V.; Anderson, I.; Ivanova, N.; Reznik, G.; Los, T.; Lykidis, A.; Bhattacharyya, A.; Bartman, A.; Gardner, W.; Grechkin, G.; *et al.* Genome sequence and analysis of the oral bacterium *Fusobacterium nucleatum* strain ATCC 25586. *J. Bacteriol.* **2002**, *184*, 2005–2018.
77. Chew, J.; Zilm, P.S.; Fuss, J.M.; Gully, N.J. A proteomic investigation of *Fusobacterium nucleatum* alkaline-induced biofilms. *BMC Microbiol.* **2012**, *12*, doi:10.1186/1471-2180-12-189.
78. Skar, C.K.; Kruger, P.G.; Bakken, V. Characterisation and subcellular localisation of the GroEL-like and DnaK-like proteins isolated from *Fusobacterium nucleatum* ATCC 10953. *Anaerobe* **2003**, *9*, 305–312.
79. Oscarsson, J.; Karched, M.; Thay, B.; Chen, C.; Asikainen, S. Proinflammatory effect in whole blood by free soluble bacterial components released from planktonic and biofilm cells. *BMC Microbiol.* **2008**, *8*, doi:10.1186/1471-2180-8-206.
80. Lee, H.R.; Jun, H.K.; Kim, H.D.; Lee, S.H.; Choi, B.K. *Fusobacterium nucleatum* GroEL induces risk factors of atherosclerosis in human microvascular endothelial cells and *apoE*(−/−) mice. *Mol. Oral Microbiol.* **2012**, *27*, 109–123.
81. Shaniztki, B.; Hurwitz, D.; Smorodinsky, N.; Ganeshkumar, N.; Weiss, E.I. Identification of a *Fusobacterium nucleatum* PK1594 galactose-binding adhesin which mediates coaggregation with periopathogenic bacteria and hemagglutination. *Infect. Immun.* **1997**, *65*, 5231–5237.
82. Liu, P.F.; Shi, W.; Zhu, W.; Smith, J.W.; Hsieh, S.L.; Gallo, R.L.; Huang, C.M. Vaccination targeting surface FomA of *Fusobacterium nucleatum* against bacterial co-aggregation: implication for treatment of periodontal infection and halitosis. *Vaccine* **2010**, *28*, 3496–3505.

83. Zilm, P.S.; Mira, A.; Bagley, C.J.; Rogers, A.H. Effect of alkaline growth pH on the expression of cell envelope proteins in *Fusobacterium nucleatum*. *Microbiology* **2010**, *156*, 1783–1794.
84. Schaible, U.E.; Kaufmann, S.H. Iron and microbial infection. *Nat. Rev. Microbiol.* **2004**, *2*, 946–953.
85. Shoji, M.; Shibata, Y.; Shiroza, T.; Yukitake, H.; Peng, B.; Chen, Y.Y.; Sato, K.; Naito, M.; Abiko, Y.; Reynolds, E.C.; *et al.* Characterization of hemin-binding protein 35 (HBP35) in *Porphyromonas gingivalis*: Its cellular distribution, thioredoxin activity and role in heme utilization. *BMC Microbiol.* **2010**, *10*, doi:10.1186/1471-2180-10-152.
86. Xu, X.; Kolodrubetz, D. Construction and analysis of hemin binding protein mutants in the oral pathogen *Treponema denticola*. *Res. Microbiol.* **2002**, *153*, 569–577.
87. Rhodes, E.R.; Menke, S.; Shoemaker, C.; Tomaras, A.P.; McGillivary, G.; Actis, L.A. Iron acquisition in the dental pathogen *Actinobacillus actinomycetemcomitans*: What does it use as a source and how does it get this essential metal? *Biometals* **2007**, *20*, 365–377.
88. Al-Qutub, M.N.; Braham, P.H.; Karimi-Naser, L.M.; Liu, X.; Genco, C.A.; Darveau, R.P. Hemin-dependent modulation of the lipid A structure of *Porphyromonas gingivalis* lipopolysaccharide. *Infect. Immun.* **2006**, *74*, 4474–4485.
89. Amarasinghe, J.J.; Scannapieco, F.A.; Haase, E.M. Transcriptional and translational analysis of biofilm determinants of *Aggregatibacter actinomycetemcomitans* in response to environmental perturbation. *Infect. Immun.* **2009**, *77*, 2896–2907.
90. Amarasinghe, J.J.; Connell, T.D.; Scannapieco, F.A.; Haase, E.M. Novel iron-regulated and Fur-regulated small regulatory RNAs in *Aggregatibacter actinomycetemcomitans*. *Mol. Oral Microbiol.* **2012**, *27*, 327–349.
91. Smalley, J.W.; Birss, A.J.; McKee, A.S.; Marsh, P.D. Haemin-restriction influences haemin-binding, haemagglutination and protease activity of cells and extracellular membrane vesicles of *Porphyromonas gingivalis* W50. *FEMS Microbiol. Lett.* **1991**, *69*, 63–67.
92. Hughes, D.T.; Sperandio, V. Inter-kingdom signalling: Communication between bacteria and their hosts. *Nat. Rev. Microbiol.* **2008**, *6*, 111–120.
93. Roberts, A.; Matthews, J.B.; Socransky, S.S.; Freestone, P.P.; Williams, P.H.; Chapple, I.L. Stress and the periodontal diseases: Effects of catecholamines on the growth of periodontal bacteria *in vitro*. *Oral Microbiol. Immunol.* **2002**, *17*, 296–303.
94. Socransky, S.S.; Haffajee, A.D.; Cugini, M.A.; Smith, C.; Kent, R.L., Jr. Microbial complexes in subgingival plaque. *J. Clin. Periodontol.* **1998**, *25*, 134–144.
95. Saito, T.; Inagaki, S.; Sakurai, K.; Okuda, K.; Ishihara, K. Exposure of *P. gingivalis* to noradrenaline reduces bacterial growth and elevates ArgX protease activity. *Arch. Oral Biol.* **2011**, *56*, 244–250.
96. Mariotti, A. Sex steroid hormones and cell dynamics in the periodontium. *Crit. Rev. Oral Biol. Med.* **1994**, *5*, 27–53.
97. Armitage, G.C. Development of a classification system for periodontal diseases and conditions. *Ann. Periodontol.* **1999**, *4*, 1–6.
98. Jonsson, R.; Howland, B.E.; Bowden, G.H. Relationships between periodontal health, salivary steroids, and *Bacteroides intermedius* in males, pregnant and non-pregnant women. *J. Dent. Res.* **1988**, *67*, 1062–1069.

99. Figuero, E.; Carrillo-de-Albornoz, A.; Herrera, D.; Bascones-Martinez, A. Gingival changes during pregnancy: I. Influence of hormonal variations on clinical and immunological parameters. *J. Clin. Periodontol.* **2010**, *37*, 220–229.

100. Gursoy, M.; Gursoy, U.K.; Sorsa, T.; Pajukanta, R.; Könönen, E. High salivary estrogen and risk of developing pregnancy gingivitis. *J. Periodontol.* **2012**, doi:10.1902/jop.2012.120512.

101. Lopatin, D.E.; Kornman, K.S.; Loesche, W.J. Modulation of immunoreactivity to periodontal disease-associated microorganisms during pregnancy. *Infect. Immun.* **1980**, *28*, 713–718.

102. Lapp, C.A.; Thomas, M.E.; Lewis, J.B. Modulation by progesterone of interleukin-6 production by gingival fibroblasts. *J. Periodontol.* **1995**, *66*, 279–284.

103. Rodriguez, E.; Lopez, R.; Paez, A.; Masso, F.; Montano, L.F. 17Beta-estradiol inhibits the adhesion of leukocytes in TNF-alpha stimulated human endothelial cells by blocking IL-8 and MCP-1 secretion, but not its transcription. *Life Sci.* **2002**, *71*, 2181–2193.

104. Shu, L.; Guan, S.M.; Fu, S.M.; Guo, T.; Cao, M.; Ding, Y. Estrogen modulates cytokine expression in human periodontal ligament cells. *J. Dent. Res.* **2008**, *87*, 142–147.

105. Jensen, J.; Liljemark, W.; Bloomquist, C. The effect of female sex hormones on subgingival plaque. *J. Periodontol.* **1981**, *52*, 599–602.

106. Raber-Durlacher, J.E.; van Steenbergen, T.J.; van der Velden, U.; de Graaff, J.; Abraham-Inpijn, L. Experimental gingivitis during pregnancy and post-partum: Clinical, endocrinological, and microbiological aspects. *J. Clin. Periodontol.* **1994**, *21*, 549–558.

107. Carrillo-de-Albornoz, A.; Figuero, E.; Herrera, D.; Bascones-Martinez, A. Gingival changes during pregnancy: II. Influence of hormonal variations on the subgingival biofilm. *J. Clin. Periodontol.* **2010**, *37*, 230–240.

108. Kornman, K.S.; Loesche, W.J. Effects of estradiol and progesterone on *Bacteroides melaninogenicus* and *Bacteroides gingivalis*. *Infect. Immun.* **1982**, *35*, 256–263.

7

The Interaction of CuS and *Halothiobacillus* HT1 Biofilm in Microscale Using Synchrotron Radiation-Based Techniques

Huirong Lin [1,2], Guangcun Chen [1], Shenhai Zhu [1], Yingxu Chen [1], Dongliang Chen [3], Wei Xu [3], Xiaohan Yu [4] and Jiyan Shi [1,*]

1 Department of Environmental Engineering, Zhejiang University, Hangzhou 310058, China; E-Mails: hrlin@iue.ac.cn (H.L.); gcchen2011@sinano.ac.cn (G.C.); zhushenhai@yahoo.com.cn (S.Z.); yingxu_chen@hotmail.com (Y.C.)

2 Institute of Urban Environment, Chinese Academy of Science, Xiamen 361021, China

3 Institute of High Energy Physics, Chinese Academy of Science, Beijing Synchrotron Radiation Facility, Beijing 100049, China; E-Mails: chendl@ihep.ac.cn (D.C.); xuw@mail.ihep.ac.cn (W.X.)

4 Institute of Applied Physics, Chinese Academy of Science, Shanghai Synchrotron Radiation Facility, Shanghai 201800, China; E-Mail: yuxiaohan@sinap.ac.cn

* Author to whom correspondence should be addressed; E-Mail: shijiyan@zju.edu.cn;

Abstract: In order to investigate the microbe-mineral interaction in the micro scale, spatial distribution and speciation of Cu and S in *Halothiobacillus* HT1 biofilm formed on a CuS surface was examined using synchrotron-based X-ray techniques. Confocal laser scanning microscope (CLSM) results indicated that *Halothiobacillus* HT1 biofilm formation gave rise to distinct chemical and redox gradients, leading to diverse niches in the biofilm. Live cells were distributed at the air-biofilm and membrane-biofilm interface. CuS was oxidized by *Halothiobacillus* HT1 biofilm, and copper penetrated into the biofilm. Sulfide was oxidized to cysteine (77.3%), sulfite (3.8%) and sulfonate (18.9%). Cu-cysteine-like species were involved in the copper homeostasis. These results significantly improve our understanding of the interfacial properties of the biofilm-mineral interface.

Keywords: *Halothiobacillus* HT1 biofilm; distribution; speciation; copper; sulfur

---

1. Introduction

Microbes cycle metals through biogeochemical processes, including accumulation, transformation and biomineralization, thus leading to changes in toxicity and bioavailability [1,2]. Sulfide oxidation by microbes is used for the biological production of metals, such as copper, gold and zinc. Sulfur oxidizing bacteria contribute strongly to the biological oxidation of metal sulfide [3,4]. However, contrary to their significant role in the global sulfur cycle and the biotechnological importance, the microbial fundamentals of sulfur oxidation are incompletely understood, due to the complexity of this reaction.

The microbe-mineral interface serves as a solid phase source of electrons, and biogeochemistry of the microbe-mineral interactions have been paid great attention to during the past few decades [5–7]. Many of the critical processes occur at the biofilm-mineral interface on the molecular scale. Therefore, attention should be focused on microenvironments, where chemical transformations occur. A better understanding of the interfacial properties of the biofilms-mineral interface, especially the interfacial chemical processes at the micron and nanometer levels, is needed [6,8].

Microbes usually operate as consortia of organisms rather than as single cells. Biofilms are physiologically distinct from bacteria growing in a free-swimming planktonic state and present genetic and physiological heterogeneity [9–11]. Growth of biofilm can enhance resistance to metal toxicity. Previous report showed that the ability of biofilms to survive heavy metals stress is better than planktonic microbes [12]. The metabolic activity, microenvironment characteristics and microbial community composition of biofilm are involved in resistance to metal toxicity. Biofilms can sorb metals and retard metal diffusion, leading to protection in the interior of the biofilms [13]. The genetic basis for metal resistance in sulfur oxidizing bacteria has been studied by several investigators [14]. Basic understanding of environmental materials and processes at the molecular scale is essential for risk assessment and management and reduction of environmental pollutants. Therefore, the description of the speciation and distribution of metals in biofilms is critically important for modeling and understanding the detoxification mechanism.

The aim of this study was to investigate the interaction in sulfur oxidizing bacteria biofilm-metal sulfide at the micro scale. A heavy-metals-tolerant *Halothiobacillus* HT1 was chosen [15]. The interaction in the CuS and biofilm of the *Halothiobacillus* interface was studied. The spatial distribution and speciation of copper in *Halothiobacillus* HT1 biofilm formed on CuS was determined using synchrotron-based X-ray fluorescence microscopy (XRF) and micro-X-ray absorption near edge structure (micro-XANES) analysis. Cell viability was detected using live-dead staining. Sulfur speciation was measured using sulfur K-edge XANES.

2. Results

2.1. Spatial Distribution of Live and Dead Halothiobacillus *HT1 Cells in the Biofilms*

Halothiobacillus HT1 is a heavy-metals-tolerant sulfur oxidizing bacterium, which belongs to Gammaproteobacteria, *Halothiobacillus* (GenBank accession number GU013549). In order to analyze the interaction of *Halothiobacillus* HT1 biofilm and CuS, the spatial distribution of live and dead cells in *Halothiobacillus* HT1 biofilm sections was studied using Live/Dead staining combined with CLSM imaging. As shown in Figure 1, after 72 h cultivation, the HT1 cells presented different distributions in the biofilm formed on CuS. At the air-biofilm interface and membrane-biofilm interface, the CLSM imaging results showed green, while in the middle, the results showed red. These results indicated that there were more live cells at the air-biofilm interface and membrane-biofilm interface than in the middle.

Figure 1. Spatial distribution of live and dead cells in *Halothiobacillus* HT1 biofilm reacted with CuS after dyeing and CLSM. (**A**) Light microscope images of *Halothiobacillus* HT1 biofilm; (**B**) a composite of live and dead cells; (**C**) live cells; (**D**) dead cells. Bars = 250 μm.

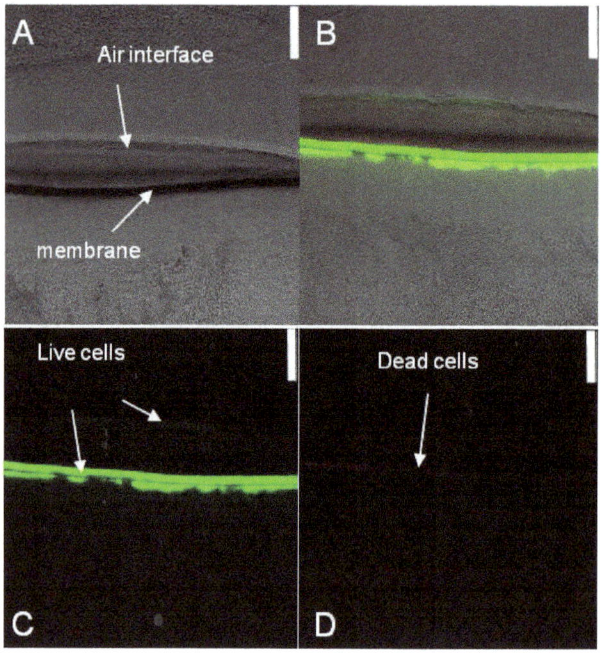

2.2. Spatial Distributions of Cu in Biofilm and Cu Speciation

XRF is considered to be a powerful tool for quantitative mapping of trace element distributions [16]. It can visualize the metal ion distribution in tissues or cells. The colony biofilms were thicker in the center and thinner at the edges. The thickness of the biofilm grown on CuS was about 150 μm, while the control was about 100 μm. Figure 2 showed the *Halothiobacillus* HT1 biofilms section, the scanning area of XRF and the distribution characters of the elements in the scanning area. The elements were not evenly distributed in the biofilms. There was a Cu accumulation layer in the middle of the *Halothiobacillus* HT1 biofilm in the presence of CuS, suggesting Cu penetrated about 100–150 μm from the membrane interface.

Compared with bulk XAFS, micro-XAFS could analyze the speciation of elements in small ozone. The Cu K-edge micro-XANES spectra were conducted to determine the speciation of Cu in the center and at the edge of the biofilm formed on CuS. Linear combination arithmetic was used to determine the probable Cu speciation in the sample as analyzed by the LSFitXAFS program. There were some differences in the XANES spectra of the center and the edge, as shown in Figure 3. As shown in Table 1, species resembling Cu-cysteine was the major species both in the center (67.8%) and at the edge (48.4%). In the center of the biofilm reacted with CuS, Cu-alginate- and Cu-citrate-like species accounted for 21.4% and 10.8%, respectively. While at the edge, Cu-histidine- (22.3%) and CuS (19.3%)-like species accounted for minor proportions, with Cu-alginate- and Cu-citrate-like species in smaller proportions (2.9% and 7.1%, respectively).

Figure 2. X-ray fluorescence microscopy (XRF) maps of qualitative spatial distributions and concentration gradients of elements in the sections of *Halothiobacillus* HT1 biofilms grown on LB medium. (**Left**) *Halothiobacillus* HT1 biofilm reacted with CuS; (**Right**) *Halothiobacillus* HT1 biofilm.

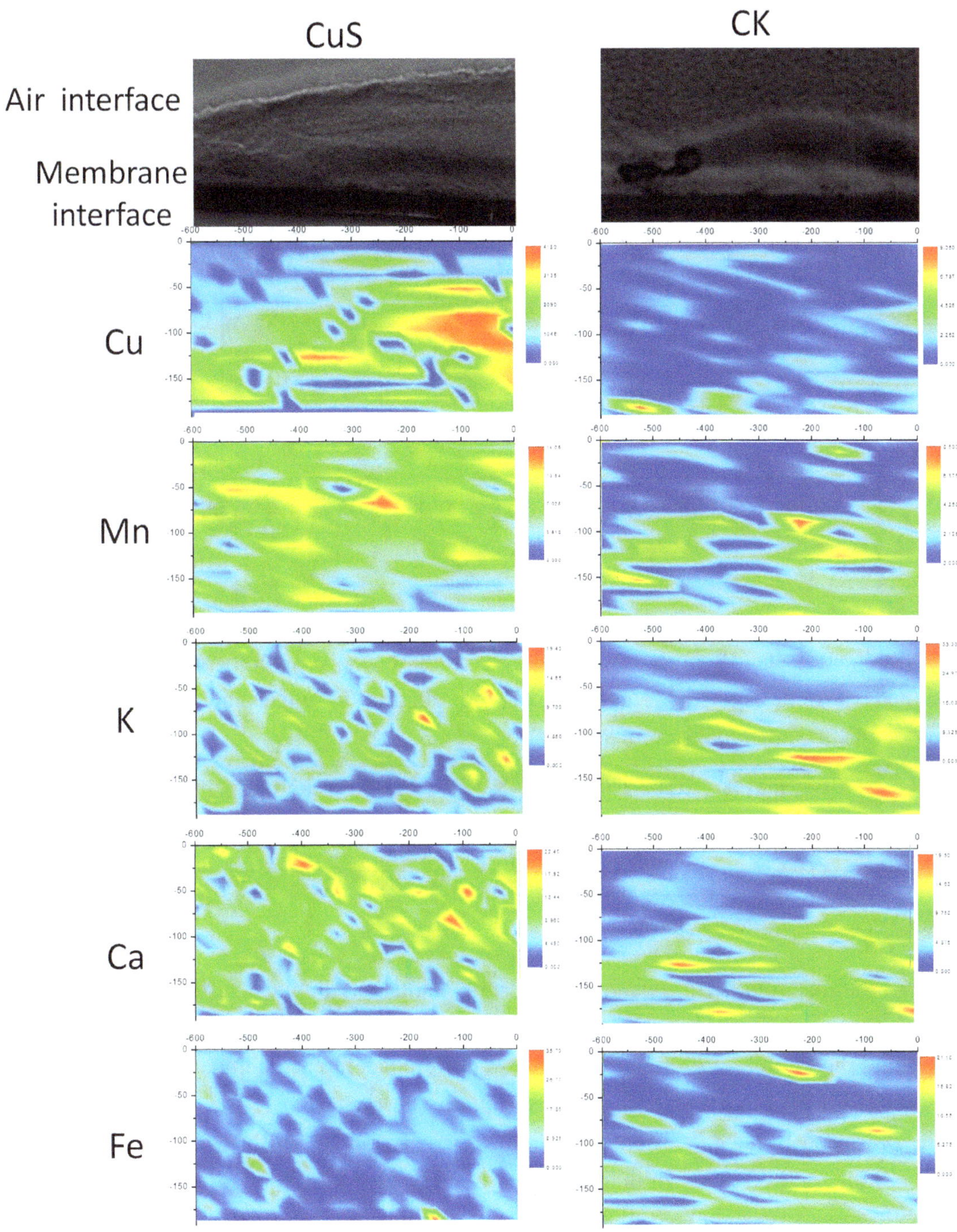

Figure 3. Comparison of copper K-edge XANES spectra of reference compounds and *Halothiobacillus* HT1 biofilm after being reacted with CuS. Black solid lines: data. Dashed lines: fits with data sets of reference compounds.

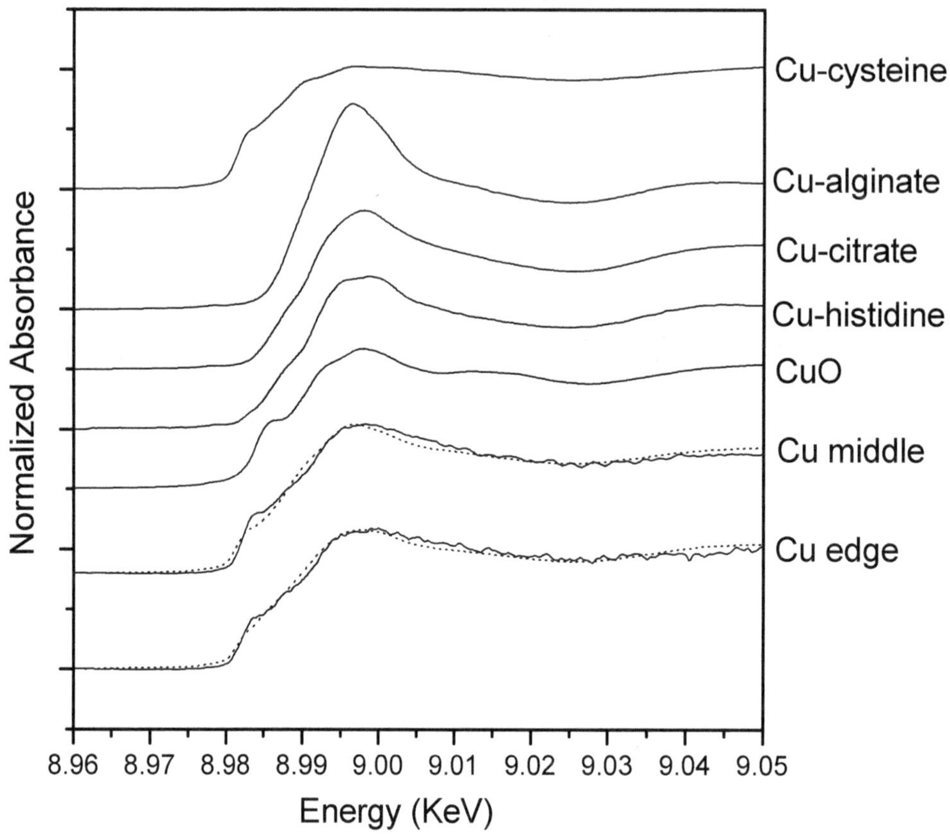

Table 1. The results of fitting the micro-XANES spectra with a linear combination of the measured data sets of representative model compounds. Copper oxidation states were assigned according to the similarity of the absorption edge energies of reference compounds. Rss: residual sum of squares.

Samples	Cu-cysteine (%)	Cu-alginate (%)	Cu-citrate (%)	Cu-histidine (%)	CuO (%)	Rss
Center	67.8	21.4	10.8	-	-	0.51
Edge	48.4	2.9	7.1	22.3	19.3	0.51

2.3. Sulfur Speciation in the Biofilms

As an essential macronutrient for microorganisms, plants and animals, sulfur (S) exists in a variety of organic and inorganic species, with oxidation states ranging from −2 to +6. The spectra of reference compounds and the samples are shown in Figure 4. Different reference compounds presented different structures of S. For example, cysteine, reduced glutathione presented R–SH, while cystine, oxidized glutathione presented R–S–S–R. There were multiple oxidations of sulfur in biofilm formed on CuS. As shown in Table 2, species resembling cysteine (R–SH) were the major sulfur species in the biofilm (77.3%), with species resembling sulfite (SO_3^{2-}) and sulfonate ($R–SO_2–O–X$) in minor proportions. While in the absence of CuS (the control sample), no oxidized sulfur was detected.

Figure 4. Comparison of sulfur K-edge XANES in *Halothiobacillus* HT1 biofilm (CK), *Halothiobacillus* HT1 biofilm reacted with CuS and reference compounds.

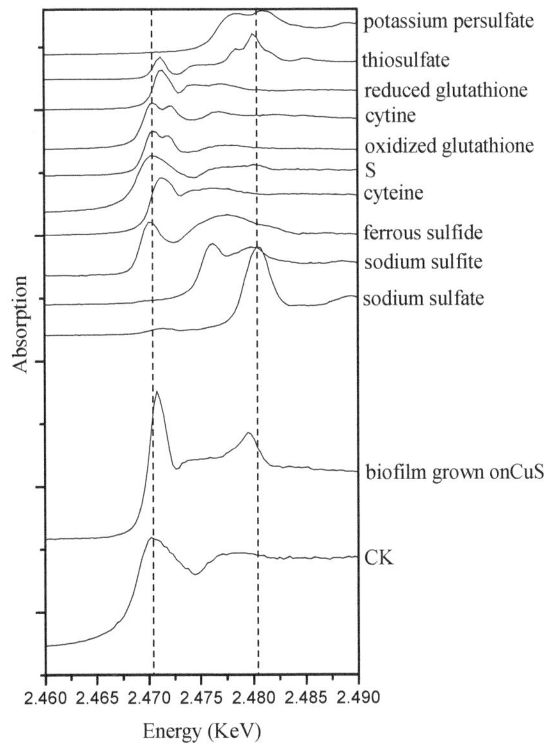

Table 2. Percentage of different S species in the biofilms as assessed by S K-edge XANES (atom% S).

Treatments	Sulfur	Cysteine	Sulfite	Sulfonate
CK	100	0	0	0
CuS	0	77.3	3.8	18.9

3. Discussion

3.1. The Relationship of Cu Distribution and Live/Dead Cells in Biofilms

Using CLSM, two active zones were found in the biofilm of HT1: the air-biofilm interface and the membrane-biofilm interface. Biofilm formation seemed to give rise to distinct chemical and redox gradients. The different distributions of the cells indicated that oxygen played an important role in the distribution of live cells [17]. At the air-biofilm interface, oxygen was enough for the growth of aerobes. As a result, *Halothiobacillus* HT1 cells in this area were active, leading to the larger quantities of live cells than in the middle. While at the membrane-biofilm interface, where the oxygen density was relatively low, more live cells were detected than in the middle. The reason might be attributed to the fact that this area played an important role in the transport of nutrition. Therefore, the cells were activated at the medium-biofilm interface. Biofilm growth of pathogenic bacteria might influence the rate of carbon and inorganic nutrient cycling [18]. The XRF results showed nonuniform distributions of the elements (Figure 4). Our results supported the previous reports that there were diverse niches in biofilms, and biofilms were physiologically distinct from bacteria in planktonic form [19–21]. The presence of gradients of dissolved gases, nutrients and solute concentrations might be responsible for the distribution of live and dead cells of *Halothiobacillus* HT1.

In addition, a better understanding of the distribution and mass transfer process within the biofilm is important for understanding their influences on biogeochemical transformation of metals. Metal might deposit as metallic oxide, phosphate, sulfate or carbonate at certain area, due to the heterogeneous distribution of pH, Eh and O_2. As a micro-analytical technique for the quantitative mapping of elemental distributions, XRF is compatible with fully hydrated biological samples [22,23]. In this study, XRF is used to detect the distribution of Cu in biofilm of *Halothiobacillus* HT1 formed on CuS. Our results indicated that CuS can be oxidized by *Halothiobacillus* HT1, as oxidized sulfur (sulfite and sulfonate) was detected using sulfur K-edge XANES analysis (Table 2). As a result, ionic copper was obtained and penetrated into the *Halothiobacillus* HT1 biofilm. As shown in Figure 2, copper accumulated in the layer near the membrane-biofilm interface. This result indicated that the CuS and *Halothiobacillus* HT1 biofilm reacted actively. This might be another reason for the larger quantity of live cells in membrane-biofilm interface than in the middle. The heterogeneous distribution of Cu found in this study was similar with the result of Templeton conducted on the distribution of Pb at biofilm-metal oxide interfaces [8]. Another study also showed that Zn equally distributed in the 12 μm surface layer of the 35 μm *E. coli* biofilm [24]. All these results confirmed the heterogeneous distribution of metal in microbe biofilms. Synchrotron-based XRF and micro-XANES spectra were successfully applied to investigate the spatial distribution of elements within HT1 biofilms and the speciation of Cu. Our results indicated that there was spatial heterogeneity vertically (Figure 2) and horizontally (Figure 3 and Table 1) within the biofilms

3.2. Effects of Halothiobacillus *HT1 Biofilms on CuS Transformation*

Some specific properties of biofilms can be related to the heterogeneity of their structure and composition. In this study, biofilms formed on CuS were about 150 μm thick, while the control was only about 100 μm, indicating that HT1 might react with the CuS, since CuS provided substrate for it. To further analyze how microbes influence the fate of metal sulfide, speciation of Cu and S was studied. Using sulfur K-edge XANES, sulfur transformation in the biofilms was found. As shown in Figure 4, sulfur was oxidized to cysteine, sulfite and sulfonate in the biofilms formed on CuS. This might be attributed to the reaction of CuS and *Halothiobacillus* HT1. Sulfur oxidizing bacteria are considered to be capable of oxidizing mineral sulfide to sulfate. In this study, oxidized sulfur detected in the biofilm indicated that CuS was oxidized biologically in the presence of *Halothiobacillus* HT1 biofilms. However, S was found to exist as diverse molecular species (cysteine, sulfite and sulfonate), which might be the intermediate products of sulfide oxidization, but not the end product (sulfate). The presence of cysteine, sulfite and sulfonate in the biofilm indicated heterogeneity in the redox state and implied that intermediate sulfur compounds were available for metabolism by *Halothiobacillus* HT1.

Intimate organic-mineral associations are often the result of bacterial activity. Attached bacteria and adsorbed organic matter may interact with sorption processes on metal sulfide surface by changing the characteristics of the electrical double layer at the solid-solution interface, blocking surface sites or providing a variety of new sites for metal binding. Extracellular polymeric substances (EPS) were considered to enclose biofilms and contain many active groups. Micro-XANES is an element-specific spectroscopic tool to identify the local environment of a target atom within a sample. Therefore, we also studied the speciation of copper in the biofilm formed on CuS by using micro-XANES. There were several active groups, such as hydroxyl (–OH), sulfhydryl (–SH) and amino ($-NH_2$), in the biofilms. The Cu-cysteine analog was the major Cu species. In correspondence, sulfur K-edge XANES results also confirmed that cysteine was the main sulfur species.

In order to survive heavy metals stress, microbes usually apply different mechanisms, such as biosorption, precipitation and complexation with sulfides, phosphates or carbonates. Active efflux or detoxification of metal ions by different transformations was found to be involved in many metal resistance systems [25]. For some bacteria, it was believed that inorganic polyphosphates and transport of metal phosphate complexes were involved in heavy metal tolerance [26–28]. Cu homeostasis proteins were also found in previous study [29]. In this study, cysteine analogs likely played an important role in the penetration of copper into the biofilm. The high percentage of Cu-cysteine-like species in the biofilm confirmed that the Cu bond to cysteine groups could be defenses against toxic copper. Complexation with –SH groups might contribute to the detoxification of Cu since –SH-containing groups were considered to be bioactive molecules, which can maintain Cu homeostasis through cation binding. These results were consistent with some previous reports concerning other bacteria, which showed that cellular proteins rich in thiol groups participated in the copper homeostasis [30].

At the biofilm-mineral interface, many critical processes occur. Our study attempted to study the interaction of CuS and *Halothiobacillus* HT1 biofilm at the micro scale. More work about the interfacial properties of the biofilm-mineral interface are needed in the future. For example, scanning transmission X-ray microscopy (STXM), which uses near-edge X-ray absorption spectroscopy (NEXAFS) as its contrast mechanism, can be further used to fully hydrate biological materials at a smaller scale (nanoscale).

4. Experimental Section

4.1. Biofilms Cultivation

Halothiobacillus HT1 was retrieved from −70 °C stock cultures by incubating on the LB agar plate at 28 °C for 48 h. Then, a single colony was suspended into 1 mL of sterile 0.9% NaCl. Zero-point-one grams of CuS (pretreated with 0.25 M EDTA, pH 8.0 for antioxidation and acetone for sulfide remover) and 1 mL sterile water were added in an EP tube to prepare the suspension. Then, 300 μL of this suspension was added on LB solid medium. Nuclepore® polyester membranes (13 mm diameter, 0.1 μm pore, 6 μm thick, Whatman) were sterilized by immersion in 70% ethanol for 2 min individually, air dried and transferred with sterile forceps to the LB solid media surface. Three membranes were spaced in an equilateral triangular pattern and 3 μL of the *Halothiobacillus* HT1 suspension were transferred to the center of each membrane and then inoculated at 28 °C for 72 h. Control treatments without adding CuS were also prepared. Each treatment was prepared in four groups for staining, XRF analysis, sulfur XANES and Cu micro-XANES analysis, respectively.

4.2. Live/Dead Cells Distribution Using CLSM Image

Biofilms were stained with the Live/Dead BacLight Kit (Molecular Probes) to determine the distribution of Live/Dead cells. Three microliters of Syto-9 and 3 μL propidium iodide were suspended in 1 mL 0.9% saline to prepare the staining agents. The biofilms were placed on the staining agent for 30 min in the dark. Live cells of *Halothiobacillus* HT1 stain green, and dead cells train orange-red.

Then, the stained biofilms were cryoembeded and cryosected as follows: biofilms were placed on the chuck. Cuts perpendicular to the plane of biofilms were made using a low profile microtome blade (Paracu™, McCormick™ Scientific). Fifty micrometer-thick cross sections from the middle of each biofilms were placed on glass coverslips and examined using microscopy to confirm the thickness and structure of biofilms sections and then stored at −20 °C for analysis. The samples were examined using a Leica TCS SP5 confocal laser scanning microscope (CLSM) to detect the distribution of live/dead cells of *Halothiobacillus* HT1.

4.3. Distribution of Cu in the Biofilm by Using Synchrotron-Based X-Ray Fluorescence

Biofilms were sectioned as described above and placed on 3M adhesive tapes (Scotch, 810, 3M Company, St. Paul, MN, USA). X-ray fluorescence (XRF) analysis of the biofilms was performed at the XRF microprobe station (beam line 4W1B) of Beijing Synchrotron Radiation Facility (BSRF), Institute of High Energy Physics of China. The electron energy in the storage ring was 2.2 GeV, with a current range from 78 to 120 mA. The size of the exciting X-ray beam was 10 μm × 20 μm. XRF spectra were collected by a PGY Si (Li) solid detector, positioned at 90° to the beam line. The scanning points of the samples were selected and observed by a microscope. Spectra data were processed by the AXIL program to integrate the area of the element excited peak. Relative contents of elements were calculated by calibrating the peak area with electron current and normalized with Compton scattering intensity. PyMca [31] and origin 8.0 was used to analyze the distribution of the elements in the biofilms.

4.4. Copper Speciation Analysis Using Cu K-Edge Micro-XANES

Two points located in the edge and center regions of the *Halothiobacillus* HT1 biofilm grown on CuS (the size of exciting X-ray beam was 40 μm × 70 μm) were selected for Cu micro-XANES analysis. A 50 μm thick cross section was obtained using cryosectioning methods described above and placed on 3 M adhesive tape and then stored at −20 °C until analysis. The Cu K-edge micro-XANES spectra were measured at beam line 15 U of Shanghai Synchrotron Radiation Facility (SSRF). It was operated at 3.5 GeV with a current of 210–150 mA in the fluorescence mode for the energy range from 8.95 to 9.06 keV. The fluorescence XANES spectra were collected for 5 s dwell times and analyzed as described by the previous description [32].

4.5. Sulfur Speciation Analysis Using Sulfur K-Edge XANES

Each of the biofilms were freeze dried, mixed and ground for XANES analysis. Sulfur speciation in the biofilms was analyzed by X-ray Absorption Spectroscopy at Beijing Synchrotron Radiation Facility (BSRF). The biofilms samples were laid on 3M tape for analysis. The storage ring was operated at the energy of 2.5 GeV with Si (111) double crystals. Spectra were recorded at 4B7A beamline (medium X-ray beamline, 2100–6000 eV) and scanned at step widths of 0.3 eV in the region between 2420 and 2520 eV. Reference compounds were ferrous sulfide, elemental sulfur, potassium persulfate, sodium thiosulfate, reduced glutathione, cystine, oxidized glutathione, cysteine, dimethyl sulfone, sodium diphenylamine sulfonate, sodium dodecyl sulfate (SDS), sodium sulfite and sodium sulfate. The X-ray energy was calibrated with reference to the spectrum of the highest resonance energy peak of Na_2SO_4 at 2,480.4 eV. The data were analyzed as described by the previous description [33].

5. Conclusions

Our results indicated that biofilm formation of *Halothiobacillus* HT1 gave rise to distinct chemical and redox gradients, leading to diverse niches in biofilms. There were two active zones in the biofilm of HT1: the air-biofilm interface and the membrane-biofilm interface. Copper was mainly distributed near the membrane-biofilm interface. *Halothiobacillus* HT1 biofilm oxidized CuS. Cysteine was synthesized for tolerance against Cu. The Cu-cysteine analog was involved in chelation within strain HT1 when it interacted with Cu to survive heavy metals stress.

Acknowledgments

This work was supported by the National Natural Science Foundation of China (11179025, 21177109, 41071302) and the Zhejiang Provincial Natural Science Foundation of China (R5110031). We gratefully thank the staff at the Beijing Synchrotron Radiation Facility (BSRF) and Shanghai Synchrotron Radiation Facility (SSRF) for their generous help.

References

1. Dynes, J.J.; Tyliszczak, T.; Araki, T.; Lawrence, J.R.; Swerhone, G.D.W.; Leppard, G.G.; Hitchcock, A.P. Speciation and quantitative mapping of metal species in microbial biofilms using scanning transmission X-ray Microscopy. *Environ. Sci. Technol.* **2006**, *40*, 1556–1565.
2. Geoffrey, M.G. Microbial influence on metal mobility and application for bioremediation. *Geoderma* **2004**, *122*, 109–119.
3. Ito, T.; Sugita, K.; Okabe, S. Isolation, characterization, and *in situ* detection of a novel chemolithoautotrophic sulfur-oxidizing bacterium in wastewater biofilms growing under microaerophilic conditions. *Appl. Environ. Microbiol.* **2004**, *70*, 3122–3129.
4. Ma, S.; Banfield, J.F. Micron-scale Fe^{2+}/Fe^{3+}, intermediate sulfur species and O_2 gradients across the biofilm-solution-sediment interface control biofilm organization. *Geochim. Cosmochim. Ac.* **2011**, *75*, 3568–3580.
5. Edwards, K.J.; Bond, P.L.; Druschel, G.K.; McGuire, M.M.; Hamers, R.J.; Banfield, J.F. Geochemical and biological aspects of sulfide mineral dissolution: Lessons from Iron Mountain, California. *Chem. Geol.* **2000**, *169*, 383–397.
6. Templeton, A.; Knowles, E. Microbial transformations of minerals and metals: Recent advances in geomicrobiology derived from synchrotron-based X-ray spectroscopy and X-ray microscopy. *Ann. Rev. Earth. Pl. Sc.* **2009**, *37*, 367–391.

7. Zachara, J.M.; Fredrickson, J.K.; Smith, S.C.; Gassman, P.L. Solubilization of Fe(III) oxide-bound trace metals by a dissimilatory Fe(III) reducing bacterium. *Geochim. Cosmochim. Ac.* **2001**, *65*, 75–93.
8. Templeton, A.S.; Trainor, T.P.; Traina, S.J.; Spormann, A.M.; Brown, G.E. Pb(II) distributions at biofilm-metal oxide interfaces. *Proc. Natl. Acad. Sci. USA* **2001**, *98*, 11897–11902.
9. Drenkard, E.; Ausubel, F.M. *Pseudomonas* biofilm formation and antibiotic resistance are linked to phenotypic variation. *Nature* **2002**, *416*, 740–743.
10. Fux, C.A.; Costerton, J.W.; Stewart, P.S.; Stoodley, P. Survival strategies of infectious biofilms. *Trends Microbiol.* **2005**, *13*, 34–40.
11. Sandt, C.; Smith-Palmer, T.; Pink, J.; Brennan, L. Pink, D. Confocal Raman microspectroscopy as a tool for studying the chemical heterogeneities of biofilms *in situ*. *J. Appl. Microbiol.* **2007**, *103*, 1808–1820.
12. Harrison, J.J.; Ceri, H.; Turner, R.J. Multimetal resistance and tolerance in microbial biofilms. *Nat. Rev. Microbiol.* **2007**, *5*, 928–938.
13. Harrison, J.J.; Ceri, H.; Stremick, C.A.; Turner, R.J. Biofilm susceptibility to metal toxicity. *Environ. Microbiol.* **2004**, *6*, 1220–1227.
14. Butcher, B.G.; Deane, S.M.; Rawlings, D.E. The chromosomal arsenic resistance genes of *Thiobacillus ferrooxidans* have an unusual arrangement and confer increased arsenic and antimony resistance to *Escherichia coli*. *Appl. Environ. Microbiol.* **2000**, *66*, 1826–1833.
15. Shi, J.Y.; Lin, H.R.; Yuan, X.F.; Zhao, Y.D. Isolation and characterization of a novel sulfur-oxidizing chemolithoautotroph *Halothiobacillus* from Pb polluted paddy soil. *Afr. J. Biotechnol.* **2011**, *10*, 4121–4126.
16. Ortega, R. Chemical elements distribution in cells. *Nucl. Instr. Meth. Phys. Res.* **2005**, *231*, 218–223.
17. Rani, S.A.; Pitts, B.; Beyenal, H.; Veluchamy, R.A.; Lewandowski, Z.; Davison, W.M.; Buckingham-Meyer, K.; Stewart, P.S. Spatial patterns of DNA replication, protein synthesis, and oxygen concentration within bacterial biofilms reveal diverse physiological states. *J. Bacteriol.* **2007**, *189*, 4223–4233.
18. Hall-Stoodley, L.; Costerton, J.W.; Stoodley, P. Bacterial biofilms: From the natural environment to infectious diseases. *Nat. Rev. Microbiol.* **2004**, *2*, 95–108.
19. Battin, T.J.; Sloan, W.T.; Kjelleberg, S.; Daims, H.; Head, I.M.; Curtis, T.P.; Eberl, L. Microbial landscapes: New paths to biofilm research. *Appl. Environ. Microbiol.* **2007**, *5*, 76–81.
20. Sauer, K.; Camper, A.K.; Ehrlich, G.D.; Costerton, J.W.; Davies, D.G. *Pseudomonas aeruginosa* displays multiple phenotypes during development as a biofilm. *J. Bacteriol.* **2002**, *184*, 1140–1154.
21. Stewart, P.S.; Franklin, M.J. Physiological heterogeneity in biofilms. *Nat. Rev. Microbiol.* **2008**, *6*, 199–210.
22. Chen, G.C.; Chen, X.C.; Yang, Y.Q.; Hay, A.G.; Yu, X.H.; Chen, Y.X. Sorption and distribution of copper in unsaturated *Pseudomonas putida* CZ1 biofilms as determined by X-ray fluorescence microscopy. *Appl. Environ. Microbiol.* **2011**, *77*, 4719–4727.

23. Fahrni, C.J. Biological applications of X-ray fluorescence microscopy: Exploring the subcellular topography and speciation of transition metals. *Curr. Opin. Chem. Biol.* **2007**, *11*, 121–127.
24. Hu, Z.Q.; Hidalgo, G.; Houston, P.L.; Hay, A.G.; Shuler, M.L.; Abruna, H.D.; Ghiorse, W.C.; Lion, L.W. Determination of spatial distributions of zinc and active biomass in microbial biofilms by two-photon laser scanning microscopy. *Appl. Environ. Microbiol.* **2005**, *71*, 4014–4021.
25. Silver, S.; Phung, L.T. Bacterial heavy metal resistance: New surprises. *Ann. Rev. Microbiol.* **1996**, *50*, 753–789.
26. Alvarez, S.; Jerez, C.A. Copper ions stimulate polyphosphate degradation and phosphate efflux in *Acidithiobacillus ferrooxidans*. *Appl. Environ. Microbiol.* **2004**, *70*, 5177–5182.
27. Kornberg, A.; Rao, N.N.; Ault-Richer, D. Inorganic polyphosphate: A molecule of many functions. *Ann. Rev. Biochem.* **1999**, *68*, 89–125.
28. Remonsellez, F.; Orell, A.; Jerez, C.A. Copper tolerance of the thermoacidophilic archaeon *Sulfolobus metallicus*: Possible role of polyphosphate metabolism. *Microbiology* **2006**, *152*, 59–66.
29. Kim, B.E.; Nevitt, T.; Thiele, D.J. Mechanisms for copper acquisition, distribution and regulation. *Nat. Chem. Biol.* **2008**, *4*, 176–185.
30. Quaranta, D.; McCarty, R.; Bandarian, V.; Rensing, C. The copper-inducible cin operon encodes an unusual methionine-rich azurin-like protein and a pre-Q0 reductase in *Pseudomonas putida* KT2440. *J. Bacteriol.* **2007**, *189*, 5361–5371.
31. PyMca. Available online: http://pymca.sourceforge.net (accessed on 15 July 2011).
32. Shi, J.Y.; Wu, B.; Yuan, X.F.; Cao, Y.Y.; Chen, X.C.; Chen, Y.X.; Hu, T.D. An X-ray absorption spectroscopy investigation of speciation and biotransformation of copper in *Elsholtzia splendens*. *Plant Soil* **2008**, *302*, 163–174.
33. Lin, H.R.; Shi, J.Y.; Chen, X.C.; Yang, J.J.; Chen, Y.X.; Zhao, Y.D.; Hu, T.D. Effects of lead upon the actions of sulfate-reducing bacteria in the rice rhizosphere. *Soil Biol. Biochem.* **2010**, *42*, 1038–1044..

8

Changing Microspatial Patterns of Sulfate-Reducing Microorganisms (SRM) during Cycling of Marine Stromatolite Mats

Alexandru I. Petrisor [1,2], Sandra Szyjka [3,†], Tomohiro Kawaguchi [3], Pieter T. Visscher [4], Robert Sean Norman [3] and Alan W. Decho [3,*]

[1] Department of Urban and Landscape Planning, School of Urban Planning, "Ion Mincu" University of Architecture and Urban Planning, str. Academiei nr. 18-20, sector 1, Bucharest 010014, Romania; E-Mail: alexandru_petrisor@yahoo.com

[2] National Institute for Research and Development in Constructions, Urbanism and Sustainable Spatial Development URBAN-INCERC, sos. Pantelimon, nr. 266, sector 2, Bucharest 021652, Romania

[3] Department of Environmental Health Sciences, Arnold School of Public Health, University of South Carolina, Columbia, SC 29208, USA; E-Mails: tkawaguchi@allenuniversity.edu (T.K.); rsnorman@mailbox.sc.edu (R.S.N.)

[4] Center for Integrative GeoSciences, University of Connecticut, 345 Mansfield Rd., U-2045 Storrs, CT 06269, USA; E-Mail: pieter.visscher@uconn.edu

[†] Present address: Department of Chemistry, University Duisburg-Essen, Universitätsstraße 2, Essen 45141, Germany; E-Mail: Sandra.szyjka@stud.uni-due.de.

* Author to whom correspondence should be addressed; E-Mail: awdecho@mailbox.sc.edu;

Abstract: Microspatial arrangements of sulfate-reducing microorganisms (SRM) in surface microbial mats (~1.5 mm) forming open marine stromatolites were investigated. Previous research revealed three different mat types associated with these stromatolites, each with a unique petrographic signature. Here we focused on comparing "non-lithifying" (Type-1) and "lithifying" (Type-2) mats. Our results revealed three major trends: (1) Molecular typing using the *dsrA* probe revealed a shift in the SRM community composition between Type-1 and Type-2 mats. Fluorescence *in-situ* hybridization (FISH) coupled to confocal scanning-laser microscopy (CSLM)-based image analyses, and

$^{35}SO_4^{2-}$-silver foil patterns showed that SRM were present in surfaces of both mat types, but in significantly ($p < 0.05$) higher abundances in Type-2 mats. Over 85% of SRM cells in the top 0.5 mm of Type-2 mats were contained in a dense 130 μm thick horizontal layer comprised of clusters of varying sizes; (2) Microspatial mapping revealed that locations of SRM and $CaCO_3$ precipitation were significantly correlated ($p < 0.05$); (3) Extracts from Type-2 mats contained acylhomoserine-lactones (C4- ,C6- ,*oxo*-C6,C7- ,C8- ,C10- ,C12- , C14-AHLs) involved in cell-cell communication. Similar AHLs were produced by SRM mat-isolates. These trends suggest that development of a microspatially-organized SRM community is closely-associated with the hallmark transition of stromatolite surface mats from a non-lithifying to a lithifying state.

Keywords: biofilms; EPS; microbial mats; microspatial; sulfate-reducing microorganisms; *dsrA* probe; chemical signals; $CaCO_3$; AHLs; $^{35}SO_4^{2-}$ silver-foil

---

Abbreviations: SRM, sulfate-reducing microorganisms; EPS, extracellular polymeric secretions; AHL, acylhomoserine lactones; QS, quorum sensing; $CaCO_3$, calcium carbonate; FISH, fluorescence *in-situ* hybridization; GIS, geographical information systems; CSLM, confocal scanning laser microscopy; daime, digital-image analysis in microbial ecology.

1. Introduction

Microbial mats exhibit dense horizontal arrays of different functional groups of bacteria and archaea living in microspatial proximity. The surface mats of open-water marine stromatolites (Highborne Cay, Bahamas) contain cyanobacteria and other common microbial functional groups such as aerobic heterotrophs, fermenters, anaerobic heterotrophs, notably sulfate reducing microbes and chemolithotrophs like sulfur oxidizing microbes [1,2]. This community cycles through three different mat types and collectively constructs organized, repeating horizontal layers of $CaCO_3$ (*i.e.*, micritic laminae and crusts), with different mineralogical features depending on community types [3,4].

Marine stromatolites represent dynamic biogeochemical systems having a long geological history. As the oldest known macrofossils on earth [5], extant marine stromatolites are still forming in isolated regions of shallow, open-water marine environments and are now known to result from microbially-mediated processes [4]. Stromatolites are ideal systems for studying microbial interactions and for examining mechanisms of organized biogeochemical precipitation of horizontal micritic crusts [4]. Interactions within and between key functional groups will be influenced, in part, by their microspatial proximities.

The surface microbial mats of Bahamian stromatolites are fueled by cyanobacterial autotrophy [6,7]. The surface communities of the mats repeatedly cycle through several distinct stages that have been termed Type-1, Type-2 and Type-3, and are categorized by characteristic changes in precipitation products, as outlined by Reid *et al.* [4]. Type-1 (binding and trapping) mats represent a non-lithifying, accretion/growth stage that possesses an abundant (and sticky) matrix of extracellular polymeric secretions (EPS) largely produced by cyanobacteria [8]. The EPS trap concentric $CaCO_3$ sediment grains called ooids, and promote an upward growth of the mats. Small microprecipitates are intermittently dispersed within the EPS [9]. This accreting community typically persists for weeks-to-months then transforms into a community that exhibits a distinct bright-green layer of cyanobacteria near the mat surface. Concurrently the surface EPS becomes a "non-sticky" gel and begins to precipitate small patches of $CaCO_3$. This morphs into the Type-2 (biofilm) community, which is visibly different from a Type-1 community in having a non-sticky mat surface and a thin, continuous (e.g., 20–50 μm) horizontal lithified layer of $CaCO_3$ (*i.e.*, micritic crust). Type-2 mats are thought to possess a more-structured microbial biofilm community of sulfate-reducing microorganisms (SRM), aerobes, sulfur-oxidizing bacteria, as well as cyanobacteria, and archaea [2]. Studies have suggested that SRM may be major heterotrophic consumers in Type-2 mats, and closely linked to the precipitation of thin laminae [1,10]. The lithifying stage sometimes further progresses into a Type-3 (endolithic) mat, which is characterized by abundant populations of endolithic coccoid cyanobacteria *Solentia* sp. that microbore, and fuse ooids through dissolution and re-precipitation of $CaCO_3$ into a thick contiguous micritized layer [4,10]. Intermittent invasions by eukaryotes can alter the development of these mat systems [11].

Over past decades a growing number of studies have shown that SRMs can exist and metabolize under oxic conditions [12–18]. Studies have shown that in marine stromatolites, the carbon products of photosynthesis are rapidly utilized by heterotrophic bacteria, including SRM [1,4,8,19]. During daylight, photosynthesis mat surface layers generate very high concentrations of molecular oxygen, mostly via cyanobacteria. Despite high O_2 levels during this time, SRM metabolic activities continue [13,16], accounting for as much as ten percent of total SRM daily carbon requirements. During darkness HS- oxidation under denitrifying conditions may lead to $CaCO_3$ precipitation [1,20]. Studies showed that concentrations of $CaCO_3$ precipitates were significantly higher in Type-2 (than in Type-1) mats [21]. Using $^{35}SO_4$ radioisotope approaches, Visscher and colleagues showed that sulfate reduction activities in Type-2 mats could be spatially aligned with precipitated lamina [10]. This has posited an important role of the SRM in the precipitation of laminae in Type-2 stromatolite mats. A similar role for SRM in precipitation of carbonate laminae has been described in lithifying hypersaline mats [22–24].

The development of a diverse, spatially-organized microbial community is often dependent upon interactions among its resident organisms and their physiochemical environment. Laboratory culture studies show that when bacteria are abundant and in spatial proximity they produce chemical signals, which are used to sense nearby cell densities and to coordinate gene expression among groups of cells in a process called quorum sensing [25]. More recently, a diverse array of chemical signals called acylhomoserine lactones (AHLs) were identified in the surface layers of stromatolite mats [26]. While quorum sensing is now a well-established process in laboratory cultures of bacteria, it is largely unexplored among the SRM [27] and its roles in natural communities are poorly understood [28,29].

Summarizing, SRM are likely to be an important regulatory component in the development and evolution of stromatolite mats [10], and in their precipitation of micritic crusts and laminae [1,22,23,30]. However, analyzing microspatial distributions of bacteria within intact microbial mats has been problematic. Here, we hypothesized that the SRM spatially organize in surface mats, communicate and coordinate activities using chemical signaling, and may be microspatially-associated with the precipitation of horizontal micritic crusts in Type-2 mats. Our study was designed to examine changes in the community, and *in situ* microspatial arrangements of SRM in non-lithifying (*i.e.*, Type-1) and lithifying (Type 2) stromatolite mats using fluorescence *in-situ* hybridization (FISH) probing coupled with confocal scanning laser microscopy (CSLM). Image-analyses, specifically using Geographical Information Systems (GIS) [31] and Digital image Analysis in Microbial Ecology (Daime; [32]) programs were employed to detect and compare changing microspatial arrangements of bacteria.

2. Results and Discussion

2.1. Overall Summary

Changes in the relative abundances and activities of specific functional groups of bacteria can be an important determinant for influencing elemental cycling and even the broader ecology of microbial mat systems [33]. In the open-water stromatolite mats of the Bahamas, the present study showed that the alternating stages of non-lithifying (Type-1) and lithifying (Type-2) surface mats possessed very different spatial distributions of bacteria, especially within the sulfate-reducing microorganism (SRM) clade. The classic Type-1 mats showed dispersions of cyanobacteria and heterotrophic bacteria, including the SRM, that were relatively random. As Type-1 mats transitioned into Type-2 mats the heterotrophic community, especially the SRM, became more abundant. Using GIS analyses, the area occupied by SRM cells in Type 2 mats was found to be double that of their Type-1 counterparts. Cells also became more microspatially-organized in Type-2 mats. This was accompanied by an increased frequency of cell clustering. Further, the relative sizes of clusters increased in Type-2 mats, eventually forming an almost contiguous thin (130 μm) horizontal layer of SRM at the uppermost mat surface. This suggested an increasing community organization. Development of this dense heterotrophic layer was concurrently associated with precipitation of $CaCO_3$, the production of cell-cell chemical signals, and a dramatic shift in the phenotypic properties of the mats.

2.2. Properties of Type-1 and Type-2 Mats

Light microscopy examinations of mat surfaces showed that Type-1 mats of stromatolites were characterized by an irregular and adherent surface (*i.e.*, Type-1 mat; Figure 1A), which collects sediment grains (*i.e.*, carbonate ooids) within a matrix of extracellular polymers (EPS). The EPS matrix is known to enhance light penetration into the mat [34]; a process that is associated with the physical stabilization of the mat since EPS often increases the cohesive properties of sediments [35]. Oxygen profiles show a diffuse zone of photosynthesis and $^{35}SO_4^{2-}$-labeled silver (Ag) foils indicated few SRM were present in the upper mm of the mat (Figure 1A, lower panel). This was followed by the appearance of a thin (30–50 μm thick) crust of $CaCO_3$ precipitate (*i.e.*, Type-2 mat; Figure 1B). The macroscopic appearance of the two types of mat surfaces was easily distinguishable under low magnification (*i.e.*, 70–150×) using a dissecting microscope.

Figure 1. Light micrographs of cross-sections showing surfaces of Type-1 and Type-2 stromatolite mats. Light micrographs of a Type-1 mat (**A**) show an irregular "sticky" EPS-laden surface that accretes ooid grains, while the Type-2 mat (**B**) is characterized by a "non-sticky, white precipitate" crust on the surface. Three ooids have been artificially placed on the Type-2 surface crust to further illustrate the precipitate. Scale bars = 500 μm. Lower panels show 2D images 1 × 1 mm in size of the surface of both mats (light grey line indicates the mat surface). Images were generated from $^{35}SO_4^{2-}$ silver (Ag) foil experiments. Mat cross-sections were incubated on silver foil impregnated with the sulfate radioisotope. SRM reduce the $^{35}SO_4^{2-}$ to $^{35}S^{2-}$, which precipitates as $Ag^{35}S$ is was visualized with radiography. Black pixels indicated areas of intense sulfate reducing activity.

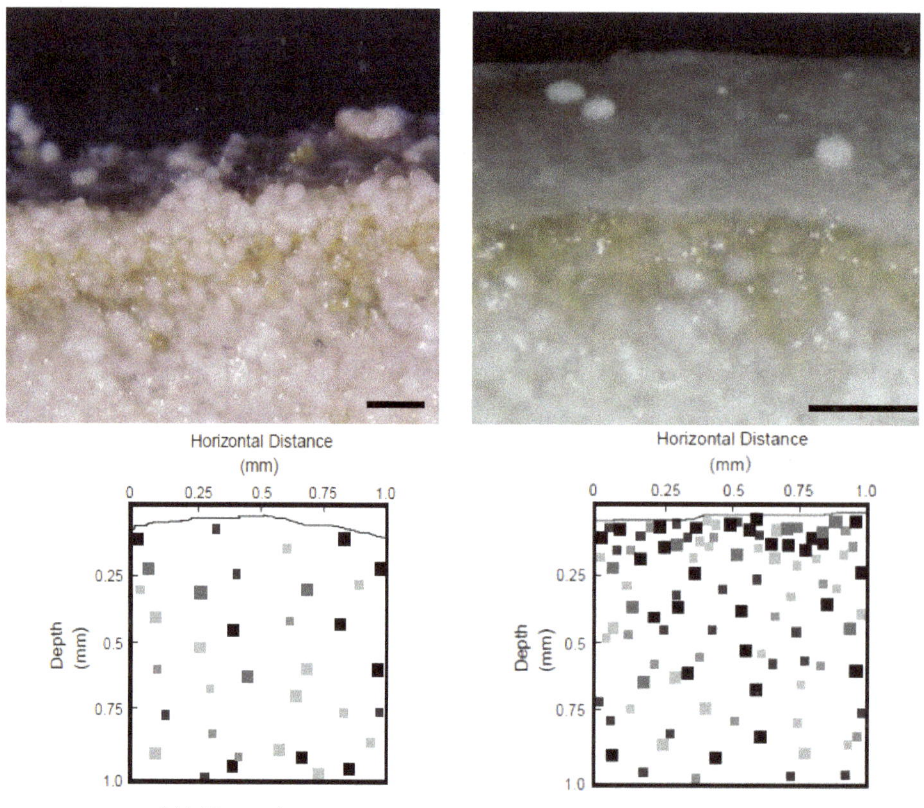

2.3. dsrA *Oligoprobing*

Our study utilized the *dsrA* oligoprobe to conservatively target SRM, including the sulfate-reducing bacteria. Sulfate reduction is known to occur in a wide range of bacteria, and some Archaea [36,37]. Through examinations of intact mat sections, and the coupling of fluorescence in situ hybridization (FISH) with confocal scanning laser microscopy (CSLM), and geographical information systems (GIS) analyses, it was possible to examine the *in situ* organization of SRM cells over microspatial scales and how the organization of this microbial functional group changed in different mat types within the stromatolite system. We showed that SRM were present in the upper-most surface layers of both Type-1 and Type-2 mats. However, within Type-1 mats, SRM cell abundances were comparatively lower, and SRM cells were relatively randomly dispersed within the EPS matrix. This was confirmed by the $^{35}SO_4^{2-}$-Ag foil observations (Figure 1B, lower panel). In contrast, distributions of cells within Type-2 mats showed that SRM became increasingly more abundant and more-clustered in their distribution, especially within the uppermost mat surface. The *dsrA* probe and $^{35}SO_4^{2-}$-Ag patterns are both in agreement for Type-2 mats as well.

The use of fluorescently-labeled rDNA oligo-probes for determinations of specific microbial cells in complex media presents several inherent obstacles [38,39]. The first relates to non-specific binding of probes in the complex media. Second, the signal intensity of a given cell is directly linked to ribosomal content and hence physiological activities of cells at the time of fixation. However, oligoprobes can be very useful for evaluation of changing spatial patterns of microorganisms [39,40]. To further examine the specificity of our *dsrA* oligoprobe, sections of Type-1 and Type-2 mats were imaged at higher magnifications (e.g., 600× to 1000×). Co-localized fluorescence of the oligoprobes (indicative of SRM cells) and also DAPI (4',6-diamidino-2-phenylindole, dihydrochloride) or PI (propidium iodide) were used to determine cell-specific binding of oligoprobes and to remove non-specific fluorescence signatures. Hence, cell areas containing both fluorescence signatures were counted as SRM cells. This allowed us to reduce the effects of non-specific binding of oligoprobes, and to digitally remove most of the non-specific binding effects in estimations of cell abundances.

2.4. Relative Abundances of SRM

Significantly ($p < 0.05$; Student's *t*-test) higher abundances of SRM cells were observed in the surfaces of Type-2 mats when compared with Type-1 mats. Using geographical information systems (GIS) analyses, abundances of cells were determined as a function of "fluorescence area" occupied by SRM cells relative to other fractions of the microbial community. Statistical analyses (Student's *t*-test) compared the portion of the total microbial community that was SRMs located within the top 130 μm of the two mat types. Appropriate transformations were made, where necessary, to normalize data for parametric tests. Relative abundances of SRMs in surfaces of Type-1 and Type-2 mats were expressed as a mean (±SE) percent (%) of total cell areas attributable to SRM within the uppermost 130 μm of the mats. Results of a student *t*-test showed the surfaces of Type-2 mats (88.0% ± 14.2%; $n = 31$ images analyzed) contained a significantly ($p < 0.0001$) higher abundance of cells (based on cell area) than Type-1 mats (39.7% ± 27.5%; $n = 21$). The results indicated that as the Type-1 community transitions into a Type-2 community, a significantly larger proportion of the total bacteria community (in Type-2 mats) were SRM.

2.4.1. SRM as Portion of Total Microbial Cells

Using direct counts of DAPI-stained cells we further confirmed that higher abundances of all microbial cells (*i.e.*, SRM, other bacteria, archaea) occurred in surfaces of Type-2 mats, when compared with Type-1 mats. The SRM comprised greater than half of the total microbial cells extractable from surface Type-2 mats. When cells were extracted from Type-2 mats and direct counts were estimated using either DAPI-staining or propidium-iodide-staining and compared to SRM cell counts using *dsrA*-staining, the SRMs represented 55.9% ± 20.0% and 56.1% ± 16.2% (mean ± SE), respectively, of the total bacteria cells detected. In contrast, SRM cells in Type-1 mats (as estimated using *dsrA*) comprised only 20.7% ± 9.3% of the total microbial cells. These observations were confirmed by the $^{35}SO_4^{2-}$-Ag foil observations that documented a 2D distribution of sulfate reducing activity (Figure 1; [10]).

Image analyses revealed interesting spatial patterns of bacteria. Images were collected from cross-sections of surface mats and focused analyses from the immediate mat surface to approximately 0.75 mm depth. Additionally, we analyzed spatial variability of the surface over a full horizontal distance of 850 μm. This allowed us to examine two-dimensional spatial patterns (e.g., horizontal layering, clustering, and dispersion) over relatively large regions of the uppermost surface of Type-1 and Type-2 mats (Figure 2A1,B1). Higher magnifications (1000×) were then used to examine smaller scale (e.g., 1 to 50 μm) patterns and clustering of cells (Figure 2A2,B2).

Figure 2. Confocal scanning laser micrographs (CSLM) illustrating relative changes microspatial distributions of SRM cells near the surface of (**A1,A2**) Type-1 (*i.e.*, relatively-scattered) and (**B1,B2**) Type-2 (*i.e.*, highly-clustered) mats. Images are cross-sections of surface mats showing SRM cells (green fluorescence; *dsrA* FISH probe), heterotrophic bacteria (red fluorescence stained with propidium-iodide (PI)) and cyanobacteria (red autofluorescence), and ooid sediment grains (artificial blue-color). Yellow circles illustrate typical clustering of SRM cells. Scale bars in **A1** and **B1** = 100 μm; in **A2** and **B2** = 10 μm.

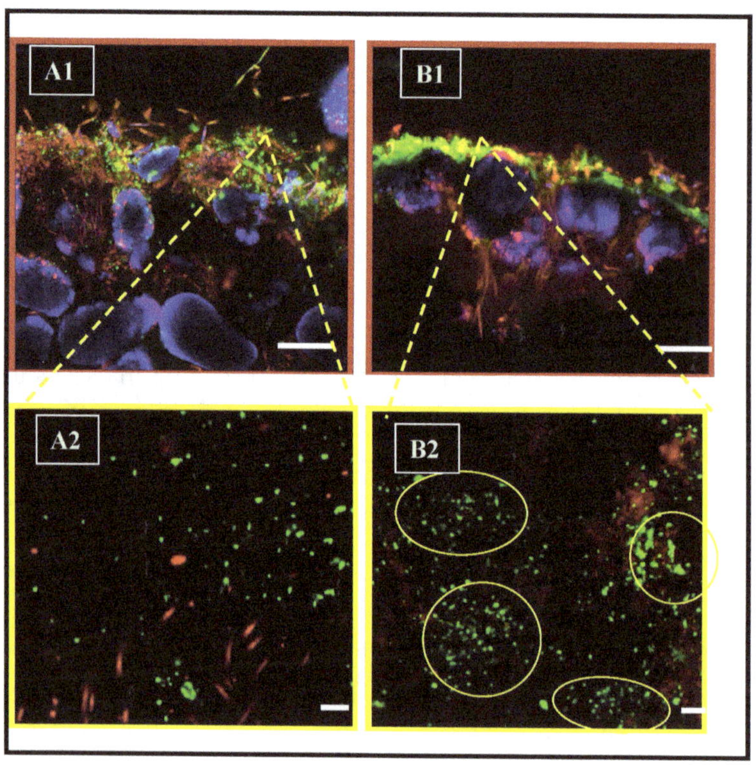

2.5. Precipitation Patterns: Microspatial Associations of SRMs and Precipitates

A highly-significant ($p < 0.05$; Student's *t*-test) statistical difference was detected in the areas occupied by precipitates. Results showed that precipitates were less abundant, in terms of area, in Type-1 mats when compared with Type-2 mats.

Based on the assumption that precipitation of $CaCO_3$ was related to SRM activities, we examined the microspatial locations of SRM cells and $CaCO_3$ precipitates within images from both Type-1 and Type-2 mats. A significant ($p < 0.05$) correlation ($r = 0.757$) was found linking SRM and $CaCO_3$ precipitates within the same image ($n = 34$). In both Type-1 and Type-2 mats, there was a close microspatial association of SRM cells and $CaCO_3$ precipitates with SRMs constituting over 80% of microbial cells that were located within a 4.4 μm distance of precipitates (Figure 3). Most of these cells occurred within a 1.1 μm distance (Table 1). This is noteworthy because although precipitates occur to a limited extent in Type-1 mats, SRM were still closely-associated with the precipitates that were present. This suggested a close relationship of SRMs and the precipitation process in both mat types.

Figure 3. Box-plot showing the percent of area occupied by all microbial cells, which were SRM. Results show that in Type-2 mats, over 80% of microbial cells (based on area occupied) were SRM. Note: Type-1 mats ($n = 21$) and Type-2 mats ($n = 31$); tails represent 95% confidence intervals (CI).

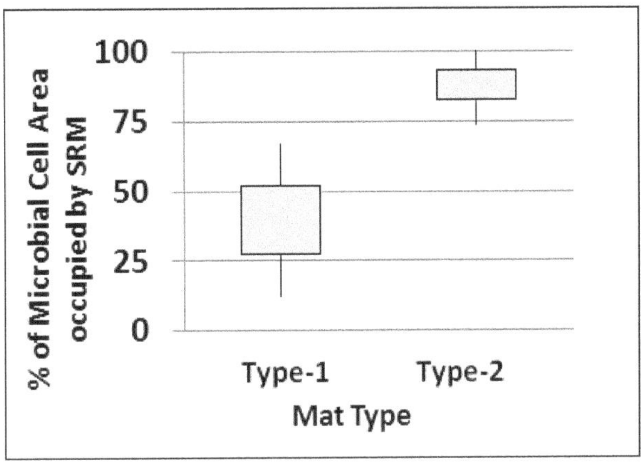

Table 1. Microspatial proximity between SRMs and $CaCO_3$ precipitates in Type-1 and Type-2 mats. Table shows percentages of total bacteria, located within 1.1, 2.2, or 4.4 μm distances from precipitates, which were SRM. Note that wherever precipitates occurred, greater than 82% of bacteria in proximity to precipitates were SRM. (n = number of samples analyzed; p-value represents results of ANOVA F-test). Type-1 mats were found to be significantly different from Type-2 ($p < 0.05$). * = designates statistical significance at $p \leq 0.05$.

% Bacteria near precipitates that were SRMs	Distance of SRM cells from $CaCO_3$ Precipitates					
	≤1.10 μm		≤2.20 μm		≤4.40 μm	
	Type-1 ($n = 12$)	Type-2 ($n = 29$)	Type-1 ($n = 12$)	Type-2 ($n = 29$)	Type-1 ($n = 12$)	Type-2 ($n = 29$)
Mean (±SE)	82.29 * ±29.92	95.51 ±7.60	82.71 * ±29.98	95.78 ±7.37	85.36 * ±25.23	96.16 ±7.11

It is important to note that in observing both Type-1 and Type-2 natural mats, variability existed over small spatial scales in the patterns of cells and precipitation products. This is likely a result of the localized interactions between bacteria and their environment. While this variability may be adaptive, in an ecological sense, it resulted in having to examine a large number of images to acquire sufficient statistical power for examination of potential differences (if present). Examination of the vertical distribution of SRMs situated within the top 500 μm indicated that the majority (over 85%) of SRM cells were located in the top 130 μm of the surface of Type-2 mats. These results suggest that SRM distributions may be used as an instrument of discrimination for categorization between Type-1 and Type-2 mats, with higher surface abundances of SRM occurring in Type-2 mats.

2.6. Phylogenetic Analysis of the dsrA *Sequences*

Phylogenetic relationships of *dsrA* gene sequences retrieved from Type-1 and Type-1-2 stromatolite mats revealed an overall low diversity (Figure 4). Type-1 *dsrA* clone sequences formed 9 different phylogenetic groups with nearly 72% of clone sequences located in a single clade most similar to *dsrA* genes of the Gram-negative delta-proteobacteria *Desulfovibrio*. Type-2 *dsrA* clones formed 6 different phylogenetic groups with nearly 83% of all clone sequences located in a single clade most similar to the delta-proteobacteria *Desulfomonile tiedjei* and other uncultured SRM capable of autotrophic growth. Most of the few remaining *dsrA* clone sequences formed monophyletic lineages that were distinct for either Type-1 or Type-2 stromatolite mats and included sequences similar to the deeply branching *Thermodesulfovibrio yellowstonii* and other uncultured sulfate-reducing bacteria. Preliminary 16S rDNA investigations of SRM diversity in a hypersaline lake with lithifying and non-lithifying mats [22], showed a dominance of delta-proteobacteria (91% and 64% of total diversity in lithifying and non-lithifying mats, respectively [2]. In this study, a wider diversity of delta-proteobacteria was observed in the lithifying mats when compared to non-lithifying mats and SRM activity was associated with the upper layer of the mats that were forming a $CaCO_3$ crust. This suggests that patterns observed in this study could apply to other lithifying systems as well.

2.7. Microspatial Clustering Analyses

Clustering, defined here as the aggregation of cells in spatial proximity, is likely an important parameter for assessing the microbial communities of stromatolites. When microbial cells are clustering together in proximity it increases their ability to interact in both positive and negative manners. Such clusters may provide a suitable proxy indicative of chemical communications, such as quorum sensing (QS) [25] and/or efficiency sensing [41]; processes that bacteria and other microorganisms likely utilize under natural conditions, especially within biofilms (e.g., microbial mats). SRM are physiologically challenged by the exposure to high O_2 levels at the surface of the mats where their activity peaks (see [2] for review). It is thought that this high activity is supported by abundant organic carbon, especially low-molecular weight compounds [8,19]. Recently QS signals have been extracted from marine stromatolite mats [26]. QS signals could be correlated with SRM and were postulated to play an important role in enabling these anaerobes to cope with O_2 concentrations that are deleterious to their physiology [42]. QS contributes to the coordination of gene expression and metabolic activities by neighboring cells, and may play important roles in the development of microbial consortia under natural conditions [42]. In other systems, QS signaling has been shown to be detectable by cells at distances extending up to 73 μm [43]. A second benefit of chemical communication resides in efficiency sensing, often considered an extended form of quorum sensing.

Efficiency sensing, however, provides cells with the ability to assess the diffusional properties of their proximal extracellular environment [41]. Finally, clustering invokes a new (and smaller) spatial scale perspective for understanding the formation of sharp geochemical gradients and the efficiency of elemental cycling that are characteristic of mats.

Figure 4. Phylogenetic tree based on translated amino acid sequences of PCR-amplified dissimilatory sulfite reductase *dsrA* genes retrieved from type I and type II stromatolites. Tree shows distributions of clones related to known sulfur-reducing bacteria and closely related sequences obtained from the GenBank database. GenBank accession numbers are shown in parentheses for non-collapsed branches and are as follows for collapsed branches: [a] AFA43406, EU127914, BAB55577, AFA43404, BAB55579, AB061543; [b] ACI31420, ABK90679; [c] ABK90745, AF334595, ABK90741, ABK90691, AAO61116, ABK90759; [d] AF271769, AF273029; [e] AF271771, AF334598; [f] AF418193, CAY20641, CAY20696; [g] YP003806924, AAK83215, AF334600; [h] AEX31202, CAJ84858, CAQ77308; [i] ACJ11472, CAJ84838, ACJ11485, ABK90809. The tree was constructed using the maximum likelihood method in MEGA 5 with values at nodes representing bootstrap confidence values with 1000 resamplings. Bootstrap values are shown for branches with more than 50% bootstrap support. Scale bar represents 0.1 substitutions per site.

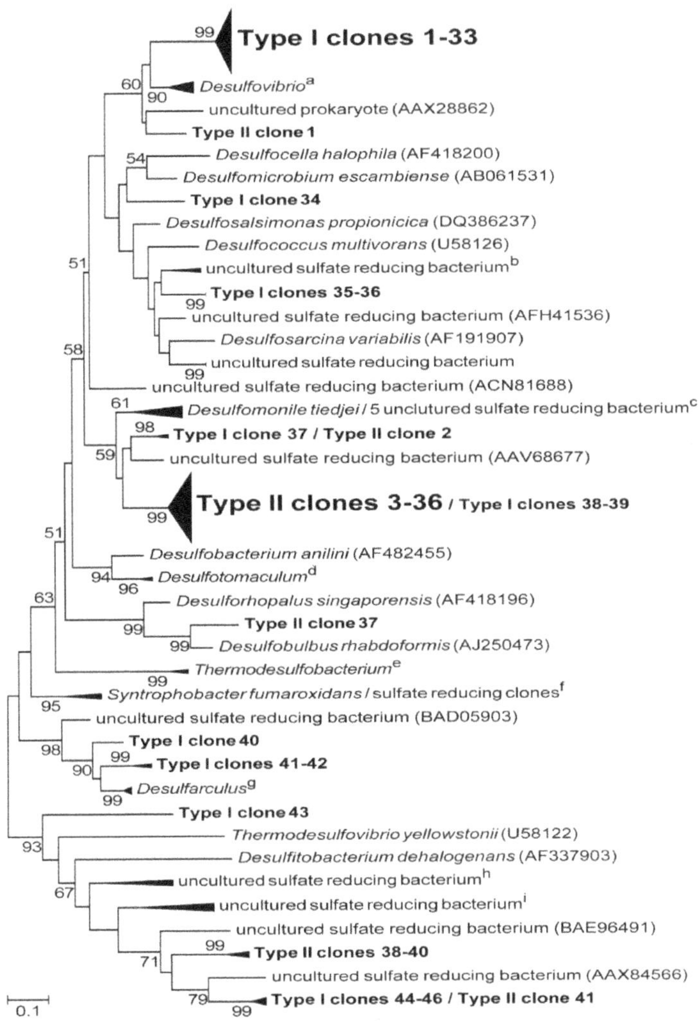

We were able to show that SRM showed little- or no-clustering in Type-1 mats but that very well-developed clustering occurred in Type-2 mats. The rapid upward growth (accreting) nature of Type-1 mats may not allow for such spatial organization to develop. The microspatial organization of cells into clusters (*i.e.*, groups of cells in proximity) was discernible at several spatial scales. Imaging using CSLM was coupled to the general labeling of cells using DAPI and PI, and more specific labeling using FISH targeting the SRM group. Using this approach, two different spatial scales of clustering became detectable. At relatively low magnifications (e.g., 200×) the distinctly higher abundances of SRMs were easily visualized near the surface of Type-2 mats (Figure 2). The non-lithifying Type-1 mats exhibited lower abundances and a relatively "random" distribution of SRM, and other bacteria, when compared with the non-random organization of bacteria in Type-2 mats. Overall differences determined by ANOVA were significant ($F = 33.55$, $p \leq 0.05$). All aposteriori specific tests (Bonferroni, and Scheffé) placed Type-1 different from the Type-2 mats, the latter of which exhibited significantly greater abundances of SRMs. At higher magnifications it became apparent that the Type-2 mat community exhibited an increase in clustering and microspatial organization, especially with regard to the SRM functional group (Figure 2). The frequency of SRM cell clusters increased, when compared with Type-1. Finally, the mean size (and variance) of clusters also increased as mats develop from a Type-1 to a Type-2 state, implying that some clusters became quite large. This occurred in the uppermost 50 μm of the surface biofilm.

These patterns were supported by image analyses using GIS [44] and Daime [32,45] programs and resulted in statistically ($p < 0.001$) higher abundances of SRM in the surfaces of Type-2 mats (when compared with Type-1). Two different, but complementary, methodological approaches (*i.e.*, Daime and GIS) were used in this study to detect microspatial clustering of cells.

2.7.1. The Daime Approach

The first approach, the Daime program [32], allowed us to examine all cell-cell distances within an image and graph the distances. Analyses of SRM spatial arrangements showed that in Type-1 mats (Figure 5A), the pair cross-correlation index $g(r)$ was close to 1 for cell-to-cell distances ranging from 0.1 to 6.44 μm, which is indicative of a relatively random distribution. A flat line ($r = 1$) was indicative of a relatively random distribution, where all cell-cell distances were equally probable. In Type-2 mats (Figure 5B), by contrast, the pair cross-correlation index was above 3 at a distance 0.36 μm, and rose to 52 at cell-cell distances of 0.03 μm. These data indicated that the SRM had a high degree of clustering, especially where cell-cell distances were very short. It can be inferred from these data that clusters were abundant in Type-2 mats and that the cells within SRM clusters were in very close proximity (*i.e.*, from 0.03 to 0.36 μm). Overall, when comparing cell distributions in Type-1 and Type-2 surface mats, there was increased clustering observed in Type-2 mats.

2.7.2. The GIS Approach

A second approach utilized GIS examined clustering of SRM cells within the surfaces of Type-1, and Type-2 mats. For each image a buffer area was created that extended from the surface of the mat to approximately 130 μm depth. Detection of SRM cells within the buffer area was based on color (as described above) using image classification of FISH-probed cells. A concentric region having a 10 μm diameter was generated around each cell. A cluster represented a group of cells having overlapping concentric regions. Subsequent statistical selection of clusters was subjectively based on cluster areas representing greater than five cells having overlapping concentric regions. The size (*i.e.*, area) of each detected cell cluster was measured. While the two methods utilize different approaches to detect clustering, both revealed a similar inference-increased clustering present in Type-2 mats.

Figure 5. Microspatial clustering arrangements of SRM cells located in the surfaces of stromatolite mats using *Daime* analyses. The graphs exhibit the pair cross-correlation function $g(r)$ for SRM cells. (**A**) In Type-1 mats, the relatively horizontal line where $g(r)$ approximates 1 indicates relatively random SRM distributions over cell-cell distances ranging from 0.1 to 6.44 μm; (**B**) In Type-2 mats, values of $g(r)$ above 1 indicate a high degree of clustering of SRM cells, especially over short (e.g., 0.03 to 0.36 μm) cell-to-cell distances. This indicates that cells in Type-2 mats are clustered closely together.

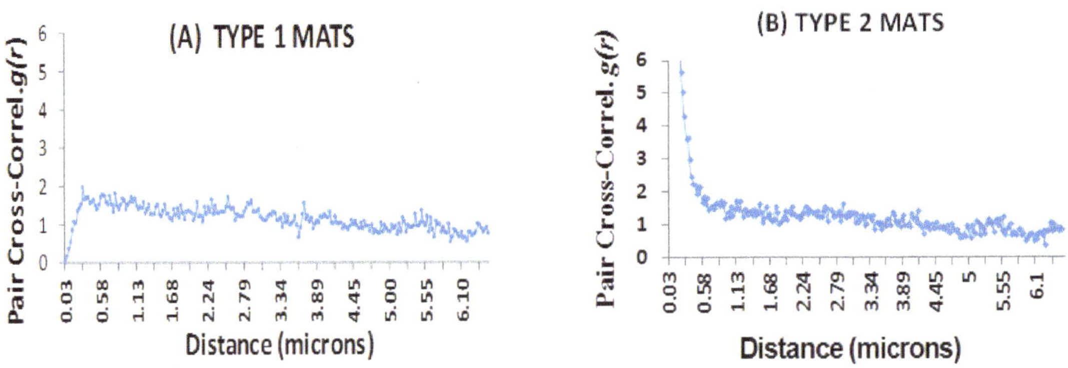

Finally, the size distribution of SRM clusters (including individual cells) was statistically analyzed using samples of 20 images that were randomly selected from microspatial regions within images from each mat type (Type-1, Type-2, and incipient Type-2) labeled with the *dsrA* oligoprobe. Type-2 exhibits the largest clusters (Figure 6). The mean cluster size was comparatively small in Type-1 mats and large in Type-2 mats. Variability followed the same pattern, increasing from Type-1 to Type-2.

2.7.3. Image Analyses

Proper image interpretation was needed to examine microscopic spatial patterns of cells within the mats. We employed GIS as a tool to decipher and interpret CSLM images collected after FISH probing, due to its power for examining spatial relationships between specific image features [46]. In order to conduct GIS interpolation of spatial relationships between different image features (e.g., groups of bacteria), it was necessary to "ground-truth" image features. This allowed for more accurate and precise quantification, and statistical comparisons of observed image features. In GIS, this is typically accomplished through "on-the-ground" sampling of the actual environment being imaged. However, in order to "ground-truth" the microscopic features of our samples (and their images) we employed separate "calibration" studies (*i.e.*, using fluorescent microspheres) designed to "ground-truth" our microscopy-based image data.

Quantitative microspatial analyses of *in-situ* microbial cells present certain logistical constraints that are not present in the analysis of dispersed cells. In the stromatolite mats, bacterial cells often occurred in aggregated groups or "clusters". Clustering of cells needed evaluation at several spatial scales in order to detect patterns of heterogeneity. Specifically, we wanted to determine if the relatively contiguous horizontal layer of dense SRM that was visible at larger spatial scales was composed of groups of smaller clusters. We employed the analysis of cell area (fluorescence) to examine *in-situ* microbial spatial patterns within stromatolites. Experimental additions of bacteria-sized (1.0 μm) fluorescent microspheres to mats (and no-mat controls) were used to assess the ability of GIS to "count cells" using cell area (based on pixels). The GIS approach (*i.e.*, cell area-derived counts) was compared with the direct counts method, and product moment correlation coefficients (r) were computed for the associations. Under these circumstances the GIS approach proved highly useful.

In the absence of mat, the correlation coefficient (r) between areas and the known concentration was 0.8054, and the correlation coefficient between direct counts and the known concentration was 0.8136. Areas and counts were also highly correlated (r = 0.9269). Additions of microspheres to natural Type-1 mats yielded a high correlation (r = 0.767) between area counts and direct counts. It is realized that extension of microsphere-based estimates to natural systems must be viewed conservatively since all microbial cells are neither spherical nor exactly 1 μm in diameter (*i.e.*, as the microspheres). Second, extraction efficiencies of microbial cells (e.g., for direct counts) from any natural matrix are uncertain, at best. Hence, the empirical estimates generated here are considered to be conservative ones. This further supports previous assertions that only relative abundances, but not absolute (*i.e.*, accurate) abundances, of cells should be estimated from complex matrices [39] such as microbial mats.

Results of microbial cell estimations derived from both direct counts and area computations, by inherent design, were subject to certain limitations. The first limitation is inherent to the process of image acquisition: many images contain only portions of items (e.g., cells or beads). In terms of counting, fragments or "small" items were summed up approximately to obtain an integer. Therefore, the solution used was the "counting rule". The problem disappears when total areas are computed. A second limitation involves image overlap [47]. This problem affects the computation of areas in the absence of a mathematical model that would account for overlapping objects. The human eye, for example, can readily distinguish between overlapping beads, and as a result traditional counting was less affected. While area computations were slightly influenced by this, the solution was approached in the same fashion as above (*i.e.*, through direct count comparisons) and the results were comparable. A third limitation relates to the three-dimensional nature of samples. Items situated slightly below the plane of focus sometimes produce residual fluorescence and appear as smaller items of the same kind or fragments. While those items might have been counted during direct counts, it was difficult to generate an objective means (*i.e.*, a systematic counting rule) to account for such items. A simple solution, however, was obtained when areas were computed during image analysis. The solution resided in the image classification process. Items situated below the plane of focus fluoresced at a lower intensity. Based on the threshold value some of them were classified as background and eliminated from computations, while others were registered as items of interest. As a result, area

Figure 6. Scheme illustrating detection of SRM clusters using GIS. (**1**) CSLM micrograph showing SRM cells labeled with *dsrA* probe with background digitally-removed, and identification of individual SRM cells (*i.e.*, black dots); (**2**) generation of artificial concentric regions with same width (10 μm) around each cell or group of cells; (**3**) identification of overlapping concentric regions; (**4**) statistical selection of clusters based on area (e.g., overlapping areas of > five cells); (**5**) Graph showing cluster sizes of SRM cells in Type-1 and Type-2 mats. Means and 95% confidence intervals are expressed as areas for SRM clusters. Note the significantly larger sizes and variability in cluster-sizes detected in Type-2 mats.

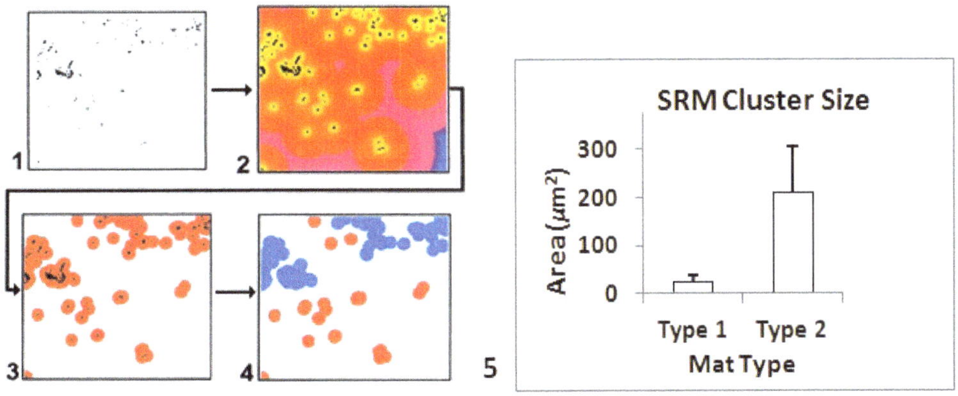

Computation incorporated a systematic approach to overcome this difficulty. Finally, the GIS-based approach was proposed as an alternative to the direct-counts method or other methods, and not as a replacement. Statistical analyses indicated that there were no significant differences between the direct counts and GIS methods when used to estimate the concentrations of microspheres, and area computations using GIS represented a successful alternative for estimating relative abundances of microbial cells in this mat system, especially at high cell abundances.

2.8. Ground-Truthing GIS at Microbial Spatial Scales

2.8.1. Fluorescent Microsphere Additions to Type 1 Mats

Results from analyses between areas of microspheres computed (via GIS) for each image individually and the total number of microspheres counted within the same image using, showed a highly-significant ($p < 0.0001$) product moment correlation coefficient ($r = 0.767$).

2.9. AHL Chemical Signals within Type-2 Mats

The high abundances of SRM cells underscore the potential impact of this clade on the mat system. The process of cell–cell chemical communication, called quorum sensing, facilitates coordination of group activities, and is now realized to play important roles in natural microbial communities [25–29]. Given the importance of sulfate reduction across many environments, it is therefore surprising that few reports exist for quorum sensing within the sulfate reducing clade, either within the delta proteobacteria [27] or the archaea. This earlier study [27] noted production of several AHLs by a stromatolite mat isolate of *Desulfovibrio* sp. (strain H2.3jlac), one of the same strains examined in this study. We examined two additional strains of SRB isolated from a Type-2 stromatolite mat: *Desulfovibrio* strain H2.3jman (isolated on mannose as the electron donor) and *Desulfovibrio* strain

H12.1lac (isolated on lactate as electron donor). Both strains also produced a wide range of AHLs (e.g., C6, C7, C8, C10) under standard culture conditions (Table 2, Figure 7). These are the same molecular congeners of AHL signals that were extracted from our natural mats, where high abundances of SRM were found.

Table 2. Summary table showing acylhomoserine lactones (AHL) extracted from the Type-2 surface mats of marine stromatolites, and from two stromatolite isolates of sulfate-reducing bacteria (SRB). AHLs were identified using mass-spectrometry, and are designated as C4-, C6-, C8-, *etc.*, based on the number of carbons in the acyl chain. An *oxo*-C6-AHL indicates a C6-AHL having an *oxo*-group at the C3-position. (* same strain used in [27]).

Sample	Strain designation	AHLs detected							
Type-2 mat extract	-	**C4-**	**C6-**	**C7-**	**C8-**	**C10-**	**C12-**	**C14-**	***oxo*-C6**
Desulfovibrio vulgaris (SRB) subsp. *oxamicus*	ATCC 33405D	**C4-**	-	-	**C8-**	-	-	-	-
SRB isolates from Type-2 mats:									
Desulfovibro strain 12.1Lac	GeneBank No. DQ822785	-	**C6-**	**C7-**	**C8-**	-	-	-	-
Desulfovibrio strain H2.3jLac *	GeneBank No. DQ822786	-	**C6-**	**C7-**	**C8-**	**C10-**	**C12-**	-	***oxo*-C6**
Desulfovibrio strain H2.3jman	-	-	**C6-**	**C7-**	**C8-**	**C10-**	-	-	-

The observed high abundances and clustering of microbial cells, coupled to the three-dimensional EPS matrix present within mats provide an ideal landscape to foster chemical communication among microbial cells, especially within Type-2 mats. The abundant SRM cell clusters, which were observed in the uppermost surfaces of the Type-2 mats using CSLM, present an ideal location for quorum sensing to occur in the mat. Under the natural conditions within microbial mats and the diffusional constraints related to EPS, quorum sensing among cells is likely to efficiently occur over relatively small spatial scales (e.g., 10's of μm). Interestingly the sizes of SRM clusters, which we measured in Type-2 mats, also occurred within this size range. It must be emphasized, however, that a single mat sample (sample core area = 5.07 cm^2) used for signal analyses contains a multitude of microbial clusters. Thus the microspatial variability of AHL signals could not be addressed here.

Figure 7. Spectra showing AHLs extracted from Type 2 mats, and AHL standards. Samples are separated using LC/MS. Peaks are shown as a relative percent (*y*-axis), while *x*-axis shows retention time (RT), expressed in minutes.

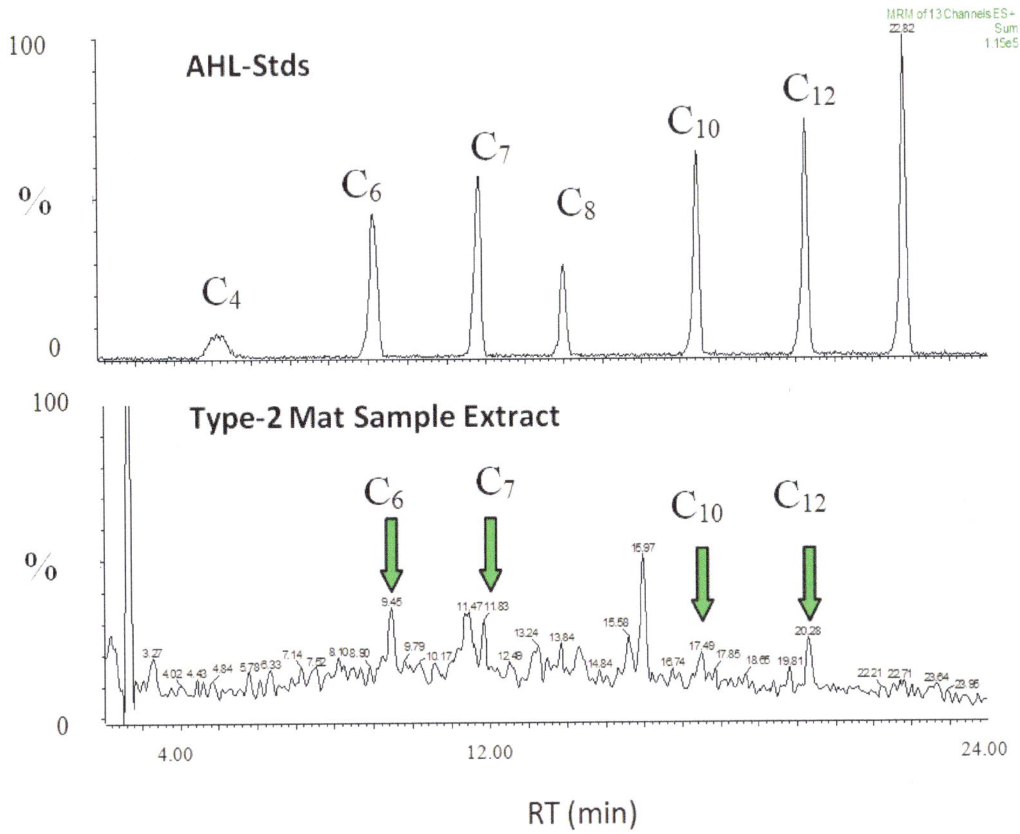

2.9.1. SRM in Oxic Environments and $CaCO_3$ Precipitation (Relevance)

Previous microelectrode studies have shown that the surfaces of both Type-1 and Type-2 mats were highly-oxygenated during daylight [10,48], with O_2 concentrations in stromatolites reaching over 600 μM during peak photosynthesis [26]. While O_2 has been classically considered to be stressful to most SRM [18], abundant populations of different SRM are now known to occur in oxygenated environments that display maximum metabolic rates under these conditions [12,14,49,50]. High abundances of SRM and sulfide-oxidizing microbes (SOM) were reported for the Highborne Cay stromatolites, and associated with this were high rates of sulfate reduction and sulfide oxidation [1]. Interestingly, this study found higher abundances and metabolic rates associated with lithifying layers (*i.e.*, Type-2 mats) than with non-lithifying layers (*i.e.*, Type-1 mats). A similar scenario was described for non-lithifying and lithifying mats in a hypersaline pond in the Bahamas, where higher cell densities and metabolic rates of sulfur-cycling organisms were associated with the mats that precipitated $CaCO_3$ [2,22]. While the SRM in the current study occurred in the uppermost surface (*i.e.*, top 130 μm) of Type-1 mats, they were significantly denser and more clustered in Type-2 mats. These data suggest that significant sulfur cycling may be occurring within the upper mm of stromatolite mats. A fundamental question guiding a theoretical understanding of stromatolite formation is: Why do SRMs tend to aggregate at the surface of Type-2 mats? Several possibilities exist to explain the occurrence of SRM at the mat surface: (1) The surface of a Type-2 mat is underlain by a dense layer of cyanobacteria, and hence, is highly-oxic during approximately half the day of each diel cycle. The SRM may receive photosynthetic excretion products from cyanobacteria on a diel basis [8]. It is postulated here that they precipitate a $CaCO_3$ cap to reduce DOC loss to the overlying water (which is oligotrophic), or to enhance efficient recycling of nutrients (e.g., N, P, Fe, *etc.*) within the mat. (2) A second possibility is that the SRM are physiologically adapted to metabolize under oxic conditions part of the time. Studies by Cyprionka [18] and others [2,51] have shown that some SRM may be physiologically adapted to cope with high O_2 levels. In this case, $CaCO_3$ precipitation could be advantageous as it produces a cement layer that increases the structural integrity of the stromatolite.

2.9.2. A Broader Role of Cell Clustering in Microbial Landscapes

Biofilms have been described as microbial landscapes owing to their physical, metabolic and functional diversity [52]. Our results emphasize that the microspatial patterns of cells within the surface biofilms of marine stromatolites may exist at several different spatial scales: (1) Micro-scale (µm) clustering, which may occur as a few (e.g., 2–5) to hundreds of cells within a single cluster. Such clustering may facilitate regulation of group activities, such as quorum sensing; (2) Aggregation of clusters: Clusters themselves may aggregate (*i.e.*, merge with adjacent cell clusters) to form a horizontal layer, within a vertical geochemical gradient region of the mat; (3) Larger mm-scale layering: The visible (to the eye) horizontal zonations, which are indicative of major functional clades within microbial mats, contribute to the exchange of autotrophically-generated DOC to heterotrophs and efficient recycling to reduce loss of DOC to overlying water. QS may be used for coordination of inter- and intra-species metabolic activities, as suggested by Decho and colleagues [42]. In the specific case of SRM, which rely on cyanobacteria for DOC but are negatively affected by the O_2 these phototrophs produce, it is of utmost importance to coordinate physiologies (including metabolisms) with other microorganisms that remove O_2 during their metabolism. This role could be fulfilled by aerobic heterotrophs and SOM, the latter benefitting from optimal SR activity to provide the substrate for sulfide oxidation. Especially noteworthy is that sulfide removal by SOM also benefits cyanobacteria, for which high concentrations of sulfide are toxic. Coordination of metabolisms may be facilitated by QS in this case. Inter-specific QS may ultimately be a key process in shaping the biofilm architecture. This is currently under investigation.

3. Experimental Section

3.1. Sampling of Intact Mats

All stromatolite sampling was conducted at a subtidal marine environment site at Highborne Cay, Exumas, Bahamas (76°51'W; 24°42'N). The site has been under long-term investigation through the Research Initiative on Bahamian Stromatolites (RIBS) project [4]. Freshly-collected intact stromatolites were dissected into working samples (approx. 2 × 2 cm), then immediately fixed (overnight, 4 °C) in a 4% paraformaldehyde (35 ppt seawater; 0.2 µm-filtered) solution. Portions of mat samples were initially trimmed into thick (approx. 2–4 mm) cross-sections using a rock saw, gently washed, and placed on glass microscope slides. Samples were then prepared for FISH. Surface mats were tentatively identified, based on light-microscopy examination of precipitation products, as either "Type-1" (*i.e.*, no visible surface precipitation), or "Type-2" (*i.e.*, crusty surface precipitation of $CaCO_3$ present) mats (Figure 1). Samples within each mat type were pooled. The samples were used to examine *in situ* distributions of cells within mats. Samples that were in-transition between full Type-1 or Type-2 were not considered further.

3.2. *Fluorescence* in-Situ *Hybridization (FISH)*

The oligodeoxynucleotide probe *dsrAB* was custom-synthesized by GeneDetect (Aukland, New Zealand) using sequences from the 16S rDNA oligonucleotide ProbeBase [53,54]. The probe *dsrAB* (GD1001-CS with GreenStar *™ FITC fluorescent labeling, Molecular Probes, Eugene, OR, USA) was used to target the dissimilatory sulfite reductase genes (*dsrAB*) of all recognized lineages of sulfate-reducing bacteria and archaea [36,38,55]. The probe was composed of a cocktail of the *DSR1F* (sequence: ACS CAC TGG AAG CACG) and the *DSR4R* (sequence: GTG TAG CAG TTA CCG CA) primers [38,56,57]. Concentrations of *dsrAB* were 5 ng per µL, and appropriate nonsense controls were used. Hybridization mixtures were removed and slides were washed for 15 min, in buffer containing 20 mM Tris-HCl (pH 7.4), 0.225 M NaCl, and 0.01% SDS. Fluorescence signals were amplified using the Alexa Fluor 488 Signal-Amplification Kit (Molecular Probes, Eugene, OR, USA) for Oregon Green Dye-Conjugated Probes (Molecular Probes, Eugene, OR, USA). DAPI (4'6'-diamidino-2-phenylindole) and PI (Molecular Probes, Eugene, OR, USA) were also used for general bacteria (DNA) staining [58,59]. FISH-probing was conducted according general methods modified from [60–62]. After fixation, intact mat samples were gently washed in phosphate-buffered saline (PBS) and stored in ethanol:PBS (1:1) at −20 °C. Samples, sliced into 2–4 mm sections on glass slides, were immersed in an ethanol series (50%, 80%, and 96%) for 3 min each. *In situ* hybridizations were performed at 50 °C overnight in a hybridization buffer containing 0.9 M NaCl, 20% formamide, 20 mM Tris-HCl (pH 7.4), and 0.01% sodium dodecyl sulfate (SDS).

3.3. *Extraction of Bacterial Cells from Mat Slurries*

Cells were extracted from the mat matrix using additional samples. This approach was conducted to determine the portion of total (extractable) cells (*i.e.*, DAPI-stained or PI-stained cells) that hybridized using the FISH probes (*i.e.*, SRM cells). Samples from the uppermost surface mats were fixed in 4% buffered paraformaldehyde overnight at 4 °C. The mat was gently homogenized into sediment slurries, then suspended in pre-filtered (0.2 µm) seawater. Cells were initially separated from sediment particulates using gentle centrifugation (1500× *g*; 2 min). Following, the cells and other organics (e.g., EPS) contained in the supernatant, were removed and subjected to repeated centrifugations (16,000× *g*; 10 min each) to pellet cells, and shear off EPS and other organics. The fixed, extracted cells were washed three times with 1× PBS (phosphate buffered saline), and stored in PBS/ethanol (1:1) at −20 °C until further processing. Cells, contained in wells on slides, were incubated at 46 °C for 90 min. in a hybridization buffer containing 0.9 M NaCl, 20% formamide, 20 mM Tris-HCl (pH 7.4), and 0.01% sodium dodecyl sulfate (SDS). The *dsrAB* probe concentration for slurry cell incubations was 1.0 ng per µL. Hybridization mixtures were removed and the slides were washed for 15 min, in buffer containing 20 mM Tris-HCl (pH 7.4), 225 mM NaCl and 0.01% SDS. Washing buffer was removed and washed with distilled water, and slides were air dried. Then, 50 μL of DAPI (or PI) was added on slides and incubated for 3 min. After washing with 80% ethanol, to remove unspecific staining, cells were rinsed in distilled H_2O and air-dried. The slides were mounted with Citifluor (Citifluor Ltd., Canterbury, UK) and the oligo-probed cells were quantitatively imaged.

3.4. Confocal Scanning Laser Microscopy (CSLM)

Images were obtained using a CSLM system (Leica TCS SP5, Leica Microsystems, Germany) equipped with a Kr-Ar laser. For CSLM imaging, three internal detectors were used, each with a 6-position emission filter wheel and a variable confocal aperture. Sample slides were viewed using 20×, 40×, 60×, or 100× objectives. The 60× and 100× objectives were used with immersion oil (Stephens Scientific Co., # M4004; Riverdale, NJ, USA; refractive index 1.515) to image individual cells. Final output was represented by colored composite images exported in a tagged image file format (TIFF). Direct counting of DAPI-stained cells and the oligoprobe-hybridized cells were performed on images of 30 independent fields using the automated image analysis software, Cell-C program [63]. In this manner, the relative proportions of SRM: total bacteria cells could be determined for each mat type using the two oligoprobes.

3.5. Image Analysis: Geographical Information Systems (GIS) Analyses

Geographical Information System (GIS) approaches [64,65] were used to analyze CSLM-generated images for spatial patterns of microbial cells and $CaCO_3$ precipitates within sections of intact surface mats. Sets of 25–30 images were sampled each from Type-1 and Type-2 mats. Briefly, images were classified using the Feature Analyst extension of ArcView GIS 3.2 [66,67]. Supervised classification was based on selecting representative pixels for each feature (e.g., SRM, cyanobacteria and bacteria). Based on these selections, the program identified all other pixels belonging to the same class. Since the fluorescence signature of cyanobacteria and bacteria was very similar, the two groups could not be separated spectrally. However, since Feature Analyst allows for the identification of linear features even when they are not continuous, all fluorescent filamentous shapes (*i.e.*, cyanobacteria) were identified. Filamentous shapes were subtracted from the image containing both cyanobacteria and other bacteria using a change-detection protocol. Following this classification, areas within images that were occupied by each feature of interest, such as SRM and other bacteria, were computed. Quantification of a given fraction of a feature that was localized within a certain delimited region was then used to examine clustering of SRM close to the mat surface, and later clustering of SRM in proximity to $CaCO_3$ precipitates.

For purposes of biological relevance, all images collected using CSLM were 512 × 512 pixels, and pixel values were converted to micrometers (*i.e.*, μm). Thus, following conversion into maps, a 512.00 × 512.00 pixel image represented an area of 682.67 × 682.67 μm. The value of 100 map pixels (approx. 130 μm) that was used to delineate abundance patterns was not arbitrary, but rather the result of analyzing sample images in search of an optimal cutoff value (rounded up to an integer expressed in pixels) for initially visualizing clustering of bacteria at the mat surface. The choice of the values used to describe the microspatial proximity of SRM to $CaCO_3$ precipitates (*i.e.*, 0.75, 1.5, and 3 pixels) was largely exploratory. Since the mechanistic relevance of these associations (e.g., diffusion distances) were not known, results were presented for three different distances in a series where each distance was double the value of the previous one. Pearson's correlation coefficients were then calculated for each putative association (see below).

3.5.1. Ground-Truthing GIS

GIS was used examine spatial relationships between specific image features such as SRM cells. In order to verify the results of GIS analyses, it was necessary to "ground-truth" image features (*i.e.*, bacteria). Therefore, separate "calibration" studies were conducted to "ground-truth" our GIS-based image data at microbial spatial scales.

3.5.2. Calibrations Using Fluorescent Microspheres

An experiment was designed to examine the correlation of "direct counts" of added spherical polymer microspheres (1.0 μm dia.) with those estimated using GIS/Image analysis approaches, which examined the total "fluorescent area" of the microspheres. The fluorescent microspheres used for these calibrations were trans-fluosphere carboxylate-modified microspheres (Molecular Probes, Molecular Probes, Eugene, OR, USA; T-8883; 1.0 μm; excit./emiss. 488/645 nm; refractive index = 1.6), and have been previously used for similar fluorescence-size calibrations [31]. Direct counts of microspheres (and later, bacteria cells) were determined [68]. Replicate serial dilutions of microspheres: *c*, *c*/2, *c*/4, *c*/8, and *c*/16, (where *c* is concentration) were homogeneously mixed in distilled water. For each dilution, five replicate slides were prepared and examined using CSLM. From each slide, five images were randomly selected. Output, in the form of bi-color images, was classified using Erdas Imagine 8.5 (Leica Geosystems AG, Heerbrugg, Switzerland). Classification was based on generating two classes ("microspheres" and background) after a maximum number of 20 iterations per pixel, and a convergence threshold of 0.95 and converted into maps. For the resulting surfaces, areas were computed in ArcView GIS 3.2. In parallel, independent direct counts of microspheres were made for each image. Statistical correlations of direct counts (of microspheres) and fluorescent image area were determined.

3.5.3. Calibrations within Intact Mats

Finally, fluorescent microspheres were added to the surface of Type-1 mats, as an external standard. Experimental additions of microspheres to Type-2 mats could not be accomplished because of the non-sticky nature of the mat surfaces. The mats were then imaged by CSLM and analyzed using the previously-described GIS-based approaches. Following image classification, the areas of microspheres were computed for each image, and correlated with the total number of microspheres counted (via direct counts approach) within the same images. This was designed to examine the ability of the image analysis approach to detect individual bacteria-sized objects (*i.e.*, 1 μm particles) within the complex matrix of natural stromatolite mats.

3.5.4. Microspatial Analyses of SRM and Microprecipitates

SRM activities have been previously implicated in the precipitation of $CaCO_3$ within the Type-2 mats of marine stromatolites [10]. Correlative microspatial associations of SRMs and $CaCO_3$

precipitates, therefore, were examined over several microspatial scales (approx. 1–5 μm distances) within Type-1 and Type-2 mats. For analyses, paired images were used of the same microspatial regions that were obtained at wavelengths specific to the FISH-probes of SRMs and $CaCO_3$ precipitates (488/550 nm = excit/emiss λ).

3.5.5. $^{35}SO_4^{2-}$-Silver Foils: 2D-Mapping of Sulfate Reducing Activity

Sulfate reducing activity was visualized using $^{35}SO_4^{2-}$-labeled Ag foil [10]. Ag foil (0.1 mm thickness, 99.99% pure; Sigma-Aldrich, St. Louis, MO, USA) was cleaned using subsequent steps of 30% *w/w* hydrogen peroxide and acetone. The foils were allowed to air dry in a class 1000 laminar flow hood. The foils were submersed in a radiolabeled sulfate ($Na_2{}^{35}SO_4$; Perkin-Elmer, Waltham, MA, USA) solution (ca. 0.1 mCi/mL) overnight and allowed to air dry. This treatment was repeated 3–4 times. $^{35}SO_4^{2-}$-Ag foils were tested for uniform distribution of the label using a BioRad Molecular Imager System GS-525 (Hercules, CA, USA). Freshly collected stromatolite samples were cut vertically and placed on the foil. After 6–9 h of incubation in the dark at 23 °C, the stromatolite mat samples were removed and the $^{35}SO_4^{2-}$ washed off the foil using distilled water. The foils (containing $^{35}SO_4^{2-}$ produced during SR) were kept in the dark and scanned using the BioRad Molecular Imager System GS-525 to visualize a 2-D $Ag^{35}SO_4^{2-}$ distribution. The individual pixels represent an area of ca. 50 × 50 μm, and darker pixels indicate a higher rate of sulfate reduction.

3.5.6. Clustering Analyses of SRMs

The microspatial arrangements of cells relative to each other (*i.e.*, clustering), and changes in relative abundances were examined by examining CSLM images of mat cross-sections. Thirty independent field images from Type-1 and Type-2 mats were examined for each mat type.

3.5.7. GIS

Clustering of SRM cells within the surfaces of Type-1 and Type-2 mats was analyzed using GIS by creating a buffer area extending from the surface of the mat to approximately 133 μm in depth. This surface region was chosen because preliminary examinations showed that most of cells appeared here. Hence our clustering analyses would examine changes in cell distributions within this surface region of the mat. Detection of SRM cells within the buffer area was based on color (as described above) using image classification of FISH-probed cells. A concentric region having a 10 μm dia. was generated around each cell. A cluster of cells represented a group of cells having overlapping concentric regions. Subsequent statistical selection of clusters was subjectively based on cluster areas representing greater than five cells. The size (*i.e.*, area) of each detected cell cluster was measured.

3.5.8. DAIME

Images collected from CSLM were also analyzed for changes in the spatial patterning of SRM cells in both Type-1 and Type-2 mats using the DAIME program [32]. Clustering within images was analysed using the Spatial:Stereology:Spatial arrangement subprogram with Daime. This calculates distances between all objects (*i.e.*, cells) within an image. Analyzed distances (*i.e.*, μm) were expressed as a pair correlation graph. Mean values of pair correlation values >1 indicated clustering at a given distance. Values approximating 1 indicated a random distribution of cells, and values <1 indicated avoidance.

3.5.9. Statistical Analyses

Following spatial analyses, the areas occupied by specific groups of bacteria (e.g., SRM, cyanobacteria) within proximity to the surface, and/or precipitates, cyanobacteria, other bacteria, and cyanobacteria) were tabulated in ArcView GIS (Environmental Systems Research Institute, Redlands, CA, USA). Data were examined using statistical analysis systems (SAS Institute Inc., Cary, NC, USA) software programs, for homogeneity of variances, then a range of statistical tests were used to examine potential differences in microspatial arrangements and associations [69,70]. Appropriate transformations were made, where necessary, to normalize data. Differences in precipitate concentrations between Type-1 and Type-2 mats were examined using a student's *t*-test. Overall differences in abundances of SRM among Type-1 and Type-2 mats were compared using analysis of variance (ANOVA). Differences in significant treatment effects were distinguished using Bonferroni and Scheffé aposteriori tests. Logistic regression analyses were used to examine clustering changes during transitions from a Type-1 to Type-2 mat. If no significant differences were detectable, mat data was pooled and analyzed as a single category. Pearson's correlation coefficient analysis was used to determine the specific correlations within given images, of areas occupied by SRM and $CaCO_3$ precipitates.

3.6. Molecular Phylogenetic Analysis of dsrA *Genes*

For molecular analysis of dissimilatory sulfite reductase *dsrA* genes, 170 mm^3 cores were removed from the surface of type I and II stromatolites. DNA was extracted from these samples using the Power Biofilm DNA Isolation Kit (MoBio Laboratories, Carlsberg, CA, USA) according to the manufacturer's protocol and used as template to generate *dsr* gene amplicons. Each PCR reaction consisted of 1.5 mM $MgCl_2$, 0.2 mM nucleotides, 0.4 uM of primers DSR1F (5'ACS(C/G)CACTGGAAGCACG-3') and DSR4R (5'GTGTAGCAGTTACCGCA3') [38], 1.25 U of Hot start polymerase (Promega), 10 ng of template DNA, and water in a 25 µL volume. PCR conditions were conducted as follows: 95 °C for 5 min, followed by 35 cycles of 95 °C for 45 s, 54 °C for 40 s, 72 °C for 2 min and a final extension at 72 °C for 10 min. PCR amplicons were purified with a QIAQuick PCR Purification Kit (Qiagen Sciences, Maryland, MD, USA) according to the manufacturer's instructions. These purified amplicons were ligated into pCR2.1-TOPO cloning vectors (Invitrogen, Carlsbad, CA, USA), and transformed into One Shot *E. coli* DH5α-T1^R competent cells following the manufacturer's protocol. Transformants were picked and grown overnight at 37 °C in LB broth containing 50 µg mL^{-1} kanamycin. Plasmids were extracted and purified using QIAprep Spin Miniprep kit (Qiagen Sciences Inc., Alameda, CA, USA), and quantified with a NanoDrop spectrophotometer (NanoDrop Technologies, Inc., Wilmington, DE, USA). Plasmid inserts were sequenced using the M13F (5'GTAAAACGACGGCCAGT3') and M13R (5'CAGGAAACAGCTATGAC3') plasmid vector primers at EnGenCore, LLC (Columbia, SC, USA) using BigDye Terminator version 3.1 cycle sequencing kit (Applied Biosystems, Warrington, UK). Resultant sequences were then searched against the GenBank database using BLASTX with default settings. Translated *dsrA* gene sequences from type I and II stromatolites were then aligned with amino acid sequences for the top BLAST hit and other characterized *dsrA* sequences using MUSCLE [71]. Next, a non-rooted phylogenetic tree was constructed using the Maximum Likelihood method based on the Whelan and Goldman model within the MEGA5 [72]. Initial tree(s) for the heuristic search were obtained by applying the Neighbor-Joining method to a matrix of pairwise distances estimated using a JTT model. A discrete Gamma distribution was used to model evolutionary rate differences among sites (5 categories (+G, parameter = 1.2797)). Tree robustness was tested using bootstrap analysis with 1000 replicates.

3.6.1. Extraction and Identification of Quorum Sensing Signals by LC/MS

Culture supernatants of SRM mat isolates were triple extracted in dichloromethane (DCM), dried under N_2 gas, and reconstituted with 50% acetonitrile, and analyzed by liquid chromatography/mass spectrometry (LC/MS) as previously described [26]. HPLC (150 mm Aquasep C_{18} column, Somerset, NJ, USA) was used to separate AHLs in samples. Detection and identification of AHLs was conducted using a Waters Premier XE triple quadrupole mass spectrometer (Milford, MA, USA) having positive-ion electrospray ionization. The MS was operated in multiple reaction monitoring mode utilizing two characteristic fragment transitions per analyte (*i.e.*, AHL). Natural mat samples, after gentle homogenization, were extracted in a similar manner to culture samples.

4. Conclusions

Abundances of SRM and their specific microspatial distributions, derived from image analyses, were used to create possible instruments of discrimination between non-lithifying Type-1 and lithifying Type-2 stromatolite mat communities. In general, Type-1 mats can be characterized as having comparatively lower abundances of SRM cells, and relatively dispersed cell distribution patterns (*i.e.*, limited-clustering of SRM cells). In contrast, Type-2 mats exhibit higher abundances and significant clustering of SRM cells within the uppermost 130 μm of the surface mat. The GIS approach may be most useful for determination of microbial cell patterns and microspatial organization (*i.e.*, areas occupied by cells) over spatial scales of tens to hundreds of microns. Once proper controls were employed, spatial relationships could be rapidly accessed.

Precipitation of micritic crusts are a characteristic feature of both fossil and present-day marine stromatolites. SRM within surface mats may play a defining role in C and S cycling processes that lead to micritic laminae formation in extant marine stromatolites. Our data suggest that development of an abundant and spatially-organized SRM community within the uppermost (oxic region) surface of stromatolite mats was closely aligned with the transition from a non-lithifying (Type-1) to a lithifying (Type-2) state. The progressive development of spatial organization (and high abundances) of SRM in surface mat layers further presents the likely possibility that quorum sensing may be involved in this transition.

Acknowledgments

This work was supported by grants from the National Sciences Foundation's BioComplexity Program (EAR—BE 0221796); Earth Sciences Program (EAR-1052974 and Environmental Genomics

Program (EF-0723707). We thank the crew of the Research Vessel Walton Smith, and the staff of the Highborne Cay Marina, for their hospitality and efficiency during field research. We thank members of the RIBS (*Research Initiative for Bahamian Stromatolites*) team for stimulating discussion in developing these ideas.

References

1. Visscher, P.T.; Reid, R.P.; Bebout, B.M.; Hoeft, S.E.; Macintyre, I.G.; Thompson, J.A., Jr. Formation of lithified micritic laminae in modern marine stromatolites (Bahamas): The role of sulfur cycling. *Am. Miner.* **1998**, *83*, 1482–1494.
2. Baumgartner, L.K.; Reid, R.P.; Dupraz, C.; Decho, A.W.; Buckley, D.H.; Spear, J.R.; Przekop, K.M.; Visscher, P.T. Sulfate reducing bacteria in microbial mats: Changing paradigms, new discoveries. *Sediment. Geol.* **2006**, *185*, 131–145.
3. Stolz, J.F.; Reid, R.P.; Visscher, P.T.; Decho, A.W.; Norman, R.S.; Aspden, R.J.; Bowlin, E.M.; Franks, J.; Foster, J.S.; Paterson, D.M.; *et al.* The microbial communities of modern marine stromatolites at Highborne Cay, Bahamas. *Atoll Res. Bull.* **2010**, *567*, 1–29.
4. Reid, R.P.; Visscher, P.T.; Decho, A.W.; Stolz, J.F.; Bebout, B.M.; Dupraz, C.; Macintyre, I.G.; Paerl, H.W.; Pinckney, J.L.; Prufert-Bebout, L.; *et al.* The role of microbes in accretion, lamination, and early lithification of modern marine stromatolites. *Nature* **2000**, *406*, 989–992.
5. Grotzinger, J.P.; Knoll, A.H. Stromatolites in PreCambrian carbonates: Evolutionary mileposts or environmental dipsticks? *Ann. Rev. Earth Planet Sci.* **1999**, *27*, 313–358.
6. Pinckney, J.L.; Reid, R.P. Productivity and community composition of stromatolitic microbial mats in the Exuma Cays, Bahamas. *Facies* **1997**, *36*, 204–207.
7. Paerl, H.W.; Steppe, T.F.; Reid, R.P. Bacterial-mediated precipitation in marine stromatolites. *Environ. Microbiol.* **2001**, *3*, 123–130.
8. Decho, A.W.; Visscher, P.T.; Reid, R.P. Production and cycling of natural microbial exopolymers (EPS) within a marine stromatolite. *Palaios* **2005**, *219*, 71–86.
9. Andres, M.S.; Sumner, D.Y.; Reid, R.P.; Swart, P.K. Isotopic fingerprints of microbial respiration in aragonite from Bahamian stromatolites. *Geology* **2006**, *34*, 973–976.
10. Visscher, P.T.; Reid, R.P.; Bebout, B.M. Microscale observations of sulfate reduction: Evidence of microbial activity forming lithified micritic laminae in modern marine stromatolites. *Geology* **2000**, *28*, 919–922.
11. Bowlin, E.M.; Klaus, J.S.; Foster, J.S.; Andres, M.S.; Custals, L.; Reid, R.P. Environmental controls on microbial community cycling in modern marine stromatolites. *Sediment. Geol.* **2012**, *263–264*, 45–55.
12. Canfield, D.E.; Des Marais, D.J. Aerobic sulfate reduction in microbial mats. *Science* **1991**, *251*, 1471–1473.
13. Visscher, P.T.; Quist, P.; van Gemerden, H. Methylated sulfur compounds in microbial mats: *In situ* concentrations and metabolism by a colorless sulfur bacterium. *Appl. Environ. Microbiol.* **1991**, *57*, 1758–1763.

14. Fründ, C.; Cohen, Y. Diurnal cycles of sulfate reduction under oxic conditions in microbial mats. *Appl. Environ. Microbiol.* **1992**, *58*, 70–77.
15. Krekeler, D.; Signalevich, P.; Teske, A.; Cypionka, H.; Cohen, Y. A sulfate-reducing bacterium from the oxic layer of a microbial mat from Solar Lake (Sinai), *Desulfovibrio oxyclinae* sp. nov. *Archiv. Microbiol.* **1997**, *176*, 69–375.
16. Visscher, P.T.; Gritzer, R.F.; Leadbetter, E.R. Low-molecular weight sulfonates, a major substrate for sulfate reducers in marine microbial mats. *Appl. Environ. Microbiol.* **1999**, *65*, 3272–3278.
17. Brune, A.; Frenzel, P.; Cypionka, H. Life at the oxic-anoxic interface: Microbial activities and adaptations. *FEMS Microbiol. Rev.* **2000**, *24*, 691–710.
18. Cypionka, H. Oxygen respiration by *Desulfovibrio* species. *Ann. Rev. Microbiol.* **2000**, *54*, 827–848.
19. Gallagher, K.L.; Kading, T.J.; Braissant, O.; Dupraz, C.; Visscher, P.T. Inside the alkalinity engine: The role of electron donors in the organomineralization potential of sulfate-reducing bacteria. *Geobiology* **2012**, *10*, 518–530.
20. Visscher, P.T.; Stolz, J.F. Microbial mats as bioreactors: Populations, processes, and products. *Palaios* **2005**, *219*, 87–100.
21. Petrisor, A.I.; Decho, A.W. Using geographical information techniques to quantify the spatial structure of endolithic boring processes within sediment grains of marine stromatolites. *J. Microbiol. Methods* **2004**, *56*, 173–180.
22. Dupraz, C.; Visscher, P.T.; Baumgartner, L.K.; Reid, R.P. Microbe-mineral interactions: Early carbonate precipitation in a hypersaline lake (Eleuthera Island, Bahamas). *Sedimentology* **2004**, *51*, 745–765.
23. Glunk, C.; Dupraz, C.; Braissant, O.; Gallagher, K.L.; Verrecchia, E.P.; Visscher, P.T. Microbially mediated carbonate precipitation in a hypersaline lake, Big Pond (Eleuthera, Bahamas). *Sedimentology* **2011**, *58*, 720–738.
24. Dupraz, C.D.; Fowler, A.; Tobias, C.; Visscher, P.T. Stromatolitic knobs in Storr's Lake (San Salvador, Bahamas): A model for formation and alteration of laminae. *Geobiology* **2013**, *11*, 527–548.
25. Fuqua, W.C.; Greenberg, E.P. Listening in on bacteria: Acylhomoserine lactone signaling. *Nat. Rev. Mol. Cell Biol.* **2002**, *3*, 685–695.
26. Decho, A.W.; Visscher, P.T.; Ferry, J.; Kawaguchi, T.; He, L.; Przekop, K.M.; Norman, R.S.; Reid, R.P. Autoinducers extracted from microbial mats reveal a surprising diversity of *N*-acylhomoserine lactones (AHLs) and abundance changes that may relate to diel pH. *Environ. Microbiol.* **2009**, *11*, 409–420.
27. Kawaguchi, T.; Chen, Y.P.; Norman, R.S.; Decho, A.W. An *in vitro* rapid detection/quantification assay for *N*-acyl homoserine lactones in environmental samples. *Appl. Environ. Microbiol.* **2008**, *74*, 3667–3671.
28. Decho, A.W.; Frey, R.L.; Ferry, J.L. Chemical challenges to bacterial AHL signaling in the environment. *Chem. Rev.* **2011**, *11*, 86–99.
29. Montgomery, K.; Charlesworth, J.C.; Lebard, R.; Visscher, P.T.; Burns, B.P. Quorum Sensing in Extreme Environments. *Life* **2013**, *3*, 131–148.

30. Petrash, D.A.; Gingras, M.K.; Lalonde, S.V.; Orange, F.; Pecoits, E.; Konhauser, K.O. Dynamic controls on accretion and lithification of modern gypsum-dominated thrombolites, Los Roques, Venezuela. *Sediment. Geol.* **2012**, doi:10.1016/j.sedgeo.2011.12.006.
31. Petrisor, A.I.; Kawaguchi, T.; Decho, A.W. Quantifying $CaCO_3$ microprecipitates within developing surface mats of marine stromatolites using GIS and digital image analysis. *Geomicrobiol. J.* **2004**, *21*, 491–496.
32. Daims, H.; Lücker, S.; Wagner, M. *daime*, a novel image analysis program for microbial ecology and biofilm research. *Environ. Microbiol.* **2006**, *8*, 200–213.
33. Des Marais, D.J. Long-Term Evolution of the Biogeochemical Carbon Cycle. In *Geomicrobiology: Interactions between Microbes and Minerals*; Banfield, J.E., Nealson, K.H., Eds.; Mineralogical Society of America Press: Washington, DC, USA, 1997; pp. 427–448.
34. Decho, A.W.; Kawaguchi, T.; Allison, M.A.; Louchard, E.M.; Reid, R.P.; Stephens, F.C.; Voss, K.J.; Wheatcroft, R.A.; Taylor, B.B. Sediment properties influencing upwelling spectral reflectance signatures: The "biofilm gel effect". *Limnol. Oceanogr.* **2003**, *48*, 431–443.
35. Paterson, D.M.; Aspden, R.J.; Visscher, P.T.; Consalvey, M.; Andres, M.S.; Decho, A.W.; Stolz, J.; Reid, R.P. Light-dependent biostabilisation of sediments by stromatolite assemblages. *PLoS One* **2008**, *3*, e3176.
36. Meyer, B.; Kuever, J. Molecular analysis of the diversity of sulfate-reducing and sulfur-oxidizing prokaryotes in the environment, using *aprA* as functional marker gene. *Appl. Environ. Microbiol.* **2007**, *73*, 7554–7679.
37. Zverlov, V.; Klein, M.; Lucker, S.; Friedrich, M.W.; Kellermann, J.; Stahl, D.A.; Loy, A.; Wagner, M. Lateral gene transfer of dissimilatory (bi)sulfite reductase revisited. *J. Bacteriol.* **2005**, *187*, 2203–2208.
38. Wagner, M.; Roger, A.J.; Flax, J.L.; Brusseau, G.A.; Stahl, D.A. Phylogeny of dissimilatory sulfite reductases supports an early origin of sulfate respiration. *J. Bacteriol.* **1998**, *180*, 2975–2982.
39. Pernthaler, J.; Glöckner, F.O.; Schönhuber, W.; Amann, R. Fluorescence *in situ* hybridization (FISH) with rRNA-targeted oligonucleotide probes. *Methods Microbiol.* **2001**, *30*, 207–226.
40. Poulsen, L.K.; Ballard, G.; Stahl, D.A. Use of rRNA fluorescence *in situ* hybridization for measuring the activity of single cells in young and established biofilms. *Appl. Environ. Microbiol.* **1993**, *59*, 1354–1360.
41. Hense, B.A.; Kuttler, C.; Müller, J.; Rothballer, M.; Hartmann, A.; Kreft, J.-U. Does efficiency sensing unify diffusion and quorum sensing. *Nat. Rev. Microbiol.* **2007**, *5*, 230–239.
42. Decho, A.W.; Norman, R.S.; Visscher, P.T. Quorum sensing in natural environments: Emerging views from microbial mats. *Trends Microbiol.* **2010**, *18*, 73–80.
43. Gantner, S.; Schmid, M.; Dürr, C.; Schuheggar, R.; Steidle, A.; Hutzler, P. *In situ* quantitation of the spatial scale of calling distances and population density-independent *N*-acylhomoserine lactone-mediated communication by rhizobacteria colonized on plant roots. *FEMS Microbiol. Ecol.* **2006**, *56*, 188–194.
44. Petrisor, A.I.; Cuc, A.; Decho, A.W. Reconstruction and computation of microscale biovolumes using geographic information systems: Potential difficulties. *Res. Microbiol.* **2004**, *155*, 447–454.
45. Daims, H.; Wagner, M. Quantification of uncultured microorganisms by fluorescence microscopy and digital image analysis. *Appl. Microbiol. Biotechnol.* **2007**, *75*, 237–248.

46. Murgante, B.; Las Casas, G. GIS and Fuzzy Sets for the Land Suitability Analysis. In *Computational Science and Its Applications, Lecture Notes in Computer Science*; Gavrilova, M.L., Kumar, V., Mun, Y., Eds.; Springer: Berlin/Heidelberg, Germany, 2004; Volume 3044, pp. 1036–1045.
47. Walker, R.F.; Ishikawa, K.; Kumagai, M. Fluorescence-assisted image analysis of freshwater microalgae. *J. Microbiol. Methods* **2002**, *51*, 149–162.
48. Visscher, P.T.; Surgeon, T.M.; Hoeft, S.E.; Bebout, B.M.; Thompson, J., Jr.; Reid, R.P. Microelectrode Studies in Modern Marine Stromatolites: Unraveling the Earth's Past? In *Electrochemical Methods for Environmental Analysis of Trace Metal Biogeochemistry*; ACS Symposium Series 881; Taillefert, M., Rozan, T., Eds.; Oxford University Press: New York, NY, USA, 2002; pp. 265–282.
49. Visscher, P.T.; van Ende, F.P.; Schaub, B.E.M.; van Gemerden, H. Competition between anoxygenic phototrophic bacteria and colorless sulfur bacteria in a microbial mat. *FEMS Microbiol. Ecol.* **1992**, *101*, 51–58.
50. Jørgensen, B.B. Sulfate reduction and thiosulfate transformations in a cyanobacterial mat during a diel oxygen cycle. *FEMS Microbiol. Ecol.* **1994**, *13*, 303–312.
51. Fournier, M.; Zhang, Y.; Wildschut, J.D.; Dolla, A.; Voordouw, J.K.; Schriemer, D.C.; Voordouw, G. Function of oxygen resistance proteins in the anaerobic, sulfate-reducing bacterium *Desulfovibrio vulgaris* Hildenborough. *J. Bacteriol.* **2003**, *185*, 71–79.
52. Battin, T.J.; Sloan, W.T.; Kjelleberg, S.; Daims, H.; Head, I.M.; Curtis, T.P.; Eberl, L. Microbial landscapes: New paths to biofilm research. *Nat. Rev. Microbiol.* **2007**, *5*, 76–81.
53. Loy, A.; Maixner, F.; Wagner, M.; Horn, H. ProbeBase—An online resource for rRNA-targeted oligonucleotide probes: New features 2007. *Nucl. Acids Res.* **2007**, *31*, 514–516.
54. Loy, A.; Küsel, K.; Lehner, A.; Drake, H.L.; Wagner, M. Microarray and functional gene analyses of sulfate-reducing prokaryotes in low sulfate, acidic fens reveal co-occurrence of recognized genera and novel lineages. *Appl. Environ. Microbiol.* **2004**, *70*, 6998–7009.
55. Klein, M.; Friedrich, M.; Roger, A.J.; Hugenholtz, P.; Fishbain, S.; Abicht, H.; Blackall, L.L.; Stahl, D.A.; Wagner, M. Multiple lateral transfers of dissimilatory sulfite reductase genes between major lineages of sulfate-reducing prokaryotes. *J. Bacteriol.* **2001**, *183*, 6028–6035.
56. Minz, D.; Fishbain, S.; Green, S.J.; Muyzer, G.; Cohen, Y.; Rittman, B.E.; Stahl, D.A. Unexpected population distribution in a microbial mat community: Sulfate-reducing bacterial localized to the highly oxic chemocline in contrast to a eukaryotic preference for anoxia. *Appl. Environ. Microbiol.* **1999**, *65*, 4659–4665.
57. Minz, D.; Flax, J.I.; Green, S.J.; Muyzer, G.; Cohen, Y.; Wagner, M.; Rittman, B.E.; Stahl, D.E. Diversity of sulfate-reducing bacteria in Oxic and Anoxic regions of a microbial mat characterized by comparative analysis of dissimilatory sulfite reductase genes. *Appl. Environ. Microbiol.* **1999**, *65*, 4666–4671.
58. Wagner, M.; Erhart, R.; Manz, W.; Amann, R.; Lemmer, H.; Wedi, D.; Schleifer, K.-H. Development of an rRNA-targeted oligonucleotide probe specific for the genus *Acinetobacter* and its application for *in situ* monitoring in activated sludge. *Appl. Environ. Microbiol.* **1994**, *60*, 792–800.
59. Wagner, M.; Amann, R.; Kämpfer, P.; Assmus, B.; Hartmann, A.; Hutzler, P.; Springer, N.; Schleifer, K.-H. Identification and *in-situ* detection of Gram-negative filamentous bacteria in activated sludge. *Syst. Appl. Microbiol.* **1994**, *17*, 405–417.

60. Snaidr, J.; Amann, R.; Huber, I.; Ludwig, W.; Schleifer, K.-H. Phylogenetic analysis and *in situ* identification of bacteria in activated sludge. *Appl. Environ. Microbiol.* **1997**, *63*, 2884–2896.
61. Llobet-Brossa, E.; Rossello-Mora, R.; Amann, R. Microbial community composition of Wadden Sea sediments as revealed by fluorescence *in situ* hybridization. *Appl. Environ. Microbiol.* **1998**, *64*, 2691–2696.
62. Amann, R.I, Schleifer, K.-H. Nucleic Acid Probes and Their Application in Environmental Microbiology. In *Bergey's Manual of Systematic Bacteriology*; Garrity, G.M., Ed.; Springer: New York, NY, USA, 2001; pp. 67–82.
63. Selinummi, J.; Seppälä, J.; Yli-Harja, O.; Puhakka, J.A. Software for quantification of labeled bacteria from digital microscope images by automated image analysis. *Biotechniques* **2005**, *39*, 859–863.
64. Fedra, K.; Feoli, E. GIS technology and spatial analysis in coastal zone management. *EEZ Technol.* **1998**, *3*, 171–179.
65. Jensen, J.R. *Remote Sensing of the Environment. An Earth Resource Perspective*; Prentice Hall: Upper Saddle River, NJ, USA, 2000.
66. ERDA Imagine Support (2003). Available online: http://support.erdas.com (accessed on 24 November 2003).
67. Feature Analyst. 2003. Available online: http://www.featureanalyst.com/index.html/support/support/Support.html (accessed on 25 November 2003).
68. Kirchman, D.L. Statistical Analysis of Direct Counts of Microbial Abundance. In *Handbook of Methods in Aquatic Microbial Ecology*; Kemp, P.F., Sherr, B.F., Sherr, E.B., Cole, J.J., Eds.; Lewis Publishers: Boca Raton, FL, USA, 1993; pp. 117–120.
69. Cody, R.P.; Smith, J.K. *Applied Statistics and the SAS® Programming Language*; Prentice Hall: Upper Saddle River, NJ, USA, 1997; pp. 138–141.
70. SAS. *Statistical Analysis Systems*; SAS Inst. Inc.: Cary, NC, USA, 2007.
71. Edgar, R.C. MUSCLE: A multiple sequence alignment method with reduced time and space complexity. *BMC Bioinforma.* **2004**, *5*, 113.
72. Tamura, K.; Peterson, D.; Peterson, N.; Stecher, G.; Nei, M.; Kumar, S. MEGA5: Molecular evolutionary genetics analysis using maximum likelihood, evolutionary distance, and maximum parsimony methods. *Mol. Biol. Evol.* **2011**, *28*, 2731–2739.

9

Systematic Exploration of Natural and Synthetic Flavonoids for the Inhibition of *Staphylococcus aureus* Biofilms

Suvi Manner [1], Malena Skogman [1], Darla Goeres [2], Pia Vuorela [1,3] and Adyary Fallarero [1,*]

[1] Pharmaceutical Sciences, Department of Biosciences, Abo Akademi University, Artillerigatan 6A, 3rd Floor, Biocity, Turku FI-20520, Finland; E-Mails: smanner@abo.fi (S.M.); malena.skogman@abo.fi (M.S.); pia.vuorela@abo.fi (P.V.)

[2] Center for Biofilm Engineering, Montana State University, Bozeman, MT 59717, USA; E-Mail: darla_g@erc.montana.edu

[3] Division of Pharmaceutical Biology, Faculty of Pharmacy, University of Helsinki, Viikinkaari 5E, P.O. Box 56, Helsinki FI-00014, Finland

* Author to whom correspondence should be addressed; E-mail: adyary.fallarero@abo.fi;

Abstract: When single-cell (or suspended) bacteria switch into the biofilm lifestyle, they become less susceptible to antimicrobials, imposing the need for anti-biofilms research. Flavonoids are among the most extensively studied natural compounds with an unprecedented amount of bioactivity claims. Most studies focus on the antibacterial effects against suspended cells; fewer reports have researched their anti-biofilm properties. Here, a high throughput phenotypic platform was utilized to screen for the inhibitory activity of 500 flavonoids, including natural and synthetic derivatives, against *Staphylococcus aureus* biofilms. Since discrepancies among results from earlier antibacterial studies on flavonoids had been noted, the current study aimed to minimize sources of variations. After the first screen, flavonoids were classified as inactive (**443**), moderately active (**47**) or highly active (**10**). Further, exclusion criteria combining bioactivity and selectivity identified two synthetic flavans as the most promising. The body of data reported here serves three main purposes. First, it offers an improved methodological workflow for anti-biofilm screens of chemical libraries taking into account the (many times ignored) connections between anti-biofilm and antibacterial properties. This is particularly relevant for the study of flavonoids and other natural products. Second, it provides a large and freely available anti-biofilm bioactivity dataset that expands the knowledge on flavonoids and paves the way for future structure-activity relationship studies and structural optimizations. Finally, it identifies two new flavans that can successfully act on biofilms, as well as on suspended bacteria and represent more feasible antibacterial candidates.

Keywords: flavonoids; flavanones; flavans; *Staphylococcus aureus*; biofilms; antibacterial; bioactivity; chalcones; flavones

1. Introduction

One of the most ground-breaking advancements of the microbiological research during the past 40 years has been the recognition of bacterial biofilms as the predominant bacterial lifestyle instead of bacterial suspensions [1,2]. Bacterial biofilms is the term used to describe the surface-attached bacterial lifestyle. Cells in biofilms grow as communities, surrounded by a self-produced thick layer of extracellular polymeric substances (EPS, also known as matrix or slime). Biofilms are structurally and also functionally different from single-cell (suspended) bacteria [3]. The presence of the EPS protects cells in biofilms from the detrimental effects of chemical insults and harsh environmental conditions. Moreover, in the complex tri-dimensional architecture of biofilms, subpopulations of cells co-exist in all stages of growth, including a fraction of dormant cells that are not metabolically active. This combination of factors helps to explain why biofilms possess much lower susceptibility to antimicrobial therapy or biocides, when compared to suspended cells [4–6]. Biofilms are involved in a wide range of infections, such as chronic wounds [7,8], otitis [9], cystic fibrosis [10] and those associated with medical devices [11], and they are claimed to be responsible for an overwhelming proportion of persistent, antibiotic-resistant infections.

Current challenges faced within the anti-biofilms field are enormous and are present at different stages in the drug discovery process. From a pre-clinical perspective, a limited repertoire of molecules has been reported that can act *in vitro* on existing biofilms at low concentrations, especially in the case of those formed by *Staphylococcus aureus* [12]. Then, even if some compounds can successfully enter the drug discovery pipeline, another more serious challenge is the absence of a clearly-defined regulatory pathway for those products to be registered with relevant agencies, such as the European Medicines Agency, the US Food and Drug Administration or the US Environmental Protection Agency. As a result, only one disinfectant (commercially sold by Sterilex, a US company) and no antibiotics have been approved by a regulatory agency to be used specifically against bacterial biofilms. Clearly, expanding the search for anti-biofilms, as well as consistently documenting their benefits with convincing results is deemed as an imperative need of current biomedical research in the hopes of delivering tangible solutions to biofilm infections.

Flavonoids are one of the most widely recognized groups of natural products, even outside the scientific community. They are largely represented in the human diet, as they are present in plants, seeds and various foodstuffs. From a chemical viewpoint, flavonoids are phenolic compounds that consist of two benzene rings (A and B) combined with an oxygen-containing heterocyclic benzopyran ring (C). Flavonoids can be divided into different classes based on their molecular structure. The number of these classes varies according to the classification criteria. For instance, according to the position of the phenyl ring (B) relative to the benzopyran moiety, they can be classified as flavonoids (2-phenyl-benzopyrans), isoflavonoids (3-phenyl-benzopyrans) and neoflavonoids (4-phenyl-benzopyrans) [13]. Oxidation and saturation status in the heterocyclic ring also enables division of flavonoids into flavans, flavanones, dihydroflavonols, flavonols, flavones, flavone-3-ols and flavone-3,4-diols, while, depending upon the type, number and arrangement of substituents, flavonoids can be further divided into other groups, such as anthocyanidins and chalcones [14]. So far, more than 6500 flavonoids have been discovered, and a total of 14 classes have been proposed by Cushnie and Lamb [15]. From a biological perspective, flavonoids are one of the most extensively studied types of natural compounds, with a massive amount of research published supporting an unprecedented amount of bioactivity claims. In the field of anti-infectives, flavonoids have been reported to display antibacterial, antiviral, antiprotozoan and antifungal properties [16,17]. Most of the studies performed against bacterial infections have focused on suspended cells, with flavones, chalcones, flavonols, flavan-3-ols, flavanones and flavolans as the most successful flavonoid classes [15,18,19].

This article focuses on the use of flavonoids as a representative chemical template to perform a large screening in search for compounds that can counteract *S. aureus* biofilms. The screened collection consisted of a commercial set of 500 natural and synthetic flavonoids that covers various structural classes (Figure 1) and includes many well-known dietary flavonoids. To the best of our knowledge, the current contribution represents the largest systematic screening study performed with flavonoids for anti-biofilm effects. Compounds per class present in the library is indicated between parentheses. Flavanones, isoflavonoids, neoflavonoids and dihydroflavonols are also included.

Figure 1. The most important flavonoid classes represented in the flavonoids chemical library. The amount of compounds per class present in the library is indicated between parentheses. Flavanones, isoflavonoids, neoflavonoids and dihydroflavonols are also included. * The numbering of the chemical structure of the flavonols and flavans is similar to the flavones, and it follows the criteria of [15].

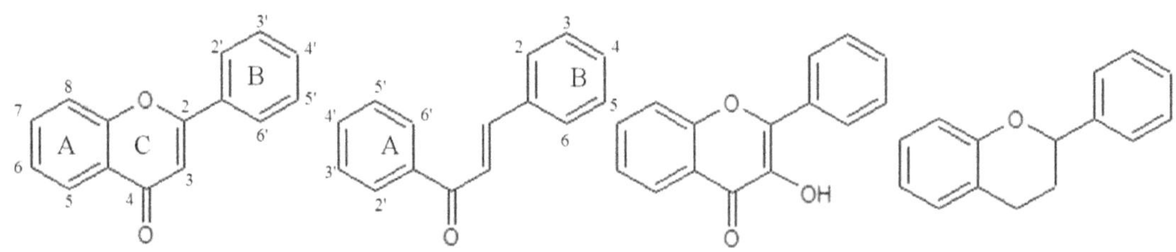

Flavones (205)* **Chalcones** (131) **Flavonols** (90)* **Flavans** (48)*

2. Results and Discussion

2.1. Design of the Flavonoid Anti-Biofilm Screening for More Meaningful Data Generation

Earlier studies dealing with the antibacterial properties of flavonoids performed by different groups have generated conflicting data. However, a seminal contribution by Cushnie and Lamb in 2005 [18] elucidated the major reasons that could explain such contradictions, namely, differences in the types of assays used, a lack of optimization or even a failure to enumerate the starting bacterial concentrations, differences in the solvents used to dissolve the flavonoids, as well as in the origin of the compounds (*i.e.*, if they are obtained from a commercial or a natural source). Taking this into account, one of the aims of this study was to design chemical screening to minimize these sources of variation.

Here, the strategy that was adopted for anti-biofilm screening against *Staphylococcus aureus* spp. includes a parallel exposure scheme in which bacteria are treated with the compounds before biofilm formation takes place (prior-to-exposure) and once biofilms are formed (post-exposure paradigm). This strategy has been described in the earlier contributions of our group [20–23], and a detailed scheme is presented in [22]. It solely relies on optimized methods to quantify biofilm viability (with resazurin staining) and biomass (with crystal violet staining) under repeatable conditions that have been summarized here and are also detailed in earlier contributions [20–23]. Key conditions, such as biofilm formation time, starting bacterial concentrations or available surface for biofilm formation, are comparable to those that have been applied in anti-biofilm studies on 96-microwell plates [24,25], making inter-laboratory comparisons feasible. Special attention was paid to control the initial concentration of added bacteria, as it is one of the most critical factors in antimicrobial assays [25], as well as to maintain it throughout the screening and follow-up studies. Unlike the antibacterial assays against planktonic bacteria for which straightforward guidelines exist from the Clinical and Laboratory Standards Institute, as well as from the European Committee on Antimicrobial Susceptibility Testing (EUROCAST), only one standard for an anti-biofilm screening assay in microwell plates has been accepted so far by the American Society for Testing of Materials (ASTM E2799-12), and it does not apply to *Staphylococcus aureus* biofilms.

To avoid identifying only strain-specific hits, the initial screening was performed simultaneously against two clinical *S. aureus* strains, and only those flavonoids highly active against both strains and in both exposure paradigms (prior-to and post-biofilm formation) were chosen for further study (selection criteria are discussed in Section 2.2). Moreover, all the tested compounds were obtained from the same company (TimTec, see Section 3.1), to avoid differences due to the source. In Table S1, a list of the flavonoids library is included. Their structures can be retrieved with the company codes using the freely available TimTec database (http://www.echemstore.com/). Trivial names of the natural flavonoids, as well as short names of the natural derivatives were also compiled and included in this Table. All the 500 flavonoids used in this study were prepared in 20 mM in dimethylsulfoxide (DMSO), as this is the most widely used solvent for the solubilization of large chemical libraries. Precipitation upon dilution in DMSO was recorded for 31 flavonoids (Table S1). All the precipitated samples corresponded to being inactive, most likely because precipitation lowers the compound concentration in the solution, resulting in an underestimation of the anti-biofilm activity [26]. However, the population of non-entirely solubilized flavonoids accounted for less than 6.5% of the total collection and, therefore, has a minimal impact on the usefulness of this study.

2.2. Anti-Biofilm Screening: Inactive and Moderately Active Flavonoids

Results of the inhibitory activity of the entire flavonoids collection prior-to (a) and post-(b) biofilm formation based on viability results are shown in Figure 2. The selection process is summarized in Scheme 1, and the raw screening data are presented in Table S1.

Figure 2. Anti-biofilm effects of the flavonoids collection when added prior-to (**a**) or post-biofilm formation (**b**). Strains 1 and 2 are *S. aureus* ATCC 25923 and Newman clinical strains, respectively. Highly active flavonoids are those present in both shadowed areas. Primary screening results are presented in Table S1.

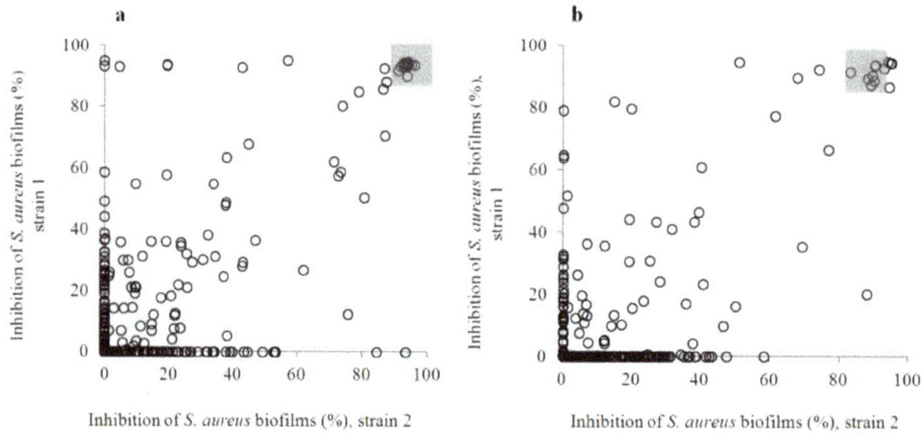

Scheme 1. Summary of the anti-biofilm screening and selection criteria applied to the collection in this study. Highly and moderately active flavonoids are listed in Tables S2 and S3, respectively.

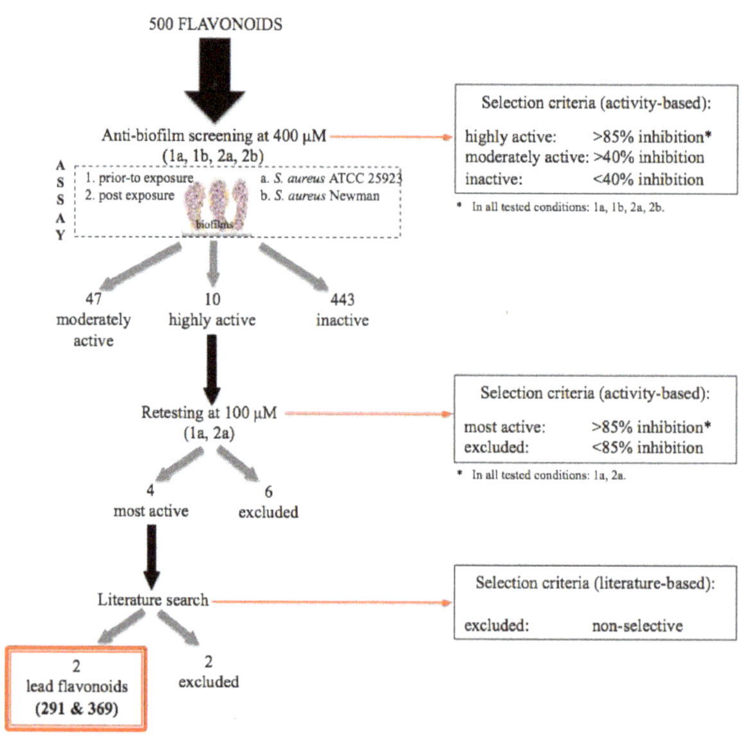

More than 80% of flavonoids (**443**) caused less than 40% inhibition of the biofilm formation by *S. aureus* (Figure 2) at 400 µM, and they were declared inactive. In general, many flavonoids had strain-specific effects, and their inhibition percentages were lower in the post-exposure paradigm. Upon closer inspection of the screening results (Table S1), ten flavonoids were tentatively identified as highly active, causing more than an 85% inhibition of biofilm viability (based on resazurin staining), when measured using both exposure assays, as well as both *S. aureus* strains. This corresponds to the compounds that are present in both of the shadowed areas marked in the Figure 2, and they are compiled in Table S2. The calculated overall hit rate, based on the amount of compounds identified as highly active, was 2%.

Between the inactive and highly active flavonoids, **47** were found to inhibit biofilm viability more than 40%, and they were classified as moderately active. Flavonoids that displayed activity in only one assay condition (for instance, compound **231**) were intentionally excluded. Thus, all the flavonoids within the moderately active group also displayed varying levels of activity in the other assay conditions. Eight of the moderately active flavonoids were only active in the pre-exposure scheme, and five showed strain-selectivity. A list of the moderately active ones is presented in Table S3.

Observations were first made on the group of inactive flavonoids (Table S1). Dietary flavonoids, present in citrus, such as kaempferol (**69**) and naringenin (**164**), have been proven to act as quorum sensing (QS) inhibitors by interfering with the interaction between acyl-homoserine lactones (AHLs, the signal molecules of Gram-negative bacteria) and their receptors, leading to inhibition of biofilm formation by *Escherichia coli* O157:H7 and *Vibrio harveyi* BB120 [15,27], but they were classified as inactive in this study. Other QS inhibitors, such as baicalein (**159**) and catechin (**160**), were found inactive, as well, when, previously, they had been shown to act on the cytoplasmic membrane-associated receptors, TraR and RhlR, respectively, in this way suppressing biofilm formation by *Pseudomonas aeruginosa* [28,29]. The lack of activity of all these compounds in *S. aureus* biofilms would thus suggest that they may preferentially act in Gram-negative bacteria.

Several flavonoids present in the moderately active group have been extensively studied and were earlier reported with different bioactivities in a variety of targets, including antimicrobial activity (Table S3). However, a total of 20 flavonoids in this group had not been reported as active in the PubChem Compound (http://www.ncbi.nlm.nih.gov/pccompound), the largest publicly available repository for small molecular weight molecules, which is connected to their bioactivities via the PubChem Bioassay project (http://www.ncbi.nlm.nih.gov/pcassay). Thus, 20 new bioactive flavonoids were identified as a result of this contribution.

Chalcones and flavones are the most highly represented flavonoid classes (Table S2) with 12 and 14 moderately active ones, respectively. This could be a reflection of the fact that these two chemical classes were also the most abundant within the collection (Figure 1). Within the chalcones, isoliquiritigenin (**372**) has been earlier shown to possess potent antibacterial properties against suspended *Ralstonia solanacearum* [30] and biofilm-forming *Porphyromonas gingivalis* [31], both Gram-negative bacteria, but no information had been published concerning the effects against Gram-positive bacteria. Chalcones typically exhibited activity prior to biofilm formation, which was not surprising, as many of them had been reported to exhibit antibacterial properties. Eight of them (**189**, **267**, **372**, **466**, **467**, **470**, **474**, **477**) are hydroxylated in position 2' in the A ring, which has earlier been shown to benefit their antibacterial effects [32,33]. Four share a hydroxylation in position 4' in the A ring (**189**, **372**, **470** and **477**), and this hydroxylation is also present in three other chalcones (**251**, **425** and **475**); thus, within the moderately active chalcones, a total of 11 compounds (out of 12) have hydroxylations in positions 2' or 4' in the A ring. In position 4 of the B ring, hydroxylation (**189**, **267** and **372**) or methoxylation (**251**, **425** and **477**) seemed to be also a relevant structural feature, but it always occurred in conjunction with the substitutions in the A ring discussed earlier. Thus, this may be important, but not sufficient.

In the case of flavones, substitutions in the A ring in positions 7 (-*O*-acyl or -*O*-alkylamino) and 5 (–OH) had been indicated as crucial for antibacterial activity [15,34], and they are indeed present in compounds **52** and **54** (7, -*O*-acyl) and in **56** and **311** (5, –OH). However, it was more commonly present in the hydroxylation of position 7 in the A ring, which occurs in **56**, **121**, **135**, **166**, **188**, **222**, **232** and **311** (eight out of 14 flavones). In two (**166** and **232**), only hydroxylations in the A ring are present (no substituents in the B ring), and one of them (**166**, 7,8-dihydroxyflavone) was found active in the two exposure conditions and in both *S. aureus* strains, with inhibition percentages ranging from 61.5% to 93.3%. The impact of substitutions in the B ring of flavones has been much less studied in the literature. Here, it was found that hydroxylation or *O*-acylation in positions 3' (**52**, **54**, **56**, **211**, **214** and **222**) or 4' (**52**, **54**, **56**, **211**, **214** and **311**) were beneficial. The essentiality of the B ring substitution is supported by flavone 214 (3',4'-dihydroxyflavone), which displayed high inhibition values of biofilm formation by *S. aureus* 25923 strain (over 64%) in the absence of any other changes in the A ring. Additionally, four moderately active flavones also share a substituent (–OH or $-OCH_3$) in 6' (**110**, **121**, **129** and **135**). These indicated motifs in the B ring could be valuable guidance for future chemical refinements on the flavones scaffold. The key structural features for chalcones and flavones, which have been discussed above, are summarized in Figure 3.

Figure 3. Key structural features present in the anti-biofilm chalcones and flavones (moderately and highly active ones).

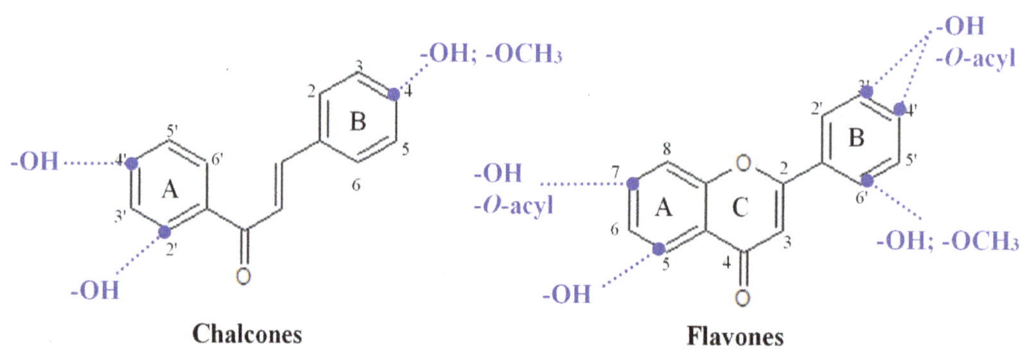

Moreover, seven moderately active flavonols (the third most represented flavonoid class after flavones and chalcones) were identified within the moderately active flavonoids. Simultaneous hydroxylations of ring A and B in various positions occurred in all flavonols, except in one (**155**), where only –OH groups were present in the A ring. Among these flavonols, quercetin (**58**) was identified. Quercetin is one of the most extensively studied dietary flavonoids and natural compounds in general. One interesting aspect is that quercetin was found slightly more effective in pre-existing biofilms. Quercetin has been reported to be antibacterial and anti-biofilm against *S. aureus*. In fact, Lee *et al.* [35] recently showed that it inhibits biofilm formation (prior-to exposure) by *S. aureus* strains, including ATCC 25923, at concentrations within the range of 17–170 μM. Thus, the activity found here is lower than the one reported earlier [35]. Unfortunately, [35] included no information on the solvent used to dissolve the quercetin, and it has been shown that quercetin is highly prone to aggregation, depending upon the vehicle, ionic strength and pH [36]. Formation of aggregates has been postulated as one possible reason for quercetin's promiscuous activity in a variety of assays [37] and may be one explanation for the discrepancy between the results of [35] and the current study. Similarly, the pentaacetate derivative of quercetin (**54**) was identified here to be moderately active (Table S1), but it was on the boundary of being classified as highly active, causing more than an 80% inhibition of biofilm formation in all the assay conditions. Lastly, fisetin (**162**) was found as a moderately active flavonol and had been earlier reported to specifically inhibit *S. aureus* 8324 biofilms at concentrations within the range of 28–56 μM with negligible effects on planktonic bacteria [38].

Novel information was also gathered for other natural flavonoids. The flavonol, galangin (**155**), and the isoflavone, osajin (**288**), were also highly effective in the pre-exposure assay and inhibited the viability of existing biofilms. In the first case, galangin (**155**) has been previously reported as a potent antibacterial against suspended *S. aureus* [26], which could reasonably explain its effectiveness in the pre-exposure assay. No prior reports of the antibacterial properties of osajin (**288**) have been found, but claims of its antioxidant activity have been supported [39].

2.3. Highly Active Flavonoids and Further Selection Process

The ten compounds that were identified as highly active, according to the initial screening results, are listed in Table 1. Their structures and bioactivities, as described in the PubChem Compound database are compiled in Table S2.

Table 1. Results from the first screening assay and reconfirmation trial for the ten flavonoids originally classified as highly active (summarized in Scheme 1).

Code	Company ID	Class	Inhibition of biofilm formation (%)					
			Primary screening at 400 μM				Reconfirmation trial at 100 μM	
			Prior-to-exposure		Post-exposure		Prior-to-exposure	Post-exposure
			strain 1	strain 2	strain 1	strain 2	strain 1	
33	ST014848	isoflavone	93.8	93.8	94.5	94.9	19.7 ± 5.6	0
59	ST024709	flavanone	94.1	92.6	94.8	94.0	95.5 ± 0.1	92.5 ± 3.2
139	ST056204	flavanone	93.8	95.1	86.9	94.3	95.0 ± 0.1	93.6 ± 1.1
291	ST075672	flavan	93.2	92.6	94.6	94.7	95.4 ± 0.1	94.1 ± 1.2
369	ST081006	flavan	93.6	96.1	93.9	90.3	95.3 ± 0.2	93.3 ± 1.7
424	ST092293	chalcone	93.8	94.3	93.7	90.2	15.9 ± 5.1	0
432	ST093738	flavan	94.1	94.0	89.4	88.1	0	0
446	ST098360	flavanone	92.0	91.2	92.9	92.9	0	0
464	ST095411	chalcone	93.7	93.7	90.5	89.4	0	0
469	ST095417	chalcone	94.8	94.1	88.8	90.0	0	0

Strains 1 and 2 refer to *S. aureus* ATCC 25923 and Newman clinical strains, respectively.

Six natural flavonoids are present in the highly active group (**59**, **139**, **424**, **446**, **464** and **469**). The first two (**59**, **139**) are discussed further in this chapter. Compound **446**, isosakuranetin, ((2*S*)-5,7-dihydroxy-2-(4-methoxyphenyl)-2,3-dihydrochromen-4-one)), has been isolated from the flowers of *Chromolaena odorata*, and it has been reported to possess moderate anti-mycobacterial activity with minimal inhibitory concentrations (MIC) of 174.8 μM [40]. On the other hand, compound **464** (2',4'-dihydroxychalcone) has been recently demonstrated as the most active constituent of the plant, *Muntingia calabura*, against suspended *S. aureus* (the same strain used in the current study) with an estimated MIC value of 50 μg/mL (208 μM) [41]. Thus, at 400 μM, the antibacterial activity of **464** on suspended cells could be one possible mechanism for the prevention of biofilm formation registered in the current study. None of these highly active flavonoids have been previously reported to possess anti-biofilm properties (Table S2). The chalcones in this group (**424**, **464** and **469**) have hydroxylation in positions 2' and 4', as previously noted in the group of moderately active ones. Because these chalcones were also effective in existing biofilms, albeit only at 400 μM, it could be that these features play a role not only for antibacterial, but also for anti-biofilm activity.

These ten highly active flavonoids were further retested at 100 μM (Scheme 1; Table 1), and the high activity was only preserved in four molecules: two flavanones (**59**, **139**) and two flavan derivatives (**291**, **369**). Their structures are shown in Figure 4.

Figure 4. The four most active anti-biofilm flavonoids identified in this contribution.

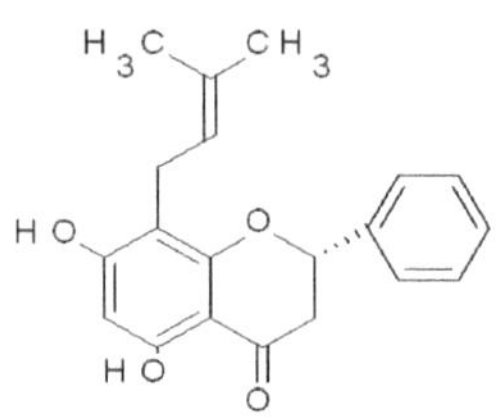
59

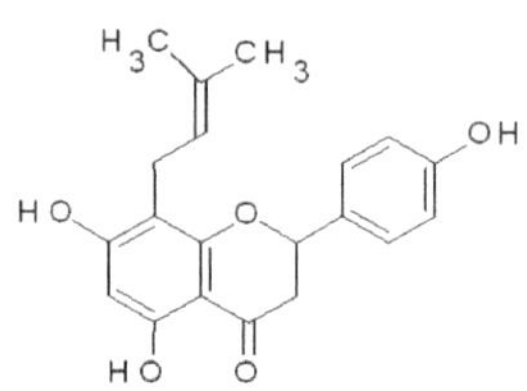
139

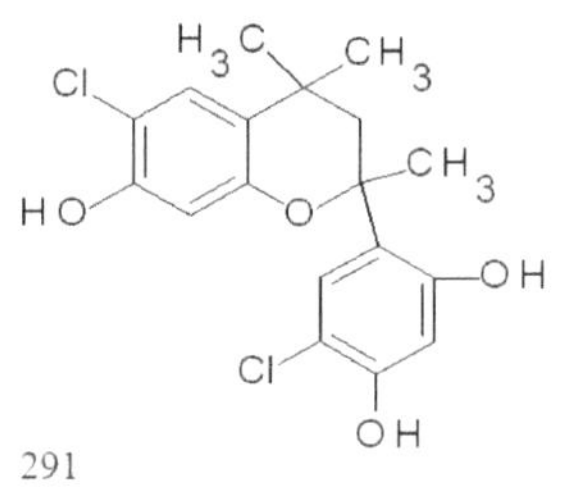
291

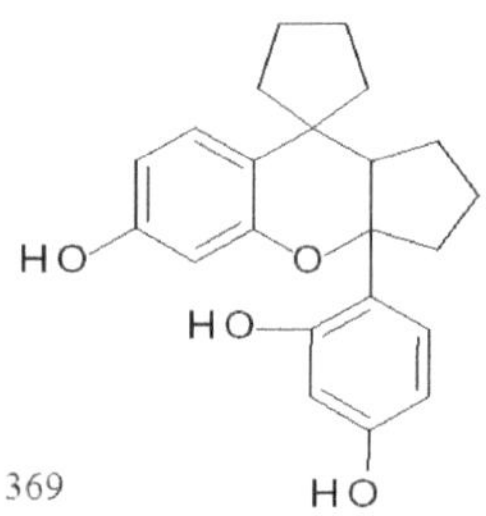
369

A literature search of flavanone glabranine **59** ((2*S*)-5,7-dihydroxy-8-(3-methylbut-2-enyl)-2-phenylchroman-4-one) promptly revealed that it is present in Mexican plants of the *Tephrosia* genus and is shown to possess inhibitory effects on the dengue virus at relatively low micromolar concentrations (25 μM) [42]. On the other hand, flavanone **139** (3,5,7-trihydroxy-2-phenylchromen-4-one), also known as the 8-prenylnaringenin, is typically found in hops and is a well-known and potent phyto-estrogen with documented *in vitro* and *in vivo* effects [43]. The side prenylation in position 8 is also present in glabranine, which suggests the importance of this side chain. The 6-prenylated naringenin (absent in this flavonoid library) has been shown to be antibacterial against suspended *S. aureus* (MIC of 18.3 μM) [44]. It has been speculated that the lipophilicity that is provided by the prenyl side chain could improve the bacterial transmembrane transport of these derivatives [45] and may tentatively facilitate the interaction of both compounds (**59** and **139**) with the biofilm matrix, in comparison to naringenin (**164**, inactive here). However, at this point, it was reasoned that compounds with other properties (such as antiviral or estrogenic) would not be selective anti-biofilms, and they may cause undesirable off-target effects. Consequently, flavanones **59** and **139** were excluded from further analysis.

Attention was then focused on the flavan derivatives, **291** (6-chloro-4-(6-chloro-7-hydroxy-2,4,4-trimethylchroman-2-yl)benzene-1,3-dioland) and **369** (4-(6-hydroxyspiro[1,2,3,3a,9a-pentahydrocyclopenta[1,2-b]chromane-9,1'-cyclopentane]-3a-yl)benzene-1,3-diol). Upon a literature search, no bioactivity data were found associated with these molecules. This search included the freely available Antimicrobial Index database (http://antibiotics.toku-e.com/), which is a very complete source of antibacterial molecules, as well as the PubChem database. One factor to keep in mind is that, in general, studies on the bioactivity of flavans appear to be more scarce in comparison to other flavonoid classes [46], and fewer screenings for their antibacterial activity have been published.

2.4. The Two Most Potent Flavonoids: Potencies, Efficacies and Mechanistic Insights

The two identified flavans prevented bacterial colonization and decreased the viability of existing biofilms (Table 2). They also caused a reduction in biofilm biomass, and the potency values quantified with the crystal violet assay were similar to the ones presented here (results not shown). The activity order was maintained in both exposure strategies, with **291** being the most active. Interestingly, the ratios between the concentrations needed to reduce 50% of the viable cells in the existing biofilm, and those needed to prevent 50% of biofilm formation varied between 2.7 and 3.4; that is a great advantage if compared with the antibiotic, penicillin G (Table 2).

Table 2. Anti-biofilm and antibacterial potencies of the top two flavonoids (flavans **291** and **369**). MIC, minimal inhibitory concentrations; MBC, minimal bactericidal concentrations.

Code	Effects on biofilms (IC_{50}, μM) (μg/mL) (95% confidence intervals)		Effects on suspended bacteria	
	Prior-to-exposure	Post-exposure	MIC, μM	MBC, μM
291	**10.2** (3.77) (8.9–11.6)	**27.9** (10.3) (22.9–33.9)	**20** (7.38)	**15** (5.54)
369	**17.7** (6.24) (12.7–24.8)	**60.5** (21.3) (49.6–73.9)	**40** (14.1)	**40** (14.1)
Penicillin G	**0.13** (0.048) (0.12–0.14)	**45.2%** *	**0.12** (0.045)	**0.13** (0.048)

* Percent inhibition at 5 mM. Penicillin fails to cause more than 50% of biofilm inhibition in the post-exposure assay, as previously shown in [22].

The same viability staining was applied to measure the efficacy these flavonoids had against suspended cells, so the comparison of the antibacterial and anti-biofilm results could be somewhat easier. The two most active flavans were shown to display a potent antibacterial activity, with similar MIC and minimal bactericidal concentrations (MBC) values quantified on the suspended cells (Table 2). MIC values, in particular, corresponding to 7.3 μg/mL and 14.1 μg/mL for compounds **291** and **369**, respectively, are indicative of a significant antibacterial activity, judging from previous indications that compounds with MICs $\leq$ 100 μg/mL are considered noteworthy and those with MICs $\leq$ 10 μg/mL are regarded to be very interesting [15,47]. Compound **291** caused 50% inhibition of biofilm formation at 10 μM, and it reached more than 90% of biofilm viability inhibition with a minor concentration increase (at 13 μM), which was very close to the MBC value quantified here (15 μM). Similarly, compound **369** had a half maximal inhibitory concentration (IC_{50}) of 17.7 μM and caused 90% inhibition of biofilm formation at 40 μM, corresponding to its MBC.

Based on how close the anti-biofilm IC_{90} and the MBC values were, it could be suggested that these flavans caused a significant killing upon direct interaction with the suspended cells, resulting in a reduction of the suspended viable cell density initially present to attach to the polystyrene surface of the microtiter well plates to begin the biofilm formation process. However, on the other hand, a former contribution demonstrated a very significant biofilm formation after 18 h, even if only a low concentration of viable bacterial cells (10^3 colony forming units, CFU/mL) was present at the beginning of the incubation period [20–23]. Thus, reducing the starting population of viable suspended bacteria even by 3-log units would still not be enough to explain the ability of these compounds to significantly inhibit bacterial colonization and biofilm formation. Consequently, it is likely that in the presence of **291** and **369**, some cells are still able to attach and form biofilms, but their maturation process is significantly hampered, resulting in the presence of less viable cells in the bacterial biofilm core, after 18 h. This possibility is supported by the fact that **291** and **369** were also potent in the post-exposure assay (when they only encounter bacterial biofilms), as previously discussed at the beginning of this section.

Next, the killing efficacy was studied. Formed *S. aureus* biofilms (18 h) were exposed to various concentrations of the two flavans during 24 h, and the density of biofilm bacteria left on the wells and bacteria in the planktonic phase were measured using agar plate counts (Table 3). The log R parameter was calculated by comparing the log counts from the treated wells to untreated controls, as in [48].

Table 3. Killing efficacy of flavans **291** and **369**.

Code	Concentration (μM)	Log Reduction (biofilm phase)	Log Reduction (planktonic phase)
291	20	0.6	0.1
	80	1.5	0.7
	200	3.5	3.5
	400	4.6	4.7
369	50	1.5	1.2
	100	3.9	4.1
	250	3.1	9.0
	400	3.9	9.0
penicillin G	400	1.0	4.0

This experiment intended to clarify the ability of these compounds to reduce the viable bacterial burden in the biofilm phase using the reference or golden standard method for bacterial quantification (the agar plate counts). Due to the laboriousness and low throughput of this method, it is only applicable for follow-up studies in screening campaigns for chemical libraries, at the point when active ones have been selected and a reasonable amount of samples are to be handled. Compounds causing at least a 3-log reduction are of greater interest, since a 3-log reduction of the biofilm burden is highly desirable to assist the immune system in clearing the remaining pathogens *in vivo*, in immunotolerant biofilm infections [49,50].

Furthermore, in this experiment, the remaining viable bacteria in the planktonic phase were quantified. Although planktonic cells are removed (and flavonoids are added), once the biofilms have been formed for 18 h, the continuation of the experiment for an additional 24 h results in space limitations for the existing *S. aureus* biofilms, due to the relatively small size of the wells. A fraction of the biofilm population then gets detached and can switch into the suspended state to maintain the stability of the established biofilms. Accordingly, at the end of the post-exposure time, a high concentration of viable biofilm bacteria is typically recovered from the untreated wells (4×10^9 CFU/mL, corresponding to 5×10^8 CFU/cm^2), as well as of suspended bacteria from the same wells (6×10^9 CFU/mL).

Flavans **291** and **369** caused more than 3-log reduction of the biofilm viable core at concentrations of 200 and 100 μM, respectively, indicative of 99.9% killing. Applying flavan **291** resulted in a similar log reduction of biofilm and planktonic cells throughout the entire concentration range, confirming that this compound has both antibacterial and anti-biofilm activity and that it can be equally effective at targeting both, under the same experimental conditions. Flavan **369**, on the other hand, was more effective in reducing the viable suspended bacteria (achieving the maximum log R values with no detectable living colonies) than the viable core of bacterial cells residing on the *S. aureus* biofilms. This suggests instead that **369** is a flavan behaving more as an antibacterial than an anti-biofilm compound.

If only the antibacterial activity is to be taken into account, **291** ranks within the top 10 most active flavonoids reported in the literature (reviewed by Cushnie and Lamb [15]). However, it is the fact that **291**, as well as **369** are also potent anti-biofilm agents that makes them highly interesting. Bacteria are switching between single-cell and biofilm states in host organisms, and the predominant lifestyle can dynamically change depending on various complex factors. Thus, compounds that can successfully act on both at fairly similar concentrations are the most promising starting point for more feasible candidates for antibacterials.

3. Experimental Section

3.1. Compounds Collection

The studied collection consisted of 500 synthetic and natural flavonoids that are commercially sold by TimTec (www.timtec.net). Structures of all compounds can be retrieved from the vendor's database (http://www.echemstore.com/). Flavonoids with 9 different basic cores are present in the library (flavanones, flavones, chalcones, flavonols, dihydroflavonols, flavans, anthocyanins, isoflavonoids and neoflavonoids). The list of all tested compounds is presented in Table S1, as well as their trivial and short names, when available. The average molecular weight of the collection is 342.8 g/mol, and the averages of predicted log$P$ and log$S$ are 3.49 and −4.46, respectively, as provided by the supplier. Compound identities were confirmed by the supplier, using Nuclear Magnetic Resonance spectroscopy (NMR) (300 MHz or higher) and Liquid Chromatography-Mass Spectrometry (LC/MS). Compound purities were measured using High-Performance Liquid Chromatography (HPLC) and ensured to be over 95%. All compounds were dissolved in dry DMSO to a concentration of 20 mM and were maintained in matrix storage tubes (Thermo Fisher Scientific, Waltham, MA, USA) at −20 °C. From those stocks, the compounds were added directly into the assay plates containing the culture media or further dilutions in dry DMSO were made for the bioactivity follow-ups with the most active compounds. Additional amounts of flavans **291** (company ID: ST075672) and **369** (company ID: ST081006) were purchased from TimTec for the characterization studies. Structural validation for these two compounds is presented in Table S4.

3.2. Biofilm Assay

S. aureus (ATCC 25923 and Newman, both clinical strains of human origin) were used for the primary screening of the flavonoids collection. Bacteria were cultured in 30 g/L tryptic soy broth (TSB, Fluka Biochemika, Buchs, Switzerland) under aerobic conditions at 37 °C, 200 rpm for 4 h to reach exponential growth. Biofilms were formed according to the conditions previously described [23]. Briefly, exponentially grown bacteria (10^6 CFU/mL, 200 μL) were added to flat-bottomed 96-well microplates (Nunclon Δ surface) (Nunc, Roskilde, Denmark), and biofilms were formed at 37 °C, 200 rpm for 18 h in 30 g/L TSB.

3.3. Exposure to Flavonoids

The anti-biofilm effects of the flavonoids were examined prior to biofilm and post-biofilm formation, as recently described in [23]. During the pre-exposure, compounds and bacterial suspension were added, and the effects were examined after incubation at 37 °C, 200 rpm for 18 h. In the post-exposure, biofilms were formed first during 18 h (37 °C, 200 rpm, as described earlier), flavonoids were added and plates were incubated for 24 h at 37 °C, 200 rpm. Penicillin G, used as the control antibiotic in all experiments, was prepared in Mueller-Hinton Broth (MHB) and tested at 400 μM. Untreated biofilms, cell-free samples with TSB and biofilms exposed to 2% DMSO were included as controls. Initial screening of the entire collection was conducted at a final compound concentration of 400 μM for both bacterial strains. The reconfirmation trial was performed for selected compounds at 100 μM.

3.4. Quantification of Biofilms

Biofilm viability and biomass were measured using resazurin and crystal violet staining assays, as previously described [23]. Biofilms were first stained with resazurin 20 μM (Sigma-Aldrich, St. Louis, MO, USA) for 20 min at room temperature (RT), (darkness, 200 rpm), followed by the measurement of the fluorescence emitted by the resorufin (the reduced form of the resazurin), at λ_{exc} = 560 nm and λ_{em} = 590 nm using a Varioskan reader (Thermo Fisher Scientific, Vantaa, Finland). The resazurin stain was then removed and replaced with crystal violet, which was added in the sample plate using a Multidrop® Combi dispenser (Thermo Fisher Scientific, Vantaa, Finland) and incubated for 5 min (RT, static conditions). After staining, biofilms were washed three times with Milli-Q (MQ) water with the aid of a Biomek 3000® liquid handling workstation (Beckman Coulter Inc., Fullerton, CA, USA). The remaining dye was recovered by the addition of 96% ethanol, and after 1 h, the absorbance was measured at λ = 595 nm using a Varioskan plate reader.

3.5. Anti-Biofilm Potency and Efficacy Testing

Anti-biofilm potencies of selected compounds (coded **291** and **369**) were measured using 18 concentration points within a concentration range of 10–400 μM in both exposure schemes. Anti-biofilm efficacies were measured using the Log Reduction (Log R) assay [48], which is based on viable cell counts in tryptic soy agar (TSA) plates. At the end of the 24 h exposure to these compounds, planktonic cells were transferred to sterile plates, while biofilms were scraped off the wells in 100 μL TSB using sterile plastic sticks and rinsed with an additional 100 μL of TSB. To disperse the bacterial aggregates, samples were immersed in a high power ultrasonic bath (Bandelin Sonorex Digitec, Zurich, Switzerland) using an in-house built-in device that allowed them to be in full contact with the water. Sonication was performed at RT (5 min, 35 kHz). The disaggregated biofilms, as well as the planktonic cells, were serially diluted, spread onto TSA plates and incubated at 37 °C overnight. The $\log_{10}$ density of viable cells (CFU/mL) in treated and control wells were determined, and Log R was calculated from the difference of the log_{10} density on the untreated biofilms (or the planktonic phase of the corresponding wells) and the log_{10} density on the biofilms treated with the flavonoids (or the planktonic phase of the corresponding wells). To confirm that all biofilms have been recovered, the scraped wells were stained with resazurin (as in Section 3.4).

3.6. Bacteriostatic and Bactericidal Effect on Planktonic Cells

Minimal inhibitory concentrations (MIC) of **291** and **369** for planktonic bacteria were first measured. After exposing biofilms to the flavonoids (at 37 °C for 18 h), the planktonic phase was transferred to sterile 96-microtiter well plates, and quantitative readouts were obtained (at λ = 620 nm) using the Varioskan Flash Multimode Plate Reader (2.4.3.37 software, Thermo Fisher Scientific Oy, Vantaa, Finland, 2004), as in [23]. The minimal bactericidal concentrations (MBC) were also measured. For this purpose, after making the MIC determinations, suspensions were mixed with resazurin 400 μM (10 μL), to reach a final concentration of 20 μM. Plates were then incubated in the darkness, RT, with 200 rpm shaking for approximately for 4 min, followed by the measurement of fluorescence (λ_{exc} = 570 nm; λ_{em} = 590 nm) using the Varioskan Flash Multimode Plate Reader (2.4.3.37 software, Thermo Fisher Scientific Oy, Vantaa, Finland, 2004).

3.7. Statistical Analysis and Data Processing

At least four replicates per treatment were included in each plate, and two separate experiments (biological replicates) were always run. Anti-biofilm activities were expressed as inhibition percentages of the untreated biofilms (Equation 1). The potencies of the anti-biofilm effects (IC_{50}) of **291** and **369** were calculated from 18 concentration points via a non-linear regression analysis (sigmoidal dose-response with variable slope), and the results are presented with 95% confidence intervals. Below, μ_{max}, and μ_{min} represent the means (average) of the signals (the fluorescence of the reduced resazurin or absorbance of the crystal violet stained biofilms) recorded in untreated biofilms and TSB controls, respectively.

$$\text{Inhibition } \% = [(\mu_{max} - \mu_{treated\ well})/(\mu_{max} - \mu_{min})] \times 100\% \qquad (1)$$

Quality control of the screening process was done via the calculation of the screening window coefficient (*Z'*) and signal-to-background (*S*/*B*) and signal-to-noise (*S*/*N*) ratios, as defined by [51,52]. Calculated parameters during the first screening behaved as follows: $Z' > 0.6$; $S/B > 16$ and $S/N > 7$. All data processing and statistical analysis was done with Microsoft Excel 2010 software and GraphPad Prism version 5.0 for Mac, GraphPad Software (San Diego, CA, USA, 2009), respectively.

4. Conclusions

The body of data produced here serves three main purposes. First, it offers an improved methodological workflow for anti-biofilm phenotypic screens of chemical libraries. By using methods optimized for biofilm quantification in the context of chemical screening and by designing the study with two exposure modes and two *Staphylococcus* spp. Strains, the prior flaws of antibacterial studies of flavonoids have been minimized. Furthermore, to improve data analysis, the probable connections between the anti-biofilm and antibacterial properties of the flavonoids were investigated, which is often times ignored in the scientific literature. This is particularly relevant for the proper interpretation of the bioactivity data obtained from flavonoids, but also from other natural products with reported antibacterial effects, for instance, phenolic acids, catechins and phytosterols, among others.

Second, it provides a large anti-biofilm bioactivity dataset for flavonoids. An effort was made to highlight relevant flavonoids and to compare current results with prior contributions, so that a better understanding of their bioactivity profiles was achieved. The dataset built here will also hopefully pave the way for future structure-activity relationship (SAR) studies and structural optimizations.

Finally, this contribution identifies two new synthetic flavans (compounds **291** and **369**) as potent antimicrobials that can successfully act on biofilms and suspended *S. aureus* and represent more feasible antibacterial candidates, to be included in a new generation of pharmaceuticals. This research may also potentially spark interest for more focused studies on this flavonoid class. More data is indeed needed on their *in vivo* bioavailabilites and safety profiles, which will benefit a better exploitation of flavans as drug leads.

Acknowledgments

The authors thank the Academy of Finland (WoodyFilm project decision 264064; Bioarmi project decision 128870), the Drug Discovery and Chemical Biology network of Biocenter Finland the Finnish Bioeconomy Cluster, the Borg Foundation and Medicinska Understödsföreningen Liv och Hälsa for financial support. The contribution of Oyvind M. Andersen (University of Bergen, Norway) during the preparation of this manuscript is also greatly appreciated.

References

1. Donlan, R.M. Biofilm formation: A clinically relevant microbiological process. *Clin. Infect. Dis.* **2001**, *33*, 1387–1392.
2. Donlan, R.M.; Costerton, J.W. Biofilms: Survival mechanisms of clinically relevant microorganisms. *Clin. Microbiol. Rev.* **2002**, *15*, 167–193.
3. Sauer, K.; Camper, A.K. Characterization of phenotypic changes in *Pseudomonas putida* in response to surface-associated growth. *J. Bacteriol.* **2001**, *183*, 6579–6589.
4. Tote, K.; vanden Berghe, D.; Deschacht, M.; de Wit, K.; Maes, L.; Cos, P. Inhibitory efficacy of various antibiotics on matrix and viable mass of *Staphylococcus aureus* and *Pseudomonas aeruginosa* biofilms. *Int. J. Antimicrob. Agents* **2009**, *33*, 525–531.
5. Landini, P.; Antoniani, D.; Burgess, J.G.; Nijland, R. Molecular mechanisms of compounds affecting bacterial biofilm formation and dispersal. *Appl. Microbiol. Biotechnol.* **2010**, *86*, 813–823.
6. Parsek, M.R.; Singh, P.K. Bacterial biofilms: An emerging link to disease pathogenesis. *Annu. Rev. Microbiol.* **2003**, *57*, 677–701.
7. Malic, S.; Hill, K.E.; Hayes, A.; Percival, S.L.; Thomas, D.W.; Williams, D.W. Detection and identification of specific bacteria in wound biofilms using peptide nucleic acid fluorescent *in situ* hybridization (PNA-FISH). *Microbiology* **2009**, *155*, 2603–2611.

8. James, G.A.; Swogger, E.; Wolcott, R.; deLancey Pulcini, E.; Secor, P.; Sestrich, J.; Costerton, J.W.; Stewart, P.S. Biofilms in chronic wounds. *Wound Repair Regener.* **2008**, *16*, 37–44.
9. Hall-Stoodley, L.; Hu, F.Z.; Gieseke, A.; Nistico, L.; Nguyen, D.; Hayes, J.; Forbes, M.; Greenberg, D.P.; Dice, B.; Burrows, A.; *et al.* Direct detection of bacterial biofilms on the middle-ear mucosa of children with chronic otitis media. *JAMA* **2006**, *296*, 202–211.
10. Singh, P.K.; Schaefer, A.L.; Parsek, M.R.; Moninger, T.O.; Welsh, M.J.; Greenberg, E.P. Quorum-sensing signals indicate that cystic fibrosis lungs are infected with bacterial biofilms. *Nature* **2000**, *407*, 762–764.
11. Adair, C.G.; Gorman, S.P.; Feron, B.M.; Byers, L.M.; Jones, D.S.; Goldsmith, C.E.; Moore, J.E.; Kerr, J.R.; Curran, M.D.; Hogg, G.; *et al.* Implications of endotracheal tube biofilm for ventilator-associated pneumonia. *Intensive Care Med.* **1999**, *25*, 1072–1076.
12. Worthington, R.J.; Richards, J.J.; Melander, C. Small molecule control of bacterial biofilms. *Org. Biomol. Chem.* **2012**, *10*, 7457–7474.
13. Samanta, A.; Das, G.; Das, S.K. Roles of flavonoids in plants. *Int. J. Pharm. Sci. Technol.* **2011**, *6*, 12–35.
14. Kumar, B.; Sandhar, H.K.; Prasher, S.; Tiwari, P.; Salhan, M.; Sharma, P. A review of phytochemistry and pharmacology of flavonoids. *Int. Pharm. Sci.* **2011**, *1*, 25–41.
15. Cushnie, T.P.; Lamb, A.J. Recent advances in understanding the antibacterial properties of flavonoids. *Int. J. Antimicrob. Agents* **2011**, *38*, 99–107.
16. Havsteen, B.H. The biochemistry and medical significance of the flavonoids. *Pharmacol. Ther.* **2002**, *96*, 67–202.
17. Alvesalo, J.; Vuorela, H.; Tammela, P.; Leinonen, M.; Saikku, P.; Vuorela, P. Inhibitory effect of dietary phenolic compounds on *Chlamydia pneumoniae* in cell cultures. *Biochem. Pharmacol.* **2006**, *71*, 735–741.
18. Cushnie, T.P.; Lamb, A.J. Antimicrobial activity of flavonoids. *Int. J. Antimicrob. Agents* **2005**, *26*, 343–356.
19. Rauha, J.P.; Remes, S.; Heinonen, M.; Hopia, A.; Kähkönen, M.; Kujala, T.; Pihlaja, K.; Vuorela, H.; Vuorela, P. Antimicrobial effects of finnish plant extracts containing flavonoids and other phenolic compounds. *Int. J. Food Microbiol.* **2000**, *56*, 3–12.
20. Sandberg, M.; Määttänen, A.; Peltonen, J.; Vuorela, P.M.; Fallarero, A. Automating a 96-well microtitre plate model for *Staphylococcus aureus* biofilms: An approach to screening of natural antimicrobial compounds. *Int. J. Antimicrob. Agents* **2008**, *32*, 233–240.
21. Sandberg, M.E.; Schellmann, D.; Brunhofer, G.; Erker, T.; Busygin, I.; Leino, R.; Vuorela, P.M.; Fallarero, A. Pros and cons of using resazurin staining for quantification of viable *Staphylococcus aureus* biofilms in a screening assay. *J. Microbiol. Methods* **2009**, *78*, 104–106.
22. Skogman, M.E.; Vuorela, P.M.; Fallarero, A. Combining biofilm matrix measurements with biomass and viability assays in susceptibility assessments of antimicrobials against *Staphylococcus aureus* biofilms. *J. Antibiot.* **2012**, *65*, 453–459.
23. Fallarero, A.; Skogman, M.; Kujala, J.; Rajaratnam, M.; Moreira, V.M.; Yli-Kauhaluoma, J.; Vuorela, P. (+)-Dehydroabietic acid, an abietane-type diterpene, inhibits *Staphylococcus aureus* biofilms *in vitro*. *Int. J. Mol. Sci.* **2013**, *14*, 12054–12072.

24. Peeters, E.; Nelis, H.J.; Coenye, T. Comparison of multiple methods for quantification of microbial biofilms grown in microtiter plates. *J. Microbiol. Methods* **2008**, *72*, 157–165.
25. Mariscal, A.; Lopez-Gigosos, R.M.; Carnero-Varo, M.; Fernandez-Crehuet, J. Fluorescent assay based on resazurin for detection of activity of disinfectants against bacterial biofilm. *Appl. Microbiol. Biotechnol.* **2009**, *82*, 773–783.
26. Cushnie, T.P.; Hamilton, V.E.; Lamb, A.J. Assessment of the antibacterial activity of selected flavonoids and consideration of discrepancies between previous reports. *Microbiol. Res.* **2003**, *158*, 281–289.
27. Vikram, A.; Jayaprakasha, G.K.; Jesudhasan, P.R.; Pillai, S.D.; Patil, B.S. Suppression of bacterial cell-cell signalling, biofilm formation and type III secretion system by citrus flavonoids. *J. Appl. Microbiol.* **2010**, *109*, 515–527.
28. Zeng, Z.; Qian, L.; Cao, L.; Tan, H.; Huang, Y.; Xue, X.; Shen, Y.; Zhou, S. Virtual screening for novel quorum sensing inhibitors to eradicate biofilm formation of *Pseudomonas aeruginosa*. *Appl. Microbiol. Biotechnol.* **2008**, *79*, 119–126.
29. Vandeputte, O.M.; Kiendrebeogo, M.; Rajaonson, S.; Diallo, B.; Mol, A.; El Jaziri, M.; Baucher, M. Identification of catechin as one of the flavonoids from *Combretum albiflorum* bark extract that reduces the production of quorum-sensing-controlled virulence factors in *Pseudomonas aeruginosa* PAO1. *Appl. Environ. Microbiol.* **2010**, *76*, 243–253.
30. Zhao, X.; Mei, W.; Gong, M.; Zuo, W.; Bai, H.; Dai, H. Antibacterial activity of the flavonoids from *Dalbergia odorifera* on *Ralstonia solanacearum*. *Molecules* **2011**, *16*, 9775–9782.
31. Feldman, M.; Santos, J.; Grenier, D. Comparative evaluation of two structurally related flavonoids, isoliquiritigenin and liquiritigenin, for their oral infection therapeutic potential. *J. Nat. Prod.* **2011**, *74*, 1862–1867.
32. Avila, H.P.; Smânia, E.F.A.; Monache, F.D.; Smânia, A. Structure-activity relationship of antibacterial chalcones. *Bioorg. Med. Chem.* **2008**, *16*, 9790–9794.
33. Alvarez, M.e.L.; Zarelli, V.E.; Pappano, N.B.; Debattista, N.B. Bacteriostatic action of synthetic polyhydroxylated chalcones against *Escherichia coli*. *Biocell* **2004**, *28*, 31–34.
34. Li, H.Q.; Shi, L.; Li, Q.S.; Liu, P.G.; Luo, Y.; Zhao, J.; Zhu, H.L. Synthesis of C(7) modified chrysin derivatives designing to inhibit beta-ketoacyl-acyl carrier protein synthase III (FABH) as antibiotics. *Bioorg. Med. Chem.* **2009**, *17*, 6264–6269.
35. Lee, J.H.; Park, J.H.; Cho, H.S.; Joo, S.W.; Cho, M.H.; Lee, J. Anti-biofilm activities of quercetin and tannic acid against *Staphylococcus aureus*. *Biofouling* **2013**, *29*, 491–499.
36. Pohjala, L.; Tammela, P. Aggregating behavior of phenolic compounds—A source of false bioassay results? *Molecules* **2012**, *17*, 10774–10790.
37. McGovern, S.L.; Shoichet, B.K. Kinase inhibitors: Not just for kinases anymore. *J. Med. Chem.* **2003**, *46*, 1478–1483.
38. Dürig, A.; Kouskoumvekaki, I.; Vejborg, R.M.; Klemm, P. Chemoinformatics-assisted development of new anti-biofilm compounds. *Appl. Microbiol. Biotechnol.* **2010**, *87*, 309–317.
39. Fránová, J.; Pavlík, M. Testing of antidiabetic and antioxidative effect of the flavonoid osajin in an experiment. *Ceska Slov. Farm.* **2007**, *56*, 200–204.

40. Suksamrarn, A.; Chotipong, A.; Suavansri, T.; Boongird, S.; Timsuksai, P.; Vimuttipong, S.; Chuaynugul, A. Antimycobacterial activity and cytotoxicity of flavonoids from the flowers of *Chromolaena odorata*. *Arch. Pharm. Res.* **2004**, *27*, 507–511.
41. Sufian, A.S.; Ramasamy, K.; Ahmat, N.; Zakaria, Z.A.; Yusof, M.I. Isolation and identification of antibacterial and cytotoxic compounds from the leaves of *Muntingia calabura* L. *J. Ethnopharmacol.* **2013**, *146*, 198–204.
42. Sánchez, I.; Gómez-Garibay, F.; Taboada, J.; Ruiz, B.H. Antiviral effect of flavonoids on the dengue virus. *Phytother. Res.* **2000**, *14*, 89–92.
43. Zanoli, P.; Zavatti, M. Pharmacognostic and pharmacological profile of *Humulus lupulus* L. *J. Ethnopharmacol.* **2008**, *116*, 383–396.
44. Mizobuchi, S.; Sato, Y. Antifungal activities of hop bitter resins and related compounds. *Agric. Biol. Chem.* **1985**, *49*, 399–403.
45. Shen, G.; Huhman, D.; Lei, Z.; Snyder, J.; Sumner, L.W.; Dixon, R.A. Characterization of an isoflavonoid-specific prenyltransferase from *Lupinus albus*. *Plant Physiol.* **2012**, *159*, 70–80.
46. Chang, S.Y.; Cheng, M.J.; Peng, C.F.; Chang, H.S.; Chen, I.S. Antimycobacterial butanolides from the root of *Lindera akoensis*. *Chem. Biodiversity* **2008**, *5*, 2690–2698.
47. Ríos, J.L.; Recio, M.C. Medicinal plants and antimicrobial activity. *J. Ethnopharmacol.* **2005**, *100*, 80–84.
48. Pitts, B.; Hamilton, M.A.; Zelver, N.; Stewart, P.S. A microtiter-plate screening method for biofilm disinfection and removal. *J. Microbiol. Methods* **2003**, *54*, 269–276.
49. Jensen, P.; Givskov, M.; Bjarnsholt, T.; Moser, C. The immune system *vs. Pseudomonas aeruginosa* biofilms. *FEMS Immunol. Med. Microbiol.* **2010**, *59*, 292–305.
50. Müller, G.; Kramer, A. Biocompatibility index of antiseptic agents by parallel assessment of antimicrobial activity and cellular cytotoxicity. *J. Antimicrob. Chemother.* **2008**, *61*, 1281–1287.
51. Zhang, J.H.; Chung, T.D.; Oldenburg, K.R. A simple statistical parameter for use in evaluation and validation of high throughput screening assays. *J. Biomol. Screening* **1999**, *4*, 67–73.
52. Bollini, S.; Herbst, J.J.; Gaughan, G.T.; Verdoorn, T.A.; Ditta, J.; Dubowchik, G.M.; Vinitsky, A. High-throughput fluorescence polarization method for identification of fkbp12 ligands. *J. Biomol. Screening* **2002**, *7*, 526–530.

10

Emulsifying Activity and Stability of a Non-Toxic Bioemulsifier Synthesized by *Microbacterium* sp. MC3B-10

Juan Carlos Camacho-Chab [1], Jean Guézennec [2], Manuel Jesús Chan-Bacab [1], Elvira Ríos-Leal [3], Corinne Sinquin [2], Raquel Muñiz-Salazar [4], Susana del C. De la Rosa-García [1], Manuela Reyes-Estebanez [1] and Benjamín Otto Ortega-Morales [1,*]

[1] Center of Environmental Microbiology and Biotechnology, Autonomous University of Campeche, Colonia Buenavista, San Francisco de Campeche 24039, Campeche, Mexico; E-Mails: juanccam@uacam.mx (J.C.C.-C.); manjchan@uacam.mx (M.J.C.-B.); scdelaro@uacam.mx (S.C.R.-G.); reyes_estebanez@hotmail.com (M.R.-E.)

[2] Ifremer Laboratoire Biotechnologies et Molécules Marines, Nantes B. P. 21105, France; E-Mails: guezennec.jean@wanadoo.fr (J.G.); Corinne.Sinquin@ifremer.fr (C.S.)

[3] Department of Biotechnology and Bioengineering, Center for Research and Advanced Studies of the National Polytechnic Institute (CINVESTAV), Mexico City 07360, Federal District, Mexico; E-Mail: erios@cinvestav.mx

[4] Laboratory of Molecular Ecology and Epidemiology, School of Health Sciences, Autonomous University of Baja California, Tijuana 22390, Baja California, Mexico; E-Mail: ramusal@uabc.edu.mx

* Author to whom correspondence should be addressed; E-Mail: beortega@uacam.mx;

Abstract: A previously reported bacterial bioemulsifier, here termed microbactan, was further analyzed to characterize its lipid component, molecular weight, ionic character and toxicity, along with its bioemulsifying potential for hydrophobic substrates at a range of temperatures, salinities and pH values. Analyses showed that microbactan is a high molecular weight (700 kDa), non-ionic molecule. Gas chromatography of the lipid fraction revealed the presence of palmitic, stearic, and oleic acids; thus microbactan may be considered a glycolipoprotein. Microbactan emulsified aromatic hydrocarbons and oils to various extents; the highest emulsification index was recorded against motor oil (96%). The stability of the microbactan-motor oil emulsion model reached its highest level (94%) at 50 °C, pH 10 and 3.5% NaCl content. It was not toxic to *Artemia salina* nauplii.

Microbactan is, therefore, a non-toxic and non-ionic bioemulsifier of high molecular weight with affinity for a range of oily substrates. Comparative phylogenetic assessment of the 16S rDNA gene of *Microbacterium* sp. MC3B-10 with genes derived from other marine *Microbacterium* species suggested that this genus is well represented in coastal zones. The chemical nature and stability of the bioemulsifier suggest its potential application in bioremediation of marine environments and in cosmetics.

Keywords: bioemulsifier; emulsifying stability; non-toxic; *Microbacterium* sp.; biotechnology

---

1. Introduction

Microorganisms produce a wide variety of high and low molecular weight biosurfactants. These active molecules include proteins, polysaccharides, lipopeptides, glycolipids, flavolipids, phospholipids and lipopolysaccharides [1]. High molecular weight biosurfactants, usually referred to as bioemulsifiers, can form and stabilize oil-in-water or water-in-oil emulsions, work at low concentrations and exhibit considerable substrate specificity [2]. The chemical diversity and functional properties of these compounds result in a broad spectrum of potential applications in sectors as diverse as agriculture, cosmetics, environmental, food, leather, paper, pharmaceutical and textile industries [2,3]. Biologically derived surface-active molecules have advantages over their synthetic counterparts, including biodegradability, lower toxicity, novel structural chemistry and high stability at extreme temperature, salinity and pH [2,4]. Some bioemulsifiers, e.g., emulsan [5], have been extensively characterized and reached commercial applications, chiefly in the bioremediation sector.

Bioprospection of various habitats, in particular marine environments [6,7], has yielded many novel biomolecules produced by microorganisms. An earlier report showed that the extracellular biopolymer produced by *Microbacterium* sp. MC3B-10 emulsified hexane, decane and hexadecane at higher efficiencies than commercial surfactants Triton® X-100 and Tween® 80 [8]. This extracellular biopolymer will be referred to as microbactan after its bacterial producer. The goal of the present study was to characterize in greater detail this extracellular biopolymer and determine the emulsifying stability as a function of key variables.

2. Results and Discussion

2.1. Characterization of Microbactan

Microbactan was produced by fermentation in shaker flasks; no attempt was made at this time to optimize its production using a bioreactor, where more strict control of variables can be achieved. The reproducibility of production conditions employed in this study was confirmed by the yield and the primary chemical profile of the biopolymer, which was comparable to the previous report [8]. Our previous work showed that microbactan was dominated by carbohydrates and proteins, and was preliminarily considered a glycoprotein. The present study also revealed the presence of lipids, shown by the peak at 2927 cm^{-1} in the FT-IR analysis, which represents the asymmetric stretch (C–H) of $-CH_2$ groups combined with that of $-CH_3$ groups in lipids [9]. Quantitative assessment by absorbance showed that the lipid signature detected by FT-IR corresponded to a lipid content of 8% ± 0.5%. The gas chromatograph analysis of the lipid fraction revealed the presence of palmitic (C-16), stearic (C-18) and oleic (C-18:1) acids. Other fatty acids, found at lower levels, were myristoleic (C-14:1) and linoleic (C-18:2). The lipid content and the previously reported chemical composition of ~90% carbohydrates and proteins [8] represented almost 100% of the total mass of the microbactan on a dry weight basis, suggesting that this biopolymer is a class of glycolipoprotein. The fatty acids detected (palmitic, stearic and oleic) have been reported in bioemulsifiers produced by *Yarrowia lipolytica* [10] and *Penicillium* sp. [11]. However, the presence of other substituents of extracellular biopolymers such as pyruvil, succinyl and sulfates, which often occur at minor levels [4,9], was not investigated.

High performance size exclusion chromatography showed that microbactan had a molecular weight (MW) approaching 700 kDa and a polydispersity index (Mw/number average molecular weight) of 1.3; a value close to 1 denotes a polymer with homogeneous monomer grouping. These results indicate that microbactan is a homogeneous high molecular weight glycolipoprotein, a finding that is consistent with previous reports showing that, in general, bioemulsifiers are high molecular weight polymers [1,12]. The commercial bioemulsifier alasan® has a molecular weight of approximately 1000 kDa [13]. Most bacterial bioemulsifiers are polymers of either carbohydrates or proteins, or even glycoproteins [1,6]. Thavasi and colleagues report marine-derived glycolipoprotein emulsifiers from *Corynebacterium kutcheri* and *Bacillus megaterium* [14,15]. The putative glycolipoprotein nature of microbactan is, therefore, to some extent unusual among marine bioemulsifiers.

Microbactan was shown by the modified double diffusion test to be a non-ionic emulsifier. The ionic character of polymers is one of the features that contribute strongly to the functional properties of emulsifiers; this property is often reported for synthetic emulsifiers or novel emulsifying formulations used for a range of applications. Despite the importance of this feature, it is rarely reported in newly discovered bioemulsifiers. The non-ionic nature of emulsifiers greatly contributes to emulsion stability, generating a number of short-range repulsive forces, such as steric, hydration, and thermal fluctuation interactions, which prevent the droplets from getting too close together [16].

2.2. Emulsifying Potential

Tables 1 and 2 show the emulsifying activity of microbactan and control commercial emulsifiers tested against different substrates at different times. Statistical analysis showed significant differences between microbactan and these controls ($F = 84.24$; $p < 0.05$). Synthetic surfactants Triton® X-100 and Tween® 80 were more efficient than microbactan and the commercial biopolymers, reaching emulsification efficiencies of 100% against oils (with the exception of motor oil), irrespective of time of incubation (24 and 96 h). Microbactan showed stronger emulsifying activities than gum arabic but was comparable to the emulsifying efficiency of xanthan gum, another bacterially-produced biopolymer. The emulsifying activity of microbactan was dependent on the type of substrate ($F = 16.2$; $p < 0.05$); this is consistent with other reports on substrate-specific hydrocarbon metabolism by marine bacteria [17,18]. Microbactan emulsions with the tested substrates were rather stable over time (Tables 1 and 2), varying only slightly between 24 and 96 h.

Table 1. Emulsifying activity of microbactan, commercial synthetic surfactants and natural biopolymer emulsifiers on various hydrophobic substrates after 24 h of evaluation.

Hydrophobic substrate	Microbactan [a]	Synthetic surfactants [a]		Biopolymers [a]	
		Tween 80	Triton-X-100	Gum arabic	Xanthan gum
Benzene	76.9 ± 2.4	98.7 ± 0.6	54.3 ± 2.0	80.7 ± 3.0	64.7 ± 1.3
Xylene	81.5 ± 3.4	94.7 ± 6.1	98.2 ± 0.4	72.6 ± 3.4	76.8 ± 1.7
Crude oil	76.5 ± 2.1	82.9 ± 0.6	100	54.7 ± 3.7	95.0 ± 0.9
Motor oil	96.3 ± 0.1	54.9 ± 1.4	68.5 ± 6	95.2 ± 1.1	89.5 ± 3.3
Sunflower oil	84.4 ± 5.4	100	100	77.8 ± 1.5	62.4 ± 2.4
Corn oil	81.1 ± 1.6	100	100	0	90.6 ± 1.9
Olive oil	76.1 ± 0.3	100	100	96.6 ± 0.1	76.6 ± 5.4
Mineral oil	0	95.9 ± 5.6	100	60.3 ± 2.8	86.5 ± 3.8

[a] Results are expressed as percentages of the total height occupied by the emulsion; values are means of at least three determinations.

Table 2. Emulsifying activity of microbactan, commercial synthetic surfactants and natural biopolymer emulsifiers on various hydrophobic substrates after 96 h of evaluation.

Hydrophobic substrate	Microbactan [a]	Synthetic surfactants [a]		Biopolymers [a]	
		Tween 80	Triton-X-100	Gum arabic	Xanthan gum
Benzene	75.3 ± 3.8	93.2 ± 5	54.3 ± 2.1	78.9 ± 0	61.7 ± 1.4
Xylene	75.4 ± 0.5	94.1 ± 5	97.9 ± 0.5	63.7 ± 3	75 ± 1.6
Crude oil	72 ± 4.8	85.5 ± 5	100 ± 0	53.2 ± 2.4	93.8 ± 2.5
Motor oil	96.3 ± 0.1	53.5 ± 2.3	59 ± 4.2	92.8 ± 1.8	83.7 ± 3.2
Sunflower oil	83.8 ± 3.4	100	100	77.8 ± 2.1	62.4 ± 2.4
Corn oil	81.1 ± 1.6	100	100	0	91.5 ± 0.8
Olive oil	76.1 ± 0.5	100	100	96.6 ± 0.6	78.2 ± 2.6
Mineral oil	0	91.2 ± 2.3	100 ± 0	56.8 ± 2.3	82.9 ± 2

[a] Results are expressed as percentages of the total height occupied by the emulsion; values are means of at least three determinations.

An established criterion for emulsion-stabilizing capacity is the ability of an emulsifier to maintain at least 50% of the original volume of the emulsion for 24 h [4,19]. Microbactan emulsions remained stable for several months, showing no sign of droplet coalescence after standing at room temperature (28 °C). This extended stability has previously been observed for glycoprotein bioemulsifiers produced by a marine *Antarctobacter* [20]. Emulsifying and surfactant activities are together responsible for important functional properties of bacterial exopolymers. Our analyses revealed that microbactan has emulsifying but not surfactant activity. Surface-active biomolecules are classified as surfactants when they lower the interfacial or surface tension and emulsifiers when they form stable emulsions [19,21]. Overall, our results indicate that microbactan is a true bioemulsifier.

2.3. Effect of Temperature, pH and NaCl on Emulsifying Activity

The effect of these three key variables that influence emulsifying activity was assessed at 24, 48, 72 and 96 h. As expected, they influenced the emulsifying activity of microbactan, but their effects occurred at different levels. A time-dependent decrease of emulsifying activity was observed at 100 °C ($F = 38.13$; $p < 0.005$). This was not the case for activities at 50 °C and 5 °C. The highest level of activity was found at 50 °C (emulsifying activity of 95.7% ± 2.5%) irrespective of time of incubation (Figure 1). The loss of emulsifying activity at 100 °C can be explained by denaturation of the protein fraction of microbactan during heating [22], as seen with other microbial biosurfactants [23]. Similarly, the activity associated with the upper limit of the range of salinities tested (3.5%, 5%, and 10%) declined as a function of time to level off at 72 and 96 h. The highest levels of activity were observed at 3.5% (Figure 2), not dissimilar to the biosurfactant produced by *Aeromonas* spp., which maintained emulsifying activity up to 5% NaCl [24]. On the other hand, no significant effect on activity was observed as a function of time with the pH values tested (Figure 3). However, slightly higher levels of emulsifying activity were recorded at acid and alkaline pH values, suggesting the ionization of functional groups that resulted in the activation of less surface-active species within the bioemulsifier matrix [25]. In comparison, biodispersan from *Acinetobacter calcoaceticus* A2 had an optimum functional pH value in the range of 9 to 12 for limestone-dispersing activity [26]. It has been shown that the emulsifying activity of certain polymers is modified at different extents when temperature and pH covariate, this is probably due to the synergistic influence of these factors on surface-active proteins, whose conformation and functional groups are influenced as a function of these factors.

Figure 1. Emulsifying activity of microbactan on motor oil at different temperatures up to 96 h. 5 °C (●), 50 °C (■) and 100 °C (▲). Values represent means ± SD ($n = 3$).

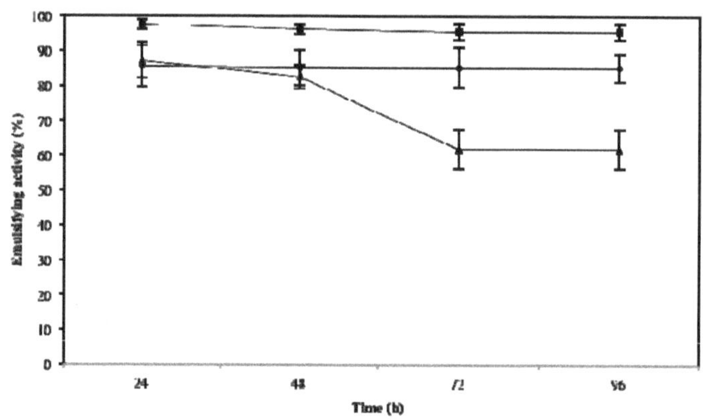

Figure 2. Emulsifying activity of microemulsan on motor oil at various salinities up to 96 h. 3.5% (●), 5% (■) and 10% (▲) NaCl concentration. Values represent means ± SD ($n = 3$).

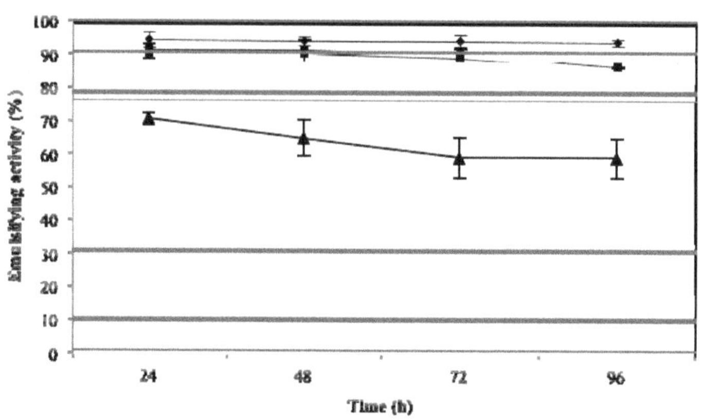

Figure 3. Emulsifying activity of microbactan on motor oil at different pH values. pH 4 (●), pH 7 (■) and pH 10 (▲).Values represent means ± SD (n = 3).

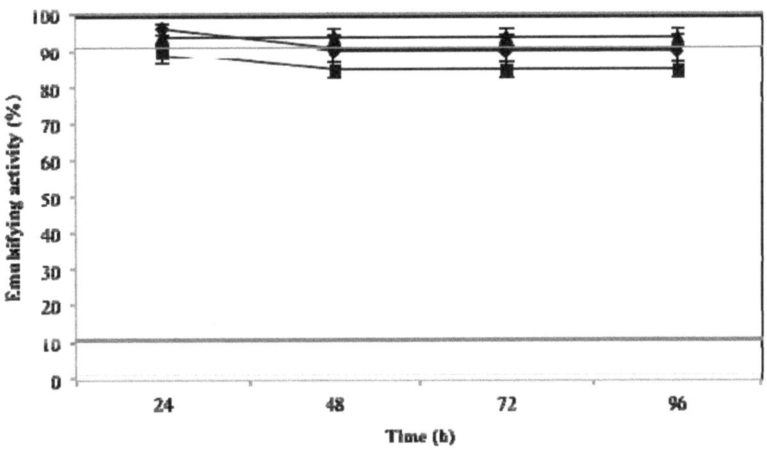

These results suggest that microbactan could find application in environmental marine processes such as enhanced oil recovery, cleaning of oil reservoirs and enhancement of biodegradation rates of spilled oils [2]. Bioemulsifiers have a wide diversity of composition and structure and are characterized by improved functionality and stability. Their potential applications include: the oil and petroleum industries, water and soil bioremediation, metal treatment and processing, detergents and laundry supplies, agriculture, textile manufacturing, pulp and paper processing, paints, cosmetics, pharmaceuticals, personal care products and food processing [4,6]. Microbactan has higher emulsifying activity at 50 °C and 3.5% NaCl, conditions typical of intertidal environments [8]. If the molecule remains stable under these conditions for long periods, it may prove specifically useful for bioremediation of polluted intertidal habitats [27]. The stability of the emulsions under diverse conditions, such as temperature, pressure, pH and ionic strength, makes this biopolymer a versatile emulsifier for use in many food and pharmaceutical formulations.

2.4. Toxicity of Microbactan

Based solely on the *Artemia salina* toxicity test, microbactan proved innocuous, as expected for a biologically derived surface-active agent; this class of compound is generally biodegradable and non-toxic [2]. Although synthetic surfactants exhibited the highest emulsifying activity in this study, Triton® X-100 (also a non-ionic surface-active), at least, proved to be toxic in our *Artemia* bioassay (Table 3). Polyoxyethylene octyl phenols (the Triton X series) are known to be highly cytotoxic, solubilizing the membrane lipid bilayer [28]. In fact, these synthetic surfactants can actually inhibit aromatic hydrocarbon biodegradation via toxic interactions, making them less suitable for bioremediation purposes [27,29]. The *Artemia salina* toxicity bioassay is a reliable primary screen, given the sensitivity of this crustacean to a wide range of biologically active compounds of diverse chemistries, including pesticide residues, mycotoxins, stream pollutants, anaesthetics, dinoflagellate toxins, morphine-like compounds and oil dispersants [30]. In addition, this bioassay has demonstrated good correlation with other cell-based tests such as tumor cell lines (e.g., KB, P-388, 388, L5178Y and L1210) and mammalian systems [31,32]. Reinforcing this finding, an experimental study on marine biofilm colonization on surfaces coated with microbactan showed that this biopolymer did not affect biofilm formation (unpublished results).

Table 3. Anti-crustacean activity of surfactants against *Artemia salina* nauplii.

Surfactant	LC_{50} (µg/mL)
Microbactan	>1000
Triton X-100	100.3 ± 3.8
Tween 80	>1000

2.5. Phylogenetic Reassessment of Microbacterium *sp. MC3B-10*

The reassessment of the phylogenetic position of *Microbacterium* sp. MC3B-10 based on its 16S rDNA gene, using a more robust bioinformatics approach, confirmed that its closest relative was *Microbacterium trichothecenolyticum*, with a similarity of 99.2% (underline, Figure 4). Stackebrandt and Ebers [33] established that a cutoff of 98.7% *16S rDNA* gene homology is appropriate for species differentiation within a genus. Following this criterion and given the fact that *Microbacterium* is a very tight genus with respect to *16S rDNA* gene homology between valid species [34], *Microbacterium* sp. MC3B-10 could be classified as *M. trichothecenolyticum.* However, definitive identification of this bacterial isolate requires a polyphasic approach including biochemical, physiological, chemotaxonomic, and nucleic acid-based methods, along with a range of microscopies, as previously shown for newly described *Microbacterium* species [35].

The *in-silica* analysis revealed novel strains of the genus *Microbacterium* (MC24 and MC60) that had been isolated from an intertidal environment in Brazil and were capable of synthesizing bioemulsifiers [36]. These isolates were also related to *Microbacterium* sp. MC3B-10, but at lower similarity levels (97%). It should be noted that, except for *M. resistens*, *M. hominis*, and *M. paraoxydans*, all microbacteria are considered environmental bacteria [34].

Certain *Microbacterium* species, such as *M. thalassium*, *M. halophilum,* and *M. phyllosphaerae*, occur in coastal habitats [35,37]. However, the biotechnological potential of this genus as a marine bioemulsifier producer has only been reported recently [8,36]. The occurrence of bioemulsifier-producing *Microbacterium* species in intertidal environments is not surprising, given the metabolic versatility exhibited by this genus; some of them can metabolize hydrocarbons, presumably through bioemulsifier synthesis [38,39]. In addition, the availability of hydrophobic substrates in coastal marine environments may select for bacteria capable of synthesizing surface-active molecules to enhance nutrient uptake [1].

Figure 4. Neighbor-joining phylogenetic tree, based on *16S rDNA* gene sequences, showing the positions of strain MC3B-10 (AY833570) relative to all known *Microbacterium* species. Accession numbers of *16S rDNA* gene sequences of reference organisms are shown in parentheses. Bootstrap values (1000 replicates, >50%) are shown above the nodes. The bar indicates the relative sequence divergence (0.01 nucleotide substitutions per site). *Curtobacterium luteum* and *C. michiganense* were used as outgroups.

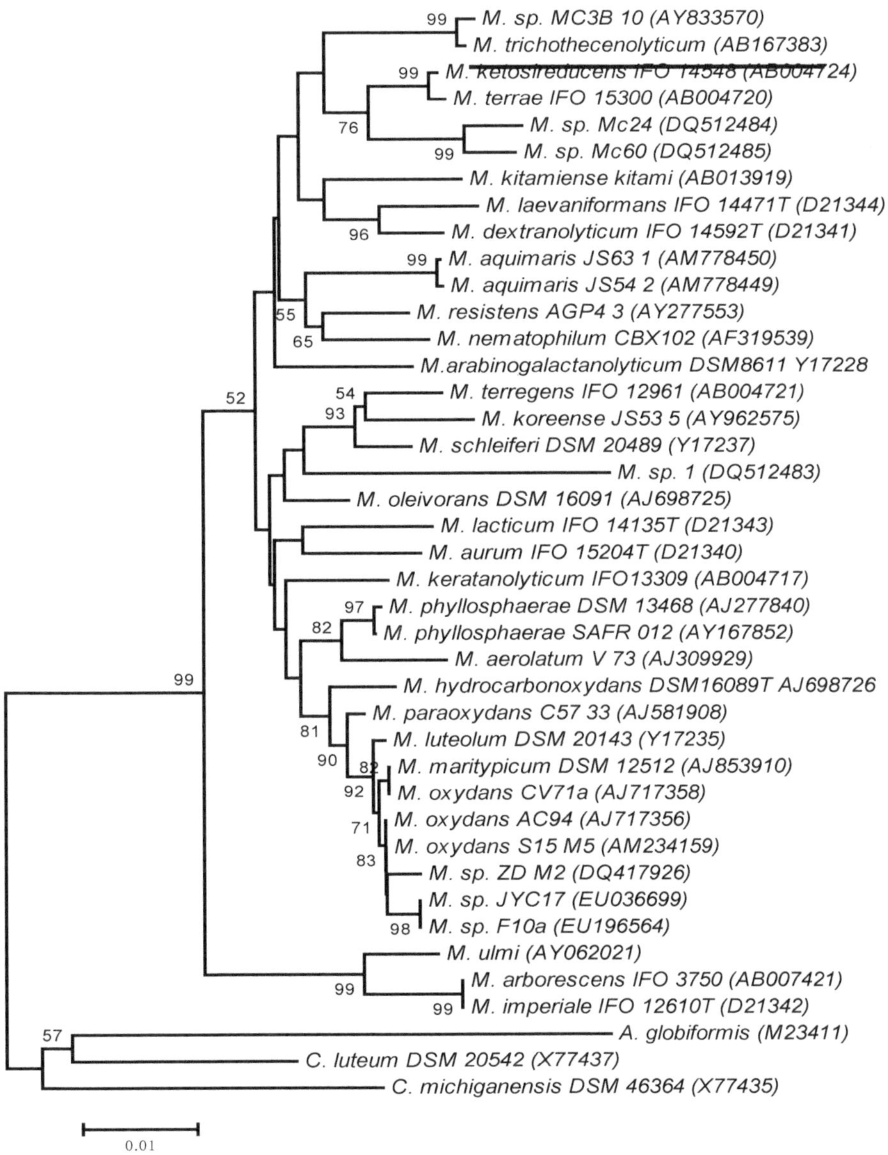

3. Experimental Section

3.1. Production of Microbactan

Microbacterium sp. MC3B-10 was originally isolated from pristine rocky intertidal shores in the state of Campeche, southern Gulf of Mexico, Mexico. Microbactan was produced following a batch shake flask fermentation of an overnight culture (50 mL) of *Microbacterium* sp. MC3B-10 as previously reported [8]. The exopolymeric material was extracted from the fermented broth, redissolved in a small volume of distilled water and then dialyzed (molecular weight cutoff 12,000–14,000 Da, Spectrum®, CA, USA) for 72 h. The resulting material was lyophilized and kept in the dark before analysis.

3.2. Lipid Analysis and Fatty Acid Composition

Lipids were determined using Fourier-transform infrared spectroscopy (FT-IR) [8] and spectrophotometry, using triolein as standard lipid [40]. Fatty acid composition was investigated as follows. After mild acid hydrolysis of the exopolymer (0.7064 g) at 80 °C for 30 min, the hydrolysate was extracted with $CHCl_3$:CH_3OH:H_2O (3:2:1) and the aqueous phase extracted two more times with $CHCl_3$ (1:1). The $CHCl_3$ fractions were combined and evaporated under reduced pressure [41]. Before GC-MS analysis, the sample was pre-treated as described previously [42]. The methylation of fatty acids was carried out with 5 mL of HCl-methanol at 80 °C for 30 min. The fatty acid methyl ethers were extracted with hexane and subjected to analysis. GC-MS was performed using helium as carrier gas on a Perkin Elmer Clarus 580 GC equipped with a Clarus SQ 8S mass spectrometer, equipped with an Elite-5 capillary column (30 m × 0.32 mm i.d., 0.25 μm film thickness).

3.3. High Performance Size Exclusion Chromatography

The molecular weight of microbactan was determined using an HPLC system Prominence Shimadzu™, a PL aquagel-OH mixed, 8 μm (Varian) guard column (U 7.5 mm × L50 mm), and a PL aquagel-OH mixed (Varian, Palo Alto, CA, USA) separation column (U 7.5 × 300 mm, operating range 10^2–10^7 g/mol). Elution was performed at 1 mL/min with 0.1 M ammonium acetate containing 0.03% NaN_3, and the eluate filtered through a 0.1 μm membrane (Durapore Membrane, PVDF, Hydrophilic type VVLP, Millipore®, Saint Quentin en Yvelines, France). A differential refractive index (RI) detector (L2490, VWR Hitachi, Fontenay sous bois, France) and a multi-angle light scattering detector (Dawn Heleos™, Wyatt, Toulouse, France) were coupled on-line and data computed with Astra software for absolute molar mass determination.

3.4. Ionic Charge

The ionic charge of microbactan was assessed using a modified double diffusion technique [43]. Briefly, two 12 mm-diameter plugs were taken from Petri dishes containing 1% agar. One of the resulting wells was filled with 200 μL of microbactan solution at 1 mg/mL and the other with commercial anionic (sodium dodecyl sulphate SDS and Teepol® 610S Sigma-Aldrich Química, S. de R.L. de C.V., Toluca, Mexico State, Mexico) or cationic (barium chloride and cetyltrimethylammonium bromide) surfactants supplied by Sigma-Aldrich. These surfactants were used at 20 mmol, except for barium chloride (50 mmol). The appearance of precipitation lines between the wells was indicative of the ionic character of microbactan. Plates were incubated at 28 °C and monitored daily.

3.5. Emulsifying Activity

The emulsifying activity of microbactan was assessed against aromatic hydrocarbons (benzene and xylene), vegetable oils (corn, olive and sunflower) and mineral, motor and crude oils, all at 1%. Oils were acquired from local commercial suppliers and the hydrocarbons were of analytical grade (J.T. Baker, Center Valley, PA, USA; E.M. Sciences, Hatfield, PA, USA). The emulsifying activity was measured by combining equal volumes of microbactan solution (1% *w*/*v*) and the target hydrophobic substrates in 12-mm-diameter glass tubes, as previously described [8]. Controls included Triton® X-100 and Tween® 80 (Research Organics, Cleveland, OH, USA), a plant-derived polysaccharide (gum arabic from Spectrum®) and xanthan gum, a bacterial polysaccharide (ICN Biomedicals, Inc, Irvine, CA, USA). All evaluations were performed in triplicate.

3.6. Effects of Temperature, Salinity and pH on Emulsifying Activity

The effect of temperature, NaCl content and pH on emulsifying activity of microbactan was tested using motor oil because this substrate yielded the highest levels of emulsification (~96%). The influence of temperature was determined by pretreating at desired temperatures (5, 50 and 100 °C) microbactan solutions that were either heated for 15 min in a water bath or cooled for 10 min in a freezer, before being assayed [23]. The effect of NaCl was investigated at three concentrations (3.5, 5 and 10% *w*/*v*) and pH values of 4, 7 and 10 were studied. Emulsification with motor oil was carried out at room temperature as in Item 3.5 [22,24].

3.7. Toxicity Test

Brine shrimp eggs (Salt Creek™, Salt Lake City, UT, USA) were hatched at 27 °C under continuous aeration and illumination in seawater prepared with sea salts (Coralife®, Rhinelander, WI, USA) at 38 g/L and supplemented with 6 mg/L of dried yeast [44]. Solutions of Triton® X-100 and Tween® 80 (Research Organics, Cleveland, OH, USA) and microbactan were dissolved in seawater at concentrations of 1000, 500, 100, 50 and 10 µg/mL [45]. The 50% Lethal Concentration (LC_{50}) was determined by counting the dead nauplii after an incubation period of 24 h. Data were analyzed with the Finney computer program as described previously [27].

3.8. Statistical Analysis

All the experiments were run in triplicate. For the emulsification assays, a two-way analysis of variance (ANOVA) was used to assess the effect of biopolymer and controls on the hydrophobic substrates. When there was a significant difference ($p \leq 0.05$) between microemulsion and controls, we applied the Tukey test aposteriori. These tests were performed with Sigma Stat software version 2.0 (1997).

3.9. Phylogenetic Reassessment

The purpose of this analysis was twofold: first, to corroborate the identification of the species of *Microbacterium* MC3B-10 based on its 16S rDNA gene (AY833570) using a more robust bioinformatics approach (see details below) with the sequences reported in [8], and second to prove if recently reported species of *Microbacterium* displaying bioemulsifying activity or originally isolated from polluted environments were phylogenetically close to *Microbacterium* sp. MC3B-10. The following additional sequences were obtained from the GenBank database, *Microbacterium* sp. Mc1 DQ512483, Mc24 DQ512484, Mc60 DQ512485, *M. aquimaris* JS54-2 AM778449, *M. aquimaris* JS63-1 AM778450, *Microbacterium* sp. F10a EU196564, *M. oleivorans* DSM 16091 AJ698726 and *M.* sp. ZD-M2 DQ417926. In total, 41 sequences (1418 bp) were aligned using the CLUSTALW algorithm implemented in MEGA 4.3 software [46]. Phylogenetic analyses were performed using the Neighbor Joining (NJ) algorithm: phylogenetic analysis using parsimony V4.0 beta 10, 2002. For NJ analyses, the evolutionary model selected was the GTR + I + G (General Time Reversible Model + Invariable sites + Gamma distribution) inferred from the program Modeltest V.3.06 [47]. Support for nodes of the NJ tree was determined by calculating bootstrap proportion values [48] based on 1000 resamplings of neighbor-joining searches. The 16S rDNA sequences of *Curtobacterium luteum* and *C. michiganense* were used as outgroup.

4. Conclusions

This study demonstrated that the novel *Microbacterium* sp. MC3B-10, probably indigenous to the marine intertidal zone, synthesizes an environmentally-friendly exopolymeric non-ionic glycolipoprotein capable of emulsifying aromatic hydrocarbons and oils. The functional stability of this bioemulsifier was retained for long periods of time and at a range of temperature, NaCl concentration and pH. The chemistry, activity and stability of microbactan make it useful for environmental and personal care applications. Overall, our results suggest that bioemulsifier-producing bacteria warrant intensified bioprospection in the intertidal zones. Additionally, this study corroborates the still untapped resource represented by marine microorganisms for new biosurfactants, bioemulsifiers and biopolymers. This work thus strengthens the notion that marine microbial diversity is a non-exhausted source of novel biomolecules and particularly emphasizes the biotechnological significance of *Microbacterium* species derived from the intertidal environment.

Acknowledgments

This research was supported by a grant Marine biofilm inhibition by bacterial exopolymeric substance-based coatings and characterization of producing isolates (F/3617-2F) 2008-2011, from the International Foundation for Science (IFS) F/3617-1, Stockholm, Sweden; Organization for the Prohibition of Chemical Weapons, The Hague, The Netherlands.

References

1. Ron, E.; Rosenberg, E. Natural roles of biosurfactants. *Environ. Microbiol.* **2001**, *3*, 229–236.

2. Banat, I.M.; Franzetti, A.; Gandolfi, I.; Bestetti, G.; Martinotti, M.; Fracchia, L.; Smyth, T.; Marchant, R. Microbial biosurfactants, production, applications and future potential. *Appl. Microbiol. Biotechnol.* **2010**, *87*, 427–444.
3. Freitas, F.; Alves, V.; Reis, M. Advances in bacterial exopolysaccharides: From production to biotechnological applications. *Trends Biotechnol.* **2011**, *29*, 388–398.
4. Freitas, F.; Alves, V.; Carvalheira, M.; Costa, N.; Oliveira, R.; Reis, M. Emulsifying behaviour and rheological properties of the extracellular polysaccharide produced by *Pseudomonas oleovorans* grown on glycerol byproduct. *Carbohyd. Polym.* **2009**, *78*, 549–556.
5. Satpute, S.; Banat, I.; Dhakephalkar, P.; Banpurkar, A.; Chopade, B. Biosurfactants, bioemulsifiers and exopolysaccharides from marine microorganisms. *Biotechnol. Adv.* **2010**, *28*, 436–450.
6. Gutiérrez, T.; Leo, V.; Walker, G.; Green, D. Emulsifying properties of a glycoprotein extract produced by a marine *Flexibacter* species strain TG382. *Enzyme Microb. Technol.* **2009**, *45*, 53–57.
7. Ortega-Morales, B.O.; Chan-Bacab, M.J.; de la Rosa-García, S.C.; Camacho-Chab, J.C. Valuable processes and products from marine intertidal microbial communities. *Curr. Opin. Biotechnol.* **2010**, *21*, 346–352.
8. Ortega-Morales, B.O.; Santiago, J.; Chan-Bacab, M.J.; Miranda, E.; Fardeau, M.; Carrero, J.; Bartolo, P.; Valadéz, A.; Guezennec, J. Characterization of extracellular polymers synthesized by tropical intertidal biofilm bacteria. *J. Appl. Microbiol.* **2007**, *102*, 254–264.
9. Beech, I.; Hanjagsit, L.; Kalaji, M.; Neal, A.L.; Zinkevich, V. Chemical and structural characterization of exopolymers produced by *Pseudomonas* sp. NCIMB 2021 in continuous culture. *Microbiology* **1999**, *145*, 1491–1497.
10. Amaral, P.; Silva, J.; Lehocky, M.; Barros-Timmons, A.; Coelho, M.; Marrucho, I.; Coutinho, J.A.P. Production and characterization of a bioemulsiftier from *Yarrowia lipolytica. Process Biochem.* **2006**, *41*, 1894–1898.
11. Luna-Velasco, M.A.; Esparza-García, F.; Cansares-Villanueva, R.O.; Rodríguez-Vázquez, R. Production and properties of a bioemulsifier synthesized by phenanthrene-degrading *Penicillum* sp. *Process Biochem.* **2007**, *42*, 310–314.
12. Maier, R.M. Biosurfactants: Evolution and diversity in bacteria. *Adv. Appl. Microbiol.* **2003**, *52*, 101–121.
13. Navon-Venezia, S.; Zosim, Z.; Gottlieb, A.; Legmann, R.; Carmeli, S.; Ron, E.Z.; Rosenberg, E. Alasan, a new bioemulsifier from *Acinetobacter radioresistens. Appl. Environ. Microb.* **1995**, *61*, 3240–3244.
14. Thavasi, R.; Jayalakshmi, S.; Balasubramanian, T.; Banat, I. Biosurfactant production by Corynebacterium kutscheri from waste motor lubricant oil and peanut oil cake. *Lett. Appl. Microbiol.* **2007**, *45*, 686–691.
15. Thavasi, R.; Jayalakshmi, S.; Balasubramanian, T.; Banat, I. Production and characterization of a glycolipid biosurfactant from *Bacillus megaterium* using economically cheaper sources. *World J. Microbiol. Biotechnol.* **2007**, *24*, 917–925.
16. Whitman, W. *Food Emulsifiers: Chemistry, Technology, Functional Properties and Applications*; Ten Alps Publishing: London, UK, 1990.
17. Kim, S.; Ravichandran, D.; Khan, S.; Kim, Y. Prospective of the cosmeceuticals derived from marine organisms. *Biotechnol. Bioprocess Eng.* **2008**, *13*, 511–523.

18. Kumar, A.; Mody, K.; Jha, B. Evaluation of biosurfactant/bioemulsifier production by a marine bacterium. *Bull. Environ. Contam. Toxicol.* **2007**, *79*, 617–621.
19. Willumsen, P.A.; Karlson, U. Screening of bacteria, isolated from PAH-contaminated soils, for production of biosurfactants and bioemulsifiers. *Biodegradation* **1997**, *7*, 415–423.
20. Gutiérrez, T.; Mulloy, B.; Bavington. C.; Black, K.; Green, D. Partial purification and chemical characterization of a glycoprotein (putative hydrocolloid) emulsifier produced by a marine bacterium *Antarctobacter*. *Appl. Microbiol. Biotechnol.* **2007**, *76*, 1017–1026.
21. Perfumo, A.; Smyth, T.J.; Marchant, R.; Banat, I.M. Production and Roles of Biosurfactants and Bioemulsifiers in Accessing Hydrophobic Substrates. In *Handbook of Hydrocarbon and Lipid Microbiology*; Timmis, K.N., Ed.; Springer-Verlag: Berlin, Germany, 2010; pp. 1502–1510.
22. Sarubbo, L.; Moura, J.; Campos-Takaki, G. Production and stability studies of the bioemulsifier obtained from a new strain of *Candida glabrata* UCP 1002. *Electron. J. Biotechnol.* **2006**, *9*, 400–406.
23. Rufino, R.; Sarubbo, A.; Campos, G. Enhancement of stability of biosurfactant produced by *Candida lipolytica* using industrial residue as substrate. *World J. Microb. Biotechnol.* **2007**, *23*, 729–734.
24. Ilori, M.O.; Amobi, C.; Odocha, A. Factors affecting biosurfactant production by oil degrading *Aeromonas* spp. isolated from a tropical environment. *Chemosphere* **2005**, *61*, 985–992.
25. Sarubbo, L.; Farias, C.; Campos-Takaki, G. Co-utilization of canola oil and glucose on the production of a surfactant by *Candida lipolytica*. *Curr. Microbiol.* **2007**, *54*, 68–73.
26. Rosenberg, E.; Rubinovitz, C.; Gottlieb, A.; Rosenhak, S.; Ron, E.Z. Production of biodispersan by *Acinetobacter calcoaceticus* A2. *Appl. Environ. Microb.* **1988**, *54*, 317–322.
27. Pelletier, E.; Delille, D.; Delille, B. Crude oil bioremediation in sub-Antarctic intertidal sediments: Chemistry and toxicity of oiled residues. *Mar. Environ. Res.* **2004**, *57*, 311–327.
28. Weyermann, J.; Lochmann, D.; Zimmer, A. A practical note on the use of cytotoxicity assays. *Int. J. Pharm.* **2005**, *288*, 369–376.
29. Meyer, B.N.; Ferrigni, N.R.; Putnam, J.E.; Jacobsen, L.B.; Nichols, D.E.; McLauglin, J.L. Brine shrimp: A convenient general bioassay for active plant constituents. *Planta. Med.* **1982**, *45*, 31–34.
30. McLaughlin, J.L. Crown-gall Tumors in Potato Discs and Brine Shrimp Lethality: Two Simple Bioassays for Higher Plant Screening and Fractionation. In *Methods in Plant Biochemistry*; Hostettmann, K., Ed.; Academic Press: London, UK, 1991; Volume 6, pp. 1–32.
31. Birndorf, H.C.; D'Alossio, J.; Bagshaw, J.C. DNA-dependant RNA-polymerases from *Artemia*, embryos. Characterization of polymerases I and II from nauplius larvae. *Dev. Biol.* **1975**, *45*, 34–43.
32. Perfeito, J.P.; Santos, M.L.; López, K.S.; Paula, J.E.; Silveira, D. Characterization and biological properties of *Pouteria torta* extracts. *Rev. Bras. Farmacogn.* **2005**, *15*, 183–186.
33. Stackebrandt, E.; Ebers, J. Taxonomic parameters revisited: Tarnished gold standards. *Microbiol. Today* **2006**, *33*, 152–155.
34. Gneiding, K.; Frodl, R.; Funke, G. Identities of *Microbacterium* spp. encountered in human clinical specimens. *J. Clin. Microbiol.* **2008**, *46*, 3646–3652.

35. Takeuchi, M.; Hatano, K. Proposal of six new species in the genus *Microbacterium* and transfer of *Flavobacterium marinotypicum* ZoBell and Upham to the genus *Microbacterium* as *Microbacterium maritypicum* comb. nov. *Int. J. Syst. Bacteriol.* **1998**, *48*, 973–982.
36. Aniszewski, E.; Silva, R.; Faria, F.; Gomes, S.; Soares, A. Bioemulsifier production by *Microbacterium sp.* strains isolated from mangrove and their application to remove cadmiun and zinc from hazardous industrial residue. *Braz. J. Microbiol.* **2010**, *41*, 235–245.
37. Burmølle, M.; Webb, J.; Rao, D.; Hansen, L.; Sørensen, S.; Kjelleberg, S. Enhanced biofilm formation and increased resistance to antimicrobial agents and bacterial invasion are caused by synergistic interactions in multispecies biofilms. *Appl. Environ. Microb.* **2006**, *72*, 3916–3925.
38. Schippers, A.; Bosecker, K.; Sproer, C.; Schumann, P. *Microbacterium oleivorans* sp. nov. and *Microbacterium hydrocarbonoxydans* sp. nov., novel crude-oil-degrading Gram-positive bacteria. *Int. J. Syst. Evol. Microbiol.* **2005**, *55*, 655–660.
39. Franzetti, A.; Gandolfi, I.; Bertolini, V.; Raimondi, C.; Piscitello, M.; Papacchini, M.; Bestetti, G. Phylogenetic characterization of bioemulsifier-producing bacteria. *Int. Biodeterior. Biodegrad.* **2011**, *65*, 1095–1099.
40. Izard, J.; Limberger, R. Rapid screening method for quantification of bacterial cell lipids from whole cells. *J. Microbiol. Meth.* **2003**, *55*, 411–418.
41. Ganfield, M.C.; Pieringer, R.A. Phosphatidylkojibiosyl diglyceride. The covalently linked lipid constituent of the membrane lipoteichoic acid from *Streptococcus faecalis* (*faecium*) ATCC 9790. *J. Biol. Chem.* **1975**, *250*, 702–709.
42. Rezanka T.; Vokoun, J.; Slavicek, J.; Podojil, M. Determination of fatty acids in algae by capillary gas chromatography-mass spectrometry. *J. Chromatogr.* **1983**, *268*, 71–78.
43. Van Oss, C.J. Specifically impermeable precipitate membranes formed through double diffusion in gels: Behavior with complex forming and with simple systems. *J. Colloid Interf. Sci.* **1968**, *27*, 684–690.
44. Ortega-Morales, B.O.; Chan-Bacab, M.J.; Miranda-Tello, E.; Fardeau M.L.; Carrero J.C.; Stein, T. Antifouling activity of sessile bacilli derived from marine surface. *J. Ind. Microbiol. Biotechnol.* **2008**, *35*, 9–15.
45. Solís, P.; Wright, C.; Anderson, M.; Gupta, M.; Phillipson, D.A. A Microwell cytotoxic using *Artemia salina* (brine shrimp). *Planta Med.* **1993**, *59*, 250–252.
46. Tamura, K.; Dudley, J.; Nei, M.; Kumar, S. MEGA4: Molecular Evolutionary Genetics Analysis (MEGA) software version 4.0. *Mol. Biol. Evol.* **2007**, *24*, 1596–1599.
47. Posada, D.; Crandall, K.A. Modeltest: Testing the model of DNA substitution. *Bioinformatics* **1998**, *14*, 817–818.
48. Felsenstein, J. Confidence limits on phylogenies: An approach using the bootstrap. *Evolution* **1985**, *39*, 783–791.

11

Enhancing Metagenomics Investigations of Microbial Interactions with Biofilm Technology

Robert J. C. McLean * and Kavita S. Kakirde

Department of Biology, Texas State University, 601 University Drive, San Marcos, TX 78666, USA

* Author to whom correspondence should be addressed; E-Mail: McLean@txstate.edu;

Abstract: Investigations of microbial ecology and diversity have been greatly enhanced by the application of culture-independent techniques. One such approach, metagenomics, involves sample collections from soil, water, and other environments. Extracted nucleic acids from bulk environmental samples are sequenced and analyzed, which allows microbial interactions to be inferred on the basis of bioinformatics calculations. In most environments, microbial interactions occur predominately in surface-adherent, biofilm communities. In this review, we address metagenomics sampling and biofilm biology, and propose an experimental strategy whereby the resolving power of metagenomics can be enhanced by incorporating a biofilm-enrichment step during sample acquisition.

Keywords: biofilm-enrichment; metagenomics; functional screening; sequence-based screening; microbial interactions

---

1. Introduction

The global distribution of microorganisms is impressive, ranging from the deep subsurface in terrestrial [1] and marine environments [2], to the upper atmosphere [3]. Although culturing techniques are improving, the vast majority of microorganisms in natural environments including soil are as yet uncultured. Estimates of microbial composition, diversity, and even ecological interactions are performed using a variety of culture-independent approaches including metagenomics [4]. One highly notable early achievement from molecular investigations was the identification of three domains of life, Archaea, Bacteria, and Eukarya [5]. The advances of sequencing technology from the traditional Sanger protocol to higher throughput, more economical approaches such as pyrosequencing and Illumina-based sequencing [6] have resulted in the generation of considerable data, and as a result these systems biology approaches require considerable bioinformatics analysis and genome sequence construction [7]. A number of highly significant results have arisen from metagenomics studies including the discovery of "Candidatus *Pelagibacter ubique*" strain HTCC1062, originally identified as clade SAR11, which is considered the most abundant microorganism in the pelagic ocean [8]. Based on genome analysis, unusual nutrient requirements for "Ca. *P. ubique*" were identified and this extreme oligotroph can now be cultured on defined media [9].

2. Experimental Strategies for Extraction of Metagenomic DNA from Soil Biofilms

Surface-adherent microbial communities (biofilms) are a common feature of microbial growth in many environments [10] including soils. In the investigation of a soil biofilm it may be of particular interest to look at specific sections that may indicate a multitude of interactions between microbial populations in the biofilm. Visualization and imaging using microscopy techniques can be used to target this subset of the entire microbial population from the sample biofilm. There are two methods for the extraction and processing of metagenomic DNA from a microbial population, direct and indirect extraction. In the direct extraction method pioneered by Ogram *et al.* [11], any extracellular DNA is first separated from the environmental sample by treating it with an alkaline buffer. The cells in the matrix are then subjected to direct mechanical (e.g., bead beating) lysis followed by extraction of DNA released from these cells. DNA recovered by centrifugation is then concentrated and purified before cloning. In contrast, the indirect method involves recovery of microbial cells from the sample. The recovered cells are subjected to cell lysis (chemical and enzymatic) followed by DNA extraction and purification [12]. Although time-consuming the indirect extraction method prevents the contamination from non-bacterial DNA [13] that may be present in the sample. Direct extraction methods provide high yield of lower size DNA fragments whereas indirect methods provide low yield of higher size DNA fragments. Both methods have distinct advantages and limitations, and the choice should be based on the intended downstream application and the objective of the study. Irrespective of the DNA extraction method, care must be taken to avoid co-isolation of organic compounds that may be present in the sample and can inhibit downstream processes. Various factors to be considered pertaining to soil metagenomics and the use of specific strategies based on the ultimate goal of the study are discussed by Kakirde *et al.* [14] and this provides a good guideline for designing a metagenomics project. Since there are multiple approaches that can be adopted at each stage of a metagenomic analysis it is important to select appropriate DNA extraction and purification methods and consider if cloning is necessary.

Direct sequencing of metagenomic DNA can be performed followed by sequence analysis. The vastly growing field of next generation sequencing technology offers a plethora of options for sequencing such as 454 Pyrosequencing and Illumina among others. Every platform offers different coverage and read length and the cost per base of sequencing is likely to become more affordable with the rapid advances in this field. The massive amount of sequence data generated by next-generation sequencers requires the use of specialized bioinformatics tools to mine and analyze the output. The sequence-only method is comparatively less time-consuming than the alternative, which is construction of metagenomic libraries and subsequent function and or sequence-based screening to identify gene products encoded by the target microbial partners. An appropriate cloning vector and a host organism should be used in capturing and cloning these genes. Depending on the desired insert size and purity, the DNA for cloning in many instances can be obtained by using commercially available kits (such as Qiagen and MoBio). Some of the methods commonly used for purification of extracted DNA are the standard phenol-chloroform extraction, cesium chloride density gradient centrifugation and chromatography. Often a combination of methods can lead to greater purity but this is also accompanied by increased DNA loss. Hence the purification protocol(s) should be selected according to the requirements of the concentration and purity of the DNA that is to be cloned. Prior to cloning DNA can be sheared using physical shearing or partial restriction digestion, size-selected by electrophoresis [15] and then electroeluted [16]. Cosmid and fosmid vectors have been used for cloning DNA from environmental samples with an insert size between 30 and 50 kb [14]. Fosmids are based on the bacterial F-factor and are stably maintained in the host due to their low copy number (1–2 copies per cell), which is tightly regulated in a host such as *E. coli*. Fosmid vectors have a higher cloning efficiency as compared to bacterial artificial chromosome (BAC) vectors. A limitation of fosmid vectors is the limited insert size. Larger inserts can be cloned by using a BAC vector, which can easily maintain fragments greater than 100 kb [17]. BAC vectors can be induced to a high copy number for increased expression and DNA yield from metagenomic clones, and can also be stably maintained at single copy [18]. In investigating specific interactions within the biofilm such as syntrophy, competition or the transfer of antibiotic resistance elements cloning would be preferable to the sequence only approach especially when looking for novel mechanisms. *E. coli* is one of the commonly used heterologous hosts in construction of metagenomic libraries since it has a high cloning efficiency and is easy to culture and work with *in vitro* [19–22]. Other heterologous hosts such as *Streptomyces* species have been used for heterologous expression of cloned metagenomic DNA in multiple studies [23,24]. The use of Archaea, specifically extreme halophiles as a host for expression of cloned DNA has been done in previous studies. The percent G + C content of the cloned genes, predominant partners (Gram positive or Gram negative) in the biofilm samples are some factors that can be considered in selecting a suitable host. Vectors systems used in the process should also be compatible with the selected host organism.

Construction of metagenomic libraries followed by a function-based screening is an excellent strategy to actually detect the gene products of the cloned inserts and could be used to identify various metabolic products, including both growth enhancing as well as antimicrobial compounds produced by microbial partners in the biofilm. The effect of these compounds on various tester microorganisms can be determined by using a bioassay method in the functional screen. Similarly the presence of specific antimicrobial resistance elements can be detected by incorporating the particular antibiotic in the bioassay during screening of the metagenomic clones. Although cost-intensive, if feasible a combined sequence and function based analysis can be very effective in determining the chemistry and basic charcteristics of the microbial partners in the biofilm interaction. The preliminary information obtained from the sequence data can be used for designing a specifically targeted function based metagenomics screen. Figure 1 summarizes the general steps of a metagenomics strategy to investigate microbial communities in environmental samples.

Figure 1. General steps in a metagenomics strategy to investigate microbial communities in environmental samples.

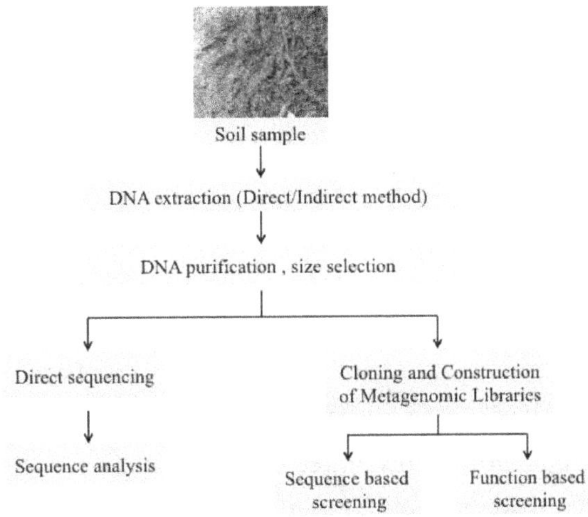

In addition to identifying genes of interest, a sequence based screening of the metagenomic libraries can be used in identification of regulatory elements that have been shown to control the formation and structure of biofilms [25]. A sequence only approach utilizing the power of the 454 sequencing technology is a good strategy for this purpose and yields good quality metagenomic sequences. These sequences can be deposited in GenBank and then referenced against available environmental databases and metagenomic datasets. The metagenomics RAST (MG-RAST) server is an excellent and free public resource that compares both protein and nucleotide databases to generate phylogenetic and functional summaries of the metagenomic sequence data [26]. MEGAN (Metagenome Analyzer), a computer program is another bioinformatics tool for analysis of high-throughput metagenomic sequence data and gene prediction that compares DNA reads against databases using comparative tools such as BLAST [27]. Metagenomic sequence analysis of microbial communities in a biofilm using the tools mentioned here can be used to identify and predict gene functions and can provide a different perspective to investigate the dynamic interactions between microbial partners within the biofilm environment.

3. Bacterial Adhesion and Biofilm Ecology

Bacterial adhesion to surfaces has been known for some time [28] but has only been recognized as a dominant mode of bacterial growth in nature in the past 20–30 years [10,29]. Surface-adherent microbial communities, now referred to as biofilms [10] are common in most environments. The prominence of biofilms is easily explained in flowing systems such as rivers [30] or pipelines [31], wherein surface adhesion enables microorganisms to persevere in spite of shear forces. Nutrients adsorb onto surfaces and microorganisms would therefore be attracted to sources of nutrition—a phenomenon sometimes referred to as the bottle effect [32]. Metabolic and genetic interactions are facilitated when organisms grow in close proximity within biofilms. Wolfaardt *et al.* [33] studied the ability of soil bacteria to grow on a commercial herbicide, diclophop methyl and found that bacteria could survive on this compound as a sole carbon source only if present as a biofilm consortium. Pure cultures of the soil isolates were unable to grow on this herbicide regardless of whether they were grown as planktonic or biofilm cultures. Similarly, mixed planktonic cultures were unable to grow on this herbicide [33].

Nitrification is another well-known biological phenomenon consisting of a two step process involving ammonia oxidation to nitrite, followed by nitrite oxidation to nitrate [34]. Ammonia oxidizing microorganisms are found in close proximity to nitrite oxidizers within nitrifying biofilms [35,36]. Syntrophic metabolism within microbial aggregates has also been reported in interspecies hydrogen transfer during anaerobic digestion of cellulose [37,38]. Biofilm growth has also been shown to promote genetic exchange through transformation [39] and conjugation [40,41] due to the close proximity of the donor and recipient organisms.

Biofilm studies with pure cultures have shown that these communities go through a developmental process [42] involving initial adhesion of microorganisms to a surface, aggregation into clumps (microcolonies), a maturation process and finally a dispersion process. In some organisms, notably *Pseudomonas aeruginosa*, *Staphylococcus aureus* and *Vibrio cholerae*, genes and mechanisms for biofilm development have been identified (reviewed in [42–44]). At the morphological level, there is evidence that similar processes occurs within mixed community biofilms, with the added complication of ecological interactions between species. In the dental field, there has been considerable work showing the population development of biofilms on teeth (dental plaque). When a hydroxyapatite tooth surface is cleaned, it becomes rapidly coated by adsorbed salivary proteins, which form a conditioning film [45]. Primary colonizing bacteria including *Streptococcus gordonii*, *Streptococcus oralis* and *Actinomyces naislundii* then attach to the conditioning film [46] and are in turn colonized by other organisms such as the cariogenic gram positive *Streptococcus mutans* [47]. Cell surface features including surface carbohydrates and carbohydrate-binding proteins (lectins), permitting the binding (coaggregation) of individual species to each other, is a major feature of population development in dental biofilms [47]. Microbial succession certainly occurs in other environments [48–50], and in biofilms associated with higher organisms, the host may play an active role in biofilm development. In the rhizosphere, plant exudates function as bacterial nutrients and play an important role in bacterial recruitment, and associated biofilm development and bacterial succession [50]. Cell signal interactions [51–53] are also important, during microbial colonization, biofilm formation and population succession. Other factors that are also important during biofilm population development include antimicrobial vesicle formation [54], antimicrobial chemicals [55] and bacteriocins [56]. At least two studies have shown that polymicrobial biofilms are more resistant to antibacterial agents and stress, than single species biofilms [57,58].

Another feature of biofilms is an indication of cell specialization. This is particularly prominent and well-described in biofilms formed by the social bacterium, *Myxococcus xanthus* in which some cells are involved in reproduction, others in nutrient acquisition, and others have structural roles [59]. Similar analogies have been shown in other organisms [43]. Certainly chemical gradients including nutrient levels, pH, and oxygen levels (in aerobic biofilms) result in a physiological gradient [60]. The structure and specialization seen within biofilms has been likened to a city [61] (Figure 2), with different physiological functions and even component species being present in clusters (microcolonies). Using the city metaphor for biofilms [60], an individual microcolony may function as one apartment building and will have ecological interactions (synergy, antagonism, synthrophic metabolism, genetic exchange, *etc.*) with neighboring microcolonies ("apartment buildings"). While biofilm structure and function is certainly complex, it largely reflects the situation in which bacteria naturally exist. As a result, broad based molecular microbial ecology studies would benefit by focusing on biofilms.

Figure 2. Biofilm enrichment strategy for metagenomics investigation. Confocal microscopy examination of a mixed population biofilm of *E. coli* and *P. aeruginosa* stained with the Live/Dead™ stain (Life Technologies, Grand Island, NY, USA) reveals microcolonies with viable (**A**) and non-viable (**B**) cells. Sampling and metagenomics analyses from these two microcolonies could suggest mechanisms underlying the loss of viability or other cellular interactions. While conventional genetic analyses can be performed for a mixed population biofilm containing known, genetically tractable organisms such as *E. coli* and *P. aeruginosa* [62], it is not practical for many naturally occurring biofilms with potentially unculturable organisms. This combination biofilm-enrichment strategy for metagenomics would be particularly useful in natural biofilms wherein the component populations may not be known.

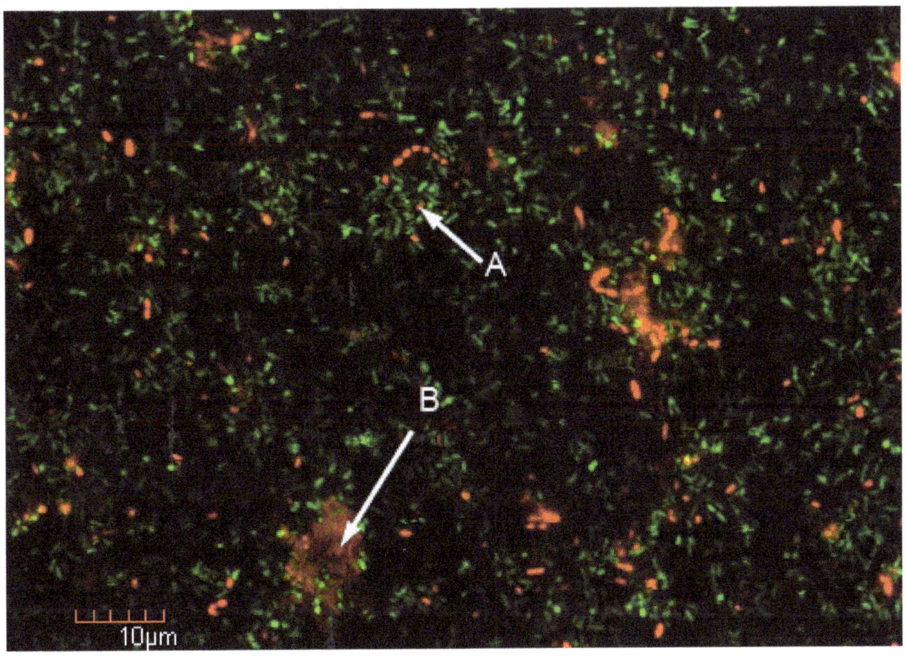

4. Biofilm Technology and Its Potential Application to Molecular Microbial Ecology

In most environments, microorganisms live as surface-adherent biofilm communities [10]. Within biofilms, many and possibly most microbial interactions and processes occur. Included in naturally occurring biofilm communities are cultivable and non-cultivable microorganisms [4]. While broad-based molecular approaches, such as metagenomics offer an invaluable insight to identifying new organisms and potential interactions, the methods commonly used to obtain the genetic material obtain samples from relatively large samples and as a result data and interpretations are based on sample averaging, which would include biofilm and planktonic populations, and likely cellular fragments and extracellular DNA. As shown in Figure 2, we propose the incorporation of biofilm technology as an experimental strategy to obtain higher resolution and more accurate investigations of microbial activities and interactions as they occur in nature.

The ideal strategy to study biofilms would be to examine samples *in situ* or alternatively those obtained directly from the field (or host if associated with a higher organism). Except for the molecular approaches used, this strategy mimics the direct morphological examinations of biofilms performed by

Zobell [28], Costerton [10] and others. In the case of easily obtained and accessible biofilms such as those associated with rock surfaces in streams ([63] or urinary catheter infections [64], access to biofilms is not an issue. Problems arise with inaccessible biofilms, particularly if these biofilms occur in the deep subsurface [2,65], or alternatively with water circulating systems in nuclear facilities [66]. While practical aspects of biofilm accessibility and data reproducibility are certainly considerations in natural samples, experimental manipulation may not be feasible. To circumvent this, a number of sampling protocols have been developed for the study of biofilms. At the simplest level, glass microscope slides or other suitable substrata may be inserted into water or soil and will be readily colonized by resident bacteria [67]. Alternatively, liquid from a pipeline or cooling system can be diverted through a biofilm sampling device [31]. An excellent three volume set of *Methods in Enzymology* [68–70] was published in 1999 and 2001, which summarizes many commonly used techniques used for biofilm research. As well, standardized biofilm growth and testing protocols for antimicrobial agent susceptibility have been developed [71–73].

As stated earlier, biofilm structure is complex and many physiological activities may change from one small population of cells (consortia) to another. Ideally, broad-based metagenomics processes to identify organisms and genes, as well as other complementary approaches such as RNA-seq [74], metabolomics [75] and proteomics [76] approaches to identify gene expression and microbial activity, could be mapped at the single cell level or within small consortia. The biofilm enrichment process for metagenomics is shown in Figure 2. Given the low (typically sub fmole) concentration of molecules in bacteria [77], analytical methods and detection limits need to be refined. As an alternative approach, broad based approaches could be used on whole biofilms and then reporter genes and chemically sensitive probes could be used to map activity using confocal microscopy [60,78]. Several fundamentally important biological issues could be addressed by this biofilm-enrichment metagenomics strategy including the mechanisms whereby microbial interactions occur in nature, do novel unrecognized interactions occur, do previously unknown organisms participate, and finally where do these interactions occur.

5. Conclusions

Direct observations of most natural environments reveal that microorganisms frequently exist within surface-adherent biofilm communities [10,43,47]. Similarly, the majority of organisms in many environments cannot be cultured but are identified through culture-independent techniques including metagenomics [3,4,6,19]. Aside from the identification of community members, culture-independent techniques are used to infer microbial interactions [58]. A number of studies using reporter gene technology and confocal microscopy reveal microbial interactions including genetic exchange, signaling, and metabolite exchange to occur between adjacent microorganisms within biofilm communities [34,36,78]. Here, we propose the use of biofilm-enrichment as an experimental strategy to enhance the resolving power of metagenomics and other culture-independent techniques to identify novel microbial interaction mechanisms.

Acknowledgments

Work in RJCM's laboratory is sponsored by a Research Enhancement Grant from Texas State

University and an endowment from the Homer Prince Foundation. We would like to dedicate this manuscript to our mentors, J.W. Costerton, T.J. Beveridge, and M.R. Liles who instilled in us a love of high quality science.

Reference

1. Balkwill, D.L.; Ghiorse, W.C. Characterization of subsurface bacteria associated with two shallow aquifers in Oklahoma. *Appl. Environ. Microbiol.* **1985**, *50*, 580–588.
2. Mason, O.U.; di Meo-Savoie, C.A.; van Nostrand, J.D.; Zhou, J.; Fisk, M.R.; Giovannoni, S.J. Prokaryotic diversity, distribution, and insights into their role in biogeochemical cycling in marine basalts. *ISME J.* **2009**, *3*, 231–242.
3. DeLeon-Rodriguez, N.; Lathem, T.L.; Rodrigues, L.M.; Barazesh, J.M.; Anderson, B.E.; Beyersdorf, A.J.; Ziemba, L.D.; Bergin, M.; Nenes, A.; Konstantinidis, K.T. Microbiome of the upper troposphere: Species composition and prevalence, effects of tropical storms, and atmospheric implications. *Proc. Natl. Acad. Sci. USA* **2013**, *110*, 2575–2580.
4. Temperton, B.; Giovannoni, S.J. Metagenomics: Microbial diversity through a scratched lens. *Curr. Opin. Microbiol.* **2012**, *15*, 605–612.
5. Woese, C.R.; Fox, G.E. Phylogenetic structure of the prokaryotic domain: The primary kingdoms. *Proc. Natl. Acad. Sci. USA* **1977**, *74*, 5088–5090.
6. Shade, A.; Gregory Caporaso, J.; Handelsman, J.; Knight, R.; Fierer, N. A meta-analysis of changes in bacterial and archaeal communities with time. *ISME J.* **2013**, *7*, 1493–1506.
7. Nagarajan, N.; Pop, M. Sequence assembly demystified. *Nat. Rev. Genet.* **2013**, *14*, 157–167.
8. Francois, P.; Tu Quoc, P.; Bisognano, C.; Kelley, W.L.; Lew, D.P.; Schrenzel, J.; Cramton, S.E.; Götz, F.; Vaudaux, P. Lack of biofilm contribution to bacterial colonisation in an experimental model of foreign body infection by *Staphylococcus aureus* and *Staphylococcus epidermidis*. *FEMS Immunol. Med. Microbiol.* **2003**, *35*, 135–140.
9. Carini, P.; Steindler, L.; Beszteri, S.; Giovannoni, S.J. Nutrient requirements for growth of the extreme oligotroph 'Candidatus *Pelagibacter*. ubique' HTCC1062 on a defined medium. *ISME J.* **2013**, *7*, 592–602.
10. Costerton, J.W.; Cheng, K.J.; Geesey, G.G.; Ladd, T.I.; Nickel, J.C.; Dasgupta, M.; Marrie, T.J. Bacterial biofilms in nature and disease. *Annu. Rev. Microbiol.* **1987**, *41*, 435–464.
11. Ogram, A.; Sayler, G.S.; Barkay, T. DNA extraction and purification from sediments. *J. Microbiol. Methods* **1987**, *7*, 57–66.
12. Holben, W.E.; Jansson, J.K.; Chelm, B.K.; Tiedje, J.M. DNA probe method for the detection of specific microorganisms in the soil bacterial community. *Appl. Environ. Microbiol.* **1988**, *54*, 703–711.
13. Osborn, A.M.; Smith, C.J. *Molecular Microbial Ecology*; Taylor and Francis: New York, NY, USA, 2005.
14. Kakirde, K.S.; Parsley, L.C.; Liles, M.R. Size does matter: Application-driven approaches for soil metagenomics. *Soil Biol. Biochem.* **2010**, *42*, 1911–1923.

15. Quaiser, A.; Ochsenreiter, T.; Klenk, H.P.; Kletzin, A.; Treusch, A.H.; Meurer, G.; Eck, J.; Sensen, C.W.; Schleper, C. First insight into the genome of an uncultivated crenarchaeote from soil. *Environ. Microbiol.* **2002**, *4*, 603–611.
16. Osoegawa, K.; Woon, P.Y.; Zhao, B. Frengen, E.; Tateno, M.; Catanese, J.J.; De Jong, P.J. An improved approach for construction of bacterial artificial chromosome libraries. *Genomics* **1998**, *52*, 1–8.
17. Shizuya, H.; Birren, B.; Kim, U.J.; Mancino, V.; Slepak, T.; Tachiiri, Y.; Simon, M. Cloning and stable maintenance of 300-kilobase-pair fragments of human DNA in *Escherichia coli* using an F-factor-based vector. *Proc. Natl. Acad. Sci. USA* **1992**, *89*, 8794–8797.
18. Wild, J.; Hradecna, Z.; Szybalski, W. Conditionally amplifiable BACs: Switching from single-copy to high-copy vectors and genomic clones. *Genome Res.* **2002**, *12*, 1434–1444.
19. Handelsman, J.; Rondon, M.R.; Brady, S.F.; Clardy, J.; Goodman, R.M. Molecular biological access to the chemistry of unknown soil microbes: A new frontier for natural products. *Chem. Biol.* **1998**, *5*, R245–R249.
20. Heath, C.; Hu, X.P.; Cary, S.C.; Cowan, D. Identification of a novel alkaliphilic esterase active at low temperatures by screening a metagenomic library from Antarctic desert soil. *Appl. Environ. Microbiol.* **2009**, *75*, 4657–4659.
21. Rondon, M.R.; August, P.R.; Bettermann, A.D.; Brady, S.F.; Grossman, T.H.; Liles, M.R.; Loiacono, K.A.; Lynch, B.A.; MacNeil, I.A.; Minor, C.; *et al.* Cloning the soil metagenome: A strategy for accessing the genetic and functional diversity of uncultured microorganisms. *Appl. Environ. Microbiol.* **2000**, *66*, 2541–2547.
22. Liles, M.R.; Williamson, L.L.; Goodman, R.M.; Handelsman, J. Isolation of High Molecular Weight Genomic DNA from Soil Bacteria for Genomic Library Construction. In *Molecular Microbial Ecology Manual*; Kowalchuk, G.A., Bruijn, F.J., Head, I.M., Akkermans, A.D.L., van Elsas, J.D., Eds.; Kluwer Academic Publishing: Dordrecht, The Netherlands, 2004; pp. 839–852.
23. Martinez, A.; Kolvek, S.J.; Hopke, J.; Yip, M.S.; Osburne, M.S. Environmental DNA fragment conferring early and increased sporulation and antibiotic production in *Streptomyces*. species. *Appl. Environ. Microbiol.* **2005**, *71*, 1638–1641.
24. King, R.W.; Bauer, J.D.; Brady, S.F. An environmental DNA-derived type II polyketide biosynthetic pathway encodes the biosynthesis of the pentacyclic polyketide erdacin. *Angew. Chem. Int. Ed.* **2009**, *48*, 6257–6261.
25. Davies, D.G.; Parsek, M.R.; Pearson, J.P.; Iglewski, B.H.; Costerton, J.W.; Greenberg, E.P. The involvement of cell-to-cell signals in the development of a bacterial biofilm. *Science* **1998**, *280*, 295–298.
26. Meyer, F.; Paarmann, D.; D'Souza, M.; Olson, R.; Glass, E.M.; Kubal, M.; Paczian, T.; Rodriguez, A.; Stevens, R.; Wilke, A.; *et al.* The metagenomics RAST server—A public resource for the automatic phylogenetic and functional analysis of metagenomes. *BMC Bioinf.* **2008**, *9*, 386.
27. Huson, D.H.; Auch, A.F.; Qi, J.; Schuster, S.C. MEGAN analysis of metagenomic data. *Genome Res.* **2007**, *17*, 377–386.
28. Zobell, C.E.; Allen, E.C. The significance of marine bacteria in the fouling of submerged surfaces. *J. Bacteriol.* **1935**, *29*, 239–251.

29. McLean, R.J.C.; Lam, J.S.; Graham, L.L. Training the biofilm generation—A tribute to JW Costerton. *J. Bacteriol.* **2012**, *194*, 6711.
30. Costerton, J.W.; Geesey, G.G.; Cheng, K.J. How bacteria stick. *Sci. Am.* **1978**, *238*, 86–95.
31. McCoy, W.F.; Bryers, J.D.; Robbins, J.; Costerton, J.W. Observations of fouling biofilm formation. *Can. J. Microbiol.* **1981**, *27*, 910–917.
32. Schmitt, J.; Nivens, D.; White, D.C.; Flemming, H.C. Changes of biofilm properties in response to sorbed substances: An FTIR-ATR study. *Water Sci. Technol.* **1995**, *32*, 149–155.
33. Wolfaardt, G.M.; Lawrence, J.R.; Robarts, R.D.; Caldwell, S.J.; Caldwell, D.E. Multicellular organization in a degradative biofilm community. *Appl. Environ. Microbiol.* **1994**, *60*, 434–446.
34. Gieseke, A.; Bjerrum, L.; Wagner, M.; Amann, R. Structure and activity of multiple nitrifying bacterial populations co-existing in a biofilm. *Environ. Microbiol.* **2003**, *5*, 355–369.
35. Egli, K.; Fanger, U.; Alvarez, P.J.; Siegrist, H.; van der Meer, J.R.; Zehnder, A.J.B. Enrichment and characterization of an anammox bacterium from a rotating biological contactor treating ammonium-rich leachate. *Arch. Microbiol.* **2001**, *175*, 198–207.
36. Schramm, A.; de Beer, D.; van den Heuvel, J.C.; Ottengraf, S.; Amann, R. Microscale distribution of populations and activities of *Nitrosospira.* and *Nitrospira.* spp. along a macroscale gradient in a nitrifying bioreactor: Quantification by *in situ* hybridization and the use of microsensors. *Appl. Environ. Microbiol.* **1999**, *65*, 3690–3696.
37. Schink, B. Synergistic interactions in the microbial world. *Antonie van Leeuwenhoek* **2002**, *81*, 257–261.
38. Thiele, J.H.; Chartrain, M.; Zeikus, J.G. Control of interspecies electron flow during anaerobic digestion: Role of floc formation in syntrophic methanogenesis. *Appl. Environ. Microbiol.* **1988**, *54*, 10–19.
39. Li, Y.H.; Lau, P.C.Y.; Lee, J.H.; Ellen, R.P.; Cvitkovitch, D.G. Natural genetic transformation of *Streptococcus mutans* growing in biofilms. *J. Bacteriol.* **2001**, *183*, 897–908.
40. Hausner, M.; Wuertz, S. High rates of conjugation in bacterial biofilms as determined by quantitative *in situ* analysis. *Appl. Environ. Microbiol.* **1999**, *65*, 3710–3713.
41. Christensen, B.B.; Sternberg, C.; Andersen, J.B.; Eberl, L.; Møller, S.; Givskov, M.; Molin, S. Establishment of new genetic traits in a microbial biofilm community. *Appl. Environ. Microbiol.* **1998**, *64*, 2247–2255.
42. Petrova, O.E.; Sauer, K. Sticky situations: Key components that control bacterial surface attachment. *J. Bacteriol.* **2012**, *194*, 2413–2425.
43. O'Toole, G.A.; Kaplan, H.B.; Kolter, R. Biofilm formation as microbial development. *Annu. Rev. Microbiol.* **2000**, *54*, 49–79.
44. Karatan, E.; Watnick, P. Signals, regulatory networks, and materials that build and break bacterial biofilms. *Microbiol. Mol. Biol. Rev.* **2009**, *73*, 310–347.
45. Marsh, P.D.; Bradshaw, D.J. Dental plaque as a biofilm. *J. Ind. Microbiol.* **1995**, *15*, 169–175.
46. Palmer, R.J., Jr.; Kazmerzak, K.; Hansen, M.C.; Kolenbrander, P.E. Mutualism *versus* independence: Strategies of mixed-species oral biofilms *in vitro* using saliva as the sole nutrient source. *Infect. Immun.* **2001**, *69*, 5794–5804.
47. Kolenbrander, P.E.; Andersen, R.N.; Kazmerzak, K.; Wu, R.; Palmer, R.J., Jr. Spatial organization of oral bacteria in biofilms. *Methods Enzymol.* **1999**, *310*, 322–332.

48. Shade, A.; McManus, P.S.; Handelsman, J. Unexpected diversity during community succession in the apple flower microbiome. *mBio* **2013**, *4*, e00602–e00612.
49. Nicol, G.W.; Tscherko, D.; Embley, T.M.; Prosser, J.I. Primary succession of soil *Crenarchaeota* across a receding glacier foreland. *Environ. Microbiol.* **2005**, *7*, 337–347.
50. Chaparro, J.M.; Badri, D.V.; Bakker, M.G.; Sugiyama, A.; Manter, D.K.; Vivanco, J.M. Root exudation of phytochemicals in *Arabidopsis* follows specific patterns that are developmentally programmed and correlate with soil microbial functions. *PLoS One* **2013**, *8*, e55731.
51. Steidle, A.; Sigl, K.; Schuhegger, R. Ihring, A.; Schmid, M.; Gantner, S.; Stoffels, M.; Riedel, K.; Givskov, M.; Hartmann, A.; *et al.* Visualization of *N*-acylhomoserine lactone-mediated cell-cell communication between bacteria colonizing the tomato rhizosphere. *Appl. Environ. Microbiol.* **2001**, *67*, 5761–5770.
52. McLean, R.J.C.; Barnes, M.B.; Windham, M.K.; Merchant, M.M.; Forstner, M.R.J.; Fuqua, C. Cell-cell influences on bacterial community development in aquatic biofilms. *Appl. Environ. Microbiol.* **2005**, *71*, 8987–8990.
53. Pacheco, A.R.; Sperandio, V. Inter-kingdom signaling: Chemical language between bacteria and host. *Curr. Opin. Microbiol.* **2009**, *12*, 192–198.
54. Mashburn, L.M.; Whiteley, M. Membrane vesicles traffic signals and facilitate group activities in a prokaryote. *Nature* **2005**, *437*, 422–425.
55. Egan, S.; James, S.; Holmstrom, C.; Kjelleberg, S. Correlation between pigmentation and antifouling compounds produced by *Pseudoalteromonas tunicata*. *Environ. Microbiol.* **2002**, *4*, 433–442.
56. Drider, D.; Fimland, G.; Héchard, Y.; McMullen, L.M.; Prévost, H. The continuing story of class IIa bacteriocins. *Microbiol. Mol. Biol. Rev.* **2006**, *70*, 564–582.
57. Whiteley, M.; Ott, J.R.; Weaver, E.A.; McLean, R.J.C. Effects of community composition and growth rate on aquifer biofilm bacteria and their susceptibility to betadine disinfection. *Environ. Microbiol.* **2001**, *3*, 43–52.
58. Burmølle, M.; Webb, J.S.; Rao, D.; Hansen, L.H.; Sørensen, S.J.; Kjelleberg, S. Enhanced biofilm formation and increased resistance to antimicrobial agents and bacterial invasion are caused by synergistic interactions in multispecies biofilms. *Appl. Environ. Microbiol.* **2006**, *72*, 3916–3923.
59. Kaplan, H.B. Multicellular development and gliding motility in *Myxococcus xanthus*. *Curr. Opin. Microbiol.* **2003**, *6*, 572–577.
60. Rani, S.A.; Pitts, B.; Beyenal, H.; Veluchamy, R.A.; Lewandowski, Z.; Davison, V.M.; Buckingham-Meyer, K.; Stewart, P.S. Spatial patterns of DNA replication, protein synthesis, and oxygen concentration within bacterial biofilms reveal diverse physiological states. *J. Bacteriol.* **2007**, *189*, 4223–4233.
61. Watnick, P.; Kolter, R. Biofilm, city of microbes. *J. Bacteriol.* **2000**, *182*, 2675–2679.
62. Chu, W.; Zere, T.R.; Weber, M.M.; Wood, T.K.; Whiteley, M.; Hidalgo-Romano, B.; Valenzuela, E., Jr.; McLean, R.J.C. Indole production promotes *Escherichia coli* mixed culture growth with *Pseudomonas aeruginosa* by inhibiting quorum signaling. *Appl. Environ. Microbiol.* **2012**, *78*, 411–419.
63. Geesey, G.G.; Richardson, W.T.; Yeomans, H.G.; Irvin, R.T.; Costerton, J.W. Microscopic examination of natural sessile bacterial populations from an alpine stream. *Can. J. Microbiol.* **1977**, *23*, 1733–1736.

64. Nickel, J.C.; Gristina, A.G.; Costerton, J.W. Electron microscopic study of an infected Foley catheter. *Can. J. Surg.* **1985**, *28*, 50–54.
65. Cusack, F.; Brown, D.R.; Costerton, J.W.; Clementz, D.M. Field and laboratory studies of microbial/fines plugging of water injection wells: Mechanism, diagnosis and removal. *J. Pet Sci. Eng.* **1987**, *1*, 39–50.
66. Santo Domingo, J.W.; Berry, C.J.; Summer, M.; Fliermans, C.B. Microbiology of spent nuclear fuel storage basins. *Curr. Microbiol.* **1998**, *37*, 387–394.
67. Marshall, K.C.; Stout, R.; Mitchell, R. Mechanisms of the initial events in the sorption of marine bacteria to solid surfaces. *J. Gen. Microbiol.* **1971**, *68*, 337–348.
68. Doyle, R.J. Biofilms. In *Methods in Enzymology*; Academic Press: San Diego, CA, USA, 1999; Volume 310, pp. 1–720.
69. Doyle, R.J. Microbial Growth in Biofilms. Part A: Developmental and Molecular Biological Aspects. In *Methods in Enzymology*; Academic Press: San Diego, CA, USA, 2001; Volume 336, pp. 1–469.
70. Doyle, R.J. Microbial Growth in Biofilms, Part B: Special Environments and Physicochemical Aspects. In *Methods in Enzymology*; Academic Press, San Diego, CA, USA, 2001; Volume 337, pp. 1–469.
71. Goeres, D.M.; Loetterle, L.R.; Hamilton, M.A.; Murga, R.; Kirby, D.W.; Donlan, R.M. Statistical assessment of a laboratory method for growing biofilms. *Microbiology* **2005**, *151*, 757–762.
72. ASTM. E2562–12: Standard Test Method for Quantification of *Pseudomonas aeruginosa* Biofilm Grown with High Shear and Continuous Flow Using CDC Biofilm Reactor. In *Annual Book of ASTM Standards*; ASTM International: West Conshohocken, PA, USA, 2012.
73. ASTM. E2871–12: Standard Test Method for Evaluating Disinfectant Efficacy against *Pseudomonas aeruginosa* Biofilm Grown in CDC Biofilm Reactor Using Single Tube Method. In *Annual Book of ASTM Standards*; ASTM International: West Conshohocken, PA, USA, 2012.
74. Westermann, A.J.; Gorski, S.A.; Vogel, J. Dual RNA-seq of pathogen and host. *Nat. Rev. Microbiol.* **2012**, *10*, 618–630.
75. Yanes, O.; Tautenhahm, R.; Patti, G.J.; Siuzdak, G. Expanding coverage of the metabolome for global metabolite profiling. *Anal. Chem.* **2011**, *83*, 2152–2161.
76. Sauer, K.; Cullen, M.C.; Rickard, A.H.; Zeef, L.A.H.; Davies, D.G.; Gilbert, P. Characterization of nutrient-induced dispersion in *Pseudomonas aeruginosa* PAO1 biofilm. *J. Bacteriol.* **2004**, *186*, 7312–7326.
77. Song, L.; Shan, D.; Zhao, M.; Pink, B.A.; Minnehan, K.A.; York, L.; Gardel, M.; Sullivan, S.; Phillips, A.F.; Hayman, R.B.; *et al.*-Direct detection of bacterial genomic DNA at sub-femtomolar concentrations using single molecule arrays. *Anal. Chem.* **2013**, *85*, 1932–1939.
78. Andersen, J.B.; Heydorn, A.; Hentzer, M.; Eberl, L.; Geisenberger, O.; Christensen, B.B.; Molin, S.; Givskov, M. *gfp*-based *N*-acyl homoserine lactone sensor systems for detection of bacterial communication. *Appl. Environ. Microbiol.* **2001**, *67*, 575–585.

12

Three-Dimensional Stratification of Bacterial Biofilm Populations in a Moving Bed Biofilm Reactor for Nitritation-Anammox

Robert Almstrand [1,†], Frank Persson [2], Holger Daims [3], Maria Ekenberg [4], Magnus Christensson [4], Britt-Marie Wilén [2], Fred Sörensson [1] and Malte Hermansson [1,*]

[1] Department of Chemistry & Molecular Biology, Microbiology, University of Gothenburg, Box 462, Göteborg SE-405 30, Sweden; E-Mail: fred.sorensson@cmb.gu.se

[2] Water Environment Technology, Department of Civil and Environmental Engineering, Chalmers University of Technology, Göteborg SE-412 96, Sweden; E-Mails: Frank.Persson@chalmers.se (F.P.); Britt-Marie.Wilen@chalmers.se (B.-M.W.)

[3] Division of Microbial Ecology, Department of Microbiology and Ecosystem Science, University of Vienna, Vienna 1090, Austria; E-Mail: daims@microbial-ecology.net

[4] AnoxKaldnes AB, Klosterängsvägen 11A, Lund SE-226 47, Sweden; E-Mails: maria.ekenberg@anoxkaldnes.com (M.E.); magnus.christensson@anoxkaldnes.com (M.C.)

[†] Present address: Civil and Environmental Engineering, Colorado School of Mines, 1500 Illinois Street, Golden, CO 80401-1887, USA; E-Mail: ralmstra@mines.edu.

* Author to whom correspondence should be addressed; E-Mail: malte.hermansson@cmb.gu.se;

Abstract: Moving bed biofilm reactors (MBBRs) are increasingly used for nitrogen removal with nitritation-anaerobic ammonium oxidation (anammox) processes in wastewater treatment. Carriers provide protected surfaces where ammonia oxidizing bacteria (AOB) and anammox bacteria form complex biofilms. However, the knowledge about the organization of microbial communities in MBBR biofilms is sparse. We used new cryosectioning and imaging methods for fluorescence *in situ* hybridization (FISH) to study the structure of biofilms retrieved from carriers in a nitritation-anammox MBBR. The dimensions of the carrier compartments and the biofilm cryosections after FISH showed good correlation, indicating little disturbance of biofilm samples by the treatment.

FISH showed that *Nitrosomonas europaea/eutropha*-related cells dominated the AOB and *Candidatus Brocadia fulgida*-related cells dominated the anammox guild. New carriers were initially colonized by AOB, followed by anammox bacteria proliferating in the deeper biofilm layers, probably in anaerobic microhabitats created by AOB activity. Mature biofilms showed a pronounced three-dimensional stratification where AOB dominated closer to the biofilm-water interface, whereas anammox were dominant deeper into the carrier space and towards the walls. Our results suggest that current mathematical models may be oversimplifying these three-dimensional systems and unless the multidimensionality of these systems is considered, models may result in suboptimal design of MBBR carriers.

Keywords: moving bed biofilm reactor (MBBR); nitritation; anammox; AOB; wastewater; biofilm; stratification

---

1. Introduction

Nitrogen removal with anaerobic ammonium oxidizing (anammox) bacteria is an emerging technology for treatment of high strength wastewater. In the process, aerobic ammonia oxidizing bacteria (AOB) oxidize half of the ammonia to nitrite, which is used by anammox bacteria as electron acceptor for oxidizing the remaining ammonia to dinitrogen gas. In the more common full-scale configurations, the processes of nitritation and anammox occur in the same reactor [1]. The AOB and anammox bacteria can reside in suspended microbial assemblages as sludge and/or granules [2], or in biofilms attached to substrata [3]. One prerequisite for nitritation-anammox is the presence of both oxic and anoxic environments for AOB and anammox bacteria, respectively. This can be attained by alternating the environmental conditions in time, such as in a sequencing batch reactor [2]. In continuous reactors on the other hand, the microbial groups need to be separated in space, with anammox bacteria residing in anoxic zones of the microbial aggregates, provided by the oxygen demanding activities of AOB. In microbial granules, distinct patterns of stratification have been predicted by modelling and observed with microbiological methods, with AOB being located in the outer aerobic zone near the water interface and with anammox bacteria in the inner anoxic parts [4–8]. For nitritation-anammox biofilms on fixed substrata, the observed localization of microorganisms has been less straightforward. In several reactors, a seemingly scattered distribution of AOB and anammox bacteria has been observed [3,9–11], while in other reactors the distribution of microorganisms has been more orderly structured [12–14], indicating that the factors determining biofilm architecture are complex. Furthermore, there is a need for quantitative image analysis to reveal the actual distribution patterns of the biofilm microorganisms. For moving bed biofilm reactors (MBBRs), which are used today in full-scale nitritation-anammox [15,16], a recent investigation has indicated a more orderly biofilm structure [17], but detailed analysis of biofilm establishment and stratification is still lacking for these systems. One probable reason for this is the challenge to retreive and analyze undisrupted biofilms from the protected carrier compartments. Furthermore, biofilm modeling is a useful tool [18,19] to improve the efficiency of nitritation-anammox MBBRs. For these models a detailed knowledge about the distribution of microorganisms is fundamental.

In this study, we have utilized a cryosectioning approach for the retrieval and study of intact biofilm of different age from protected compartments of carriers residing in a nitritation-anammox MBBR. Fluorescence *in situ* hybridization (FISH) and confocal laser scanning microscopy (CLSM) was used to identify the AOB and anammox populations. By combining FISH-CLSM with a newly developed digital image analysis method [20], the three-dimensional stratification of AOB and anammox bacteria in intact biofilms was determined and quantified. Differences between young and mature biofilms in terms of established AOB and anammox populations and biofilm structure were investigated. Finally, image analysis of total carrier compartment *i.e.*, "wall-to-wall" biofilms, allowed an assessment of problems with biofilm retrieval and shrinkage from dehydration and cryo-embedding, which has often been suggested as a serious draw-back of these biofilm research methods.

2. Results and Discussion

2.1. Reactor Conditions

The nitritation-anammox MBBR was operated for three years. Our study covers 167 days at the end of the reactor runtime, from addition of new carriers to the final sampling. Reactor conditions during this period are summarized in Table 1. Since only trace amounts of organic carbon was added to the synthetic wastewater, and even though AOB in the biofilm could be a source for organic carbon *per se* [21,22], it is unlikely that denitrification would account for any significant nitrogen removal. The ratio of nitrate production to ammonia removal was 12%, which is close to the stoichiometric value for anammox bacteria of 11%, indicating that aerobic nitrite oxidation was not important [23].

Table 1. Reactor conditions and effluent concentrations during the experimental period. Average values and standard deviation (SD). The influent (synthetic wastewater) contained 314 mg N L^{-1} as ammonium. HRT = hydraulic retention time; DO = dissolved oxygen.

	Flow ($L\ h^{-1}$)	HRT (h)	Temp (°C)	pH	DO ($mg\ L^{-1}$)	NH_4^+ ($mg\ N\ L^{-1}$)	NO_2^- ($mg\ N\ L^{-1}$)	NO_3^- ($mg\ N\ L^{-1}$)	N removal
Average	1.2	6.3	28.3	7.8	3.6	71	14	25	66%
S.D.	0.2	0.7	2.9	0.3	0.5	40	13	12	14%

2.2. Biofilm Retrieval and Sectioning

Cryo-embedding, followed by freezing and removal of the carrier plastic material enabled the retrieval of biofilm from within individual carrier compartments (for carrier details see Table S1). As shown in Figure S1, the cryosectioned biofilm maintained the square shape in the x, y dimensions. The deepest analyzed cryosections (*i.e.*, in the z dimension, going into the carrier compartment) originated from at least 800 μm (Figure 1b). This corresponds to 80% and 53% of the maximum depth of the 2 and 3 mm carriers, respectively. Deeper layers can be analyzed, but long (in the z dimension) frozen biofilm structures can easily break. Previous studies [24,25] have observed that cryo-embedding and dehydration introduce severe distortion, such as shrinkage, to biofilm structure. Since the assembled micrograph squares represent a "wall-to-wall" x, y section of the biofilm (Figure 1a), and the fact that these were of the same dimension as the actual carrier compartment (Figure S1), we suggest that for these and similar types of biofilms, significant shrinkage is not introduced by the applied cryosectioning and hybridization protocols.

Figure 1. Overview of sample retrieval and analysis. Biofilm carriers containing either young (**A1**) or mature biofilm (**A2,3**) were cryo-embedded and frozen; After removal of the carrier, biofilm from carrier compartments was frozen solid and associated with the O.C.T. compound (**A4**); Cryosectioning of the biofilm was followed by FISH and CLSM analysis of biofilm stratification in the *x*, *y* and *z* dimensions (**A5**); and vertical (*z*) distribution from one of the compartments is exemplified in (**B**), anammox (open circles), AOB (closed circles).

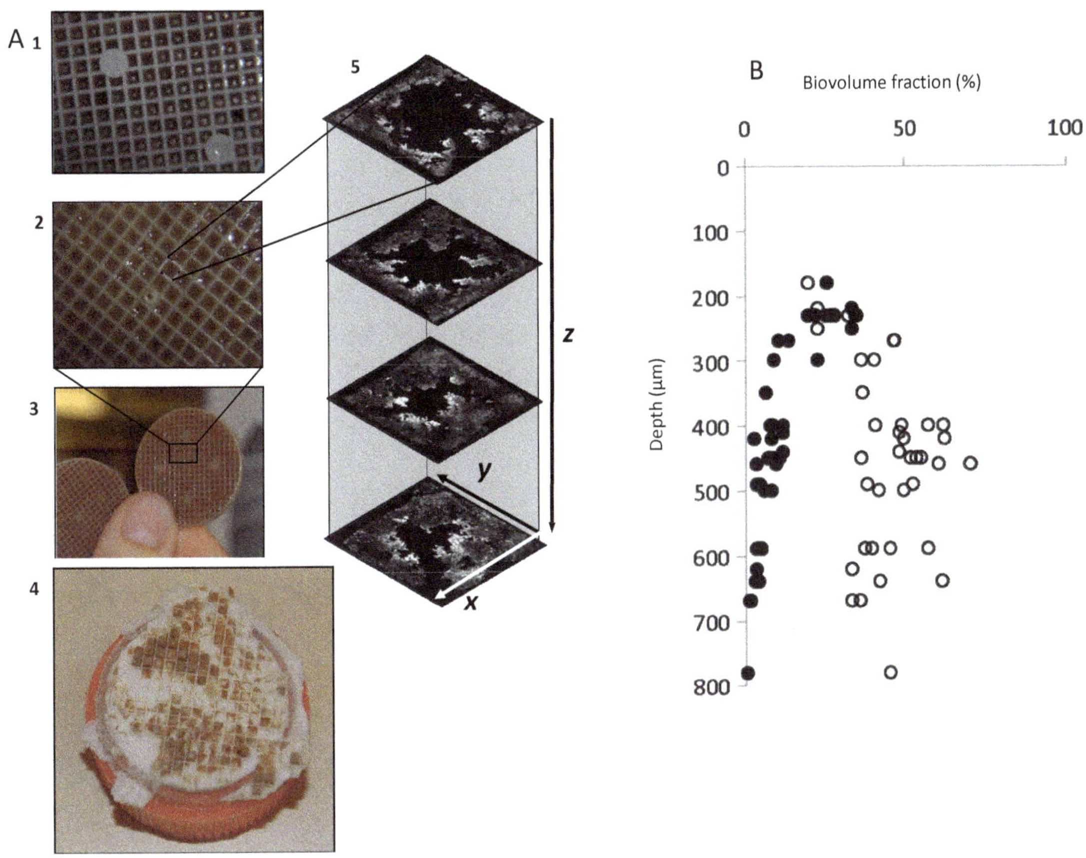

2.3. Bacterial Community Composition

A large FISH screening (Table S2) showed that the reactor community was made up of AOB in microcolonies and dense clusters of cells related to the *Nitrosomonas europaea/eutropha* lineage (cluster 7), hybridizing with the probe Nse1472. This was not unexpected considering the comparatively high ammonium concentrations in the reactor (average 314 mg N L^{-1}). *N. europaea/eutropha*-related cells are generally encountered in systems with elevated ammonium concentrations [26] and may be regarded as AOB r-strategists with a comparatively high growth rate [27,28]. Simultaneous probe hybridization with Pla46 (targeting the order *Planctomycetales*), AMX820 (targeting the genera *Ca. Brocadia* and *Ca. Kuenenia*) and Bfu613 (targeting *Ca. Brocadia fulgida*) indicated that the anammox cells were related to *Ca. Brocadia fulgida*. *Ca. Brocadia*-like cells were also dominating the anammox population on biofilm carriers in a laboratory scale CANON reactor fed with concentrated anaerobic digestion reject water [29] and are possibly r-strategists among anammox bacteria [30]. Furthermore, filamentous cells targeted by probe CF319a (phylum *Bacteroidetes* among others), were rather evenly distributed, possibly providing structural support for the biofilm, as suggested earlier [31]. Very few NOB cells were present, as targeted by the probes for *Nitrobacter*, *Nitrospira* and *Nitrotoga* (Table S2), which was in accordance with the low aerobic nitrite oxidation (see above).

2.4. Biofilm Establishment on New Carriers

Anammox cells were almost completely absent from the relatively thin biofilms on the carriers incubated for five and a half months (Figure 2c,d). A few anammox microcolonies were found, usually in the compartment corners beneath biofilm dominated by approximately 80% of *N. europaea/eutropha*. Considering the DO concentration in the reactor (on average 3.6 mg L^{-1}, Table 1), the results suggest a sequential colonization with initial formation of an AOB biofilm that with time created anaerobic microenvironments suitable for anammox bacteria in the deepest parts of the biofilm. It is likely that these environments first arose in the carrier corners, since the mature biofilms were thicker in the corners (336 ± 22 μm, average ± SD) than on the compartment sides (235 ± 19 μm). Thus, migration of intact AOB and anammox biofilm clusters alone cannot explain establishment of anammox on new carriers since AOB first needed to provide the anaerobic microenvironment. It has previously been shown that the desired stratification of AOB and anammox bacteria occurred only if anammox bacteria were established prior to AOB [13]. However, such orderly development was not necessary for stratification to develop here. The discovery of very few anammox bacteria after five and a half months suggests that the development took considerable time. Much faster establishment of anammox bacteria has been detected in anoxic reactors where AOB are not necessary to create anoxic microenvironments [32].

2.5. Biofilm Structure

As mentioned above, the biofilm was in general thicker in the corners than on the sides of the compartments. Biomass density, defined as the signal area of the Eub338 probe mix as a fraction of total compartment area (Figure S2), increased significantly with depth down to 400 μm ($R^2 = 0.4219$, $n = 17$, $\alpha = 0.05$). Thicker biofilms generally have a higher nitrogen removal, but modeling suggests improvements above a thickness of 750 μm to be small [33]. At more shallow depths (in the z dimension) erosion and detachment due to the shear forces, rather than electron donor or acceptor concentrations, may have controlled biofilm biomass. It was concluded that nitrifying biofilms on carriers fixed in a flow chamber generally showed enhanced erosion rather than growth when the flow rate increased, despite being supplied with more oxygen and substrate [19]. However, the relationship between carrier movement and the effect of flow velocity on biofilms in MBBRs is still uncertain [34].

The overall vertical distribution in the mature biofilms showed a significant decrease of AOB with carrier compartment depth (z dimension) ($R^2 = 0.80$, $n = 41$, $\alpha = 0.01$; Figure 1b). Similar to AOB, anammox cells grew in characteristic clusters or microcolonies of varying size and shape (Figure 2a,b), but instead increased significantly in relative abundance with depth from 180 μm down to 400 μm (R^2 = 0.66, n = 17, α = 0.01; Figure 1b). In the horizontal (x, y) dimension AOB dominated in the center parts and anammox in the peripheral part, close to the carrier walls (Figure 3). The least variation in relative population abundance was closer to the walls, such as for anammox in the 0–100 μm section (Figure 3c). Here erosion effects would be less severe compared with the center section where the variation of AOB was quite large between different compartments (Figure 3g). The combination of FISH, CLSM and new image analysis approaches enabled, for the first time, a detailed investigation of the three-dimensional distribution of AOB and anammox bacteria in MBBR biofilms. A uniquely detailed picture emerged, where the distinct stratification of AOB and anammox bacteria agrees well with the general pattern observed in granular biomass [4–6]. The decrease in relative abundance of AOB with carrier depth (the z dimension) is possibly caused by oxygen limitation in the deeper parts, whereas anammox bacteria increased down to 400 μm depth, below which the abundance leveled out or even decreased. This distribution pattern suggests different functional properties of the microbial communities at different depths of the biofilm in the carrier, where the outer regions contain mainly aerobic nitrite producers whereas the inner regions contain mainly anaerobic nitrite consumers. This kind of information is particularly valuable for understanding the function of MBBRs, since the continuous movement of the carriers in the reactors cause uncertainties about the flux of substrate and electron acceptors to the biofilm [34] and impede microelectrode *in situ* analysis of the gradients of substrate and electron acceptors near and in the biofilms [19].

Figure 2. Assembled FISH "wall-to-wall" micrographs of biofilm from the biofilm carrier compartments. (**A**) Mature biofilm from 230 μm depth; (**B**) Mature biofilm from 500 μm depth; (**C**) Young biofilm from 250 μm depth; and (**D**) Young biofilm from 490 μm depth. In yellow, AOB cells hybridized with probe Nse1472. In cyan, anammox cells hybridized with probe Bfu613. In green, cells hybridized with the Eub338 (I–IV) probe mix only. Scale bar = 200 μm.

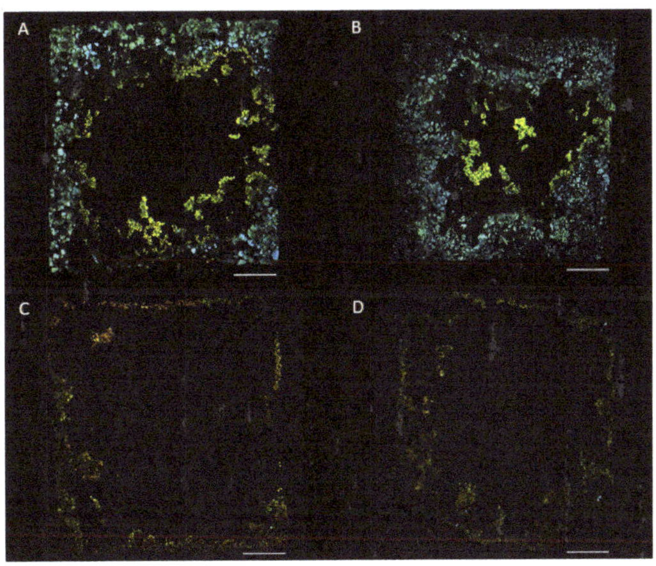

Figure 3. Assembled FISH "wall-to-wall" micrographs of biofilm from the biofilm carrier compartments in the *x*, *y* dimension. (**A**) Example of the multidirectional slicing procedure, where each colored area represents a 100 μm thick region of the biofilm from a particular distance from the compartment wall; (**B**) Schematic representation of the 100 μm thick slices, showing average biovolume fractions throughout all depths (*z*) of AOB and anammox bacteria in the carrier compartments. Average ± 95% confidence intervals, $n = 41$ ($n = 37$ for slice 400–500 μm); and (**C**–**G**) Biovolume fractions of anammox bacteria (open circles) and AOB (solid circles) in the concentric areas (colored "slices") at different depths (*z*) in the carrier compartments. The designations (**C**–**G**) refer to the concentric areas at different distance from the compartment wall, as depicted in (**B**).

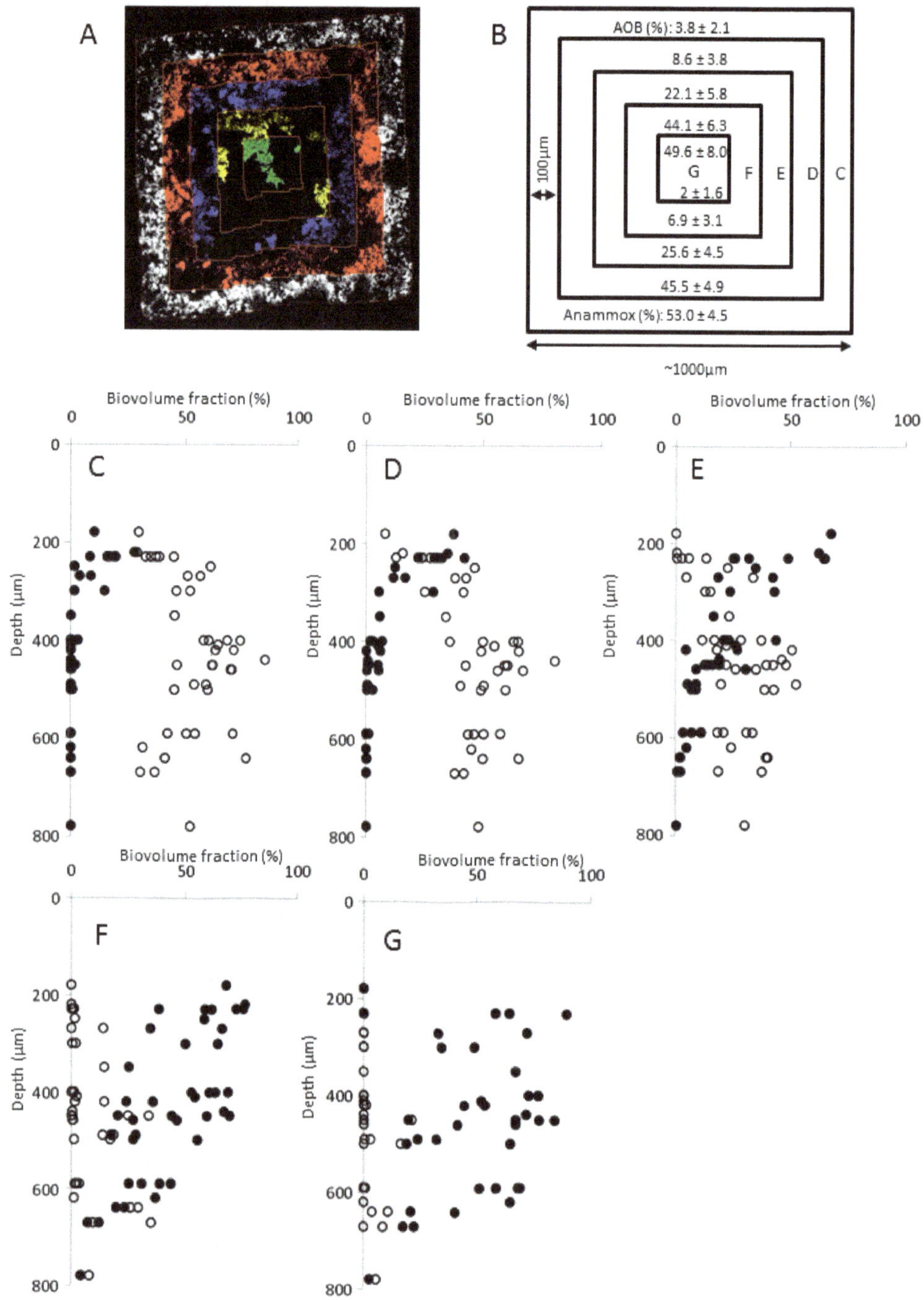

3. Experimental Procedures

3.1. Reactor Conditions

A laboratory scale (7.5 liter) MBBR reactor with biofilm carriers (Minichip, AnoxKaldnes, Lund, Sweden) fed with synthetic wastewater (see Supplementary Information) was operated for three years. The reactor conditions and process performance are shown in Table 1.

3.2. Biofilm Sampling and Preparation

Characteristics of the biofilm carriers are listed in Table S1. New carriers were added to the reactor on December 17th 2008. Carriers were sampled for initial biofilm screening on May 4th, 2009. On June 2nd 2009, carriers containing young (5.5 months old) or mature (>18 months old) biofilms were sampled from the reactor. The carriers were immediately fixed in 4% ice-cold paraformaldehyde (PFA; pH 7.2) on ice for at least 8 h. After fixation, the carriers were submerged in phosphate buffered saline (PBS, pH 7.2) for at least 20 min and thereafter cut in halves and stored in 1:1 (PBS:EtOH) at −20 °C until further use.

For screening of biofilm community composition the biofilm was brushed off from fixed halves of biofilm carriers, fixed again and homogenized [35].

For intact biofilm analyses fixed halves of biofilm carriers were embedded in Tissue-Tek® O.C.T.™ Compound (Sakura Finetek Europe B.V., Alphen aan den Rijn, The Netherlands), placed in a closed and parafilm-sealed Petri dish (Sarstedt, Helsingborg, Sweden) and incubated overnight at +4 °C. Centrifuge tubes (15 mL) were cut, placed upside down and filled to the brim with OCT compound. Incubated biofilm carriers were placed on top and covered with OCT and put in a liquid nitrogen fume chamber for approximately 60 min, until frozen solid. Tubes were fixed firmly in a vice and the carriers were removed, thereby exposing solid biofilm squares on top of the tubes (Figure 1a) which were sealed with a covering layer of OCT, frozen and stored in −70 °C. Sectioning of the biofilm from 15 compartments into 10 μm thick slices was performed in a cryotome operating at −20 °C. The produced sections were collected on SuperFrost® Plus Gold microscope slides (Menzel GmbH & KG, Braunschweig, Germany) and stored in −20 °C.

3.3. Fluorescence in Situ *Hybridization*

Screening for AOB, NOB and anammox populations was performed on homogenized biofilms, sampled on 14 May 2009, using FISH [35].

On cryosectioned samples for intact biofilm analyses, OCT was removed by a 10 min submersion in 50% EtOH. Additional fixation, to ensure that all cells in the biofilm would be accessible to the fluorescent probes and FISH, was performed as described earlier [36,37]. The fluorescent probes and unlabeled competitors were obtained from Thermo Electron (Interactiva Division, Ulm, Germany) or MWG Biotech (Ebersberg, Germany). The probes were 5' labeled with Cy3, Cy5, fluorescein or Alexa488. Hybridization was carried out at 46 °C for 2 (homogenized biofilm) or 4 h (cryosectioned biofilm), followed by washing at 48 °C for 10 min. Finally, the slides were rinsed in milli-Q water, air dried and mounted in Citifluor AF1 (Citifluor, London, UK). Each of the specific probes (Table S2) was hybridized together with the Eub338 probe mix for quantification of the relative probe signal.

3.4. Confocal Laser Scanning Microscopy

Confocal micrographs were acquired using a Bio-Rad Radiance 2000 MP microscope (Bio-Rad, Hemel Hempstead, UK) with a Nikon Plan Fluor 40×/1.40 oil objective, a Red diode laser (638 nm), a He/Ne laser (543 nm), and an Argon laser (488 nm). Images for analysis of biofilm structure were collected with the bundled software LaserSharp 2000 as 8-bit images of 1024 × 1024 pixels (resolution: 3.30 pixels/μm), and Kalman filtration ($n = 3$). For quantification of biofilm populations in homogenized samples, 8-bit images of 512 × 512 pixels (resolution: 1.65 pixels/μm), Kalman filtration ($n = 2$) were acquired. The biovolume fractions were measured in 21 microscope fields from each replicate by acquiring three images from different planes in the z dimension for each randomly selected x and y position, as previously described [38].

3.5. Digital Image Analysis

For biofilm structure analysis, the acquired images from each section were assembled manually using Photoshop CS4 extended (Adobe Systems, San Jose, CA, USA) to obtain complete coverage of the carrier compartment biofilm in each image (*i.e.*, "wall-to-wall" images). Here biofilm dimensions were measured to estimate shrinkage of the biofilm during cryosectioning, and biofilm thickness was measured (for details, see Supplementary Information). The biofilm images were then exported to the daime software version 2.0 [39] for measurements of the biovolume fractions of probe targeted bacteria in both homogenized and cryosectioned biofilm samples, and for measurements of the stratification in cryosectioned biofilms using the recently developed "Slicer" tool [20].

4. Conclusions

The applied cryosectioning and FISH protocols permitted retrieval and microscopy of intact biofilm from the MBBR carrier compartments with little biofilm distortion.

FISH data showed that *N. europaea/eutropha*-related organisms were the dominant AOB and that *Ca. Brocadia fulgida*-related organisms were the predominant anammox bacteria, suggesting a sequential colonization pattern of newly introduced carriers. Initial dominance of AOB was followed after about five and a half months by the establishment of anammox cells, probably as a result of formation of local anoxic micro-habitats by AOB.

By the use of novel digital image analysis tools and statistical analyses, substantial three-dimensional stratification of the mature biofilm was observed. In the z-dimension, the relative abundance of AOB decreased rapidly with depth, possibly due to oxygen limitation in the deeper parts of the carrier. In contrast, anammox bacteria increased in relative abundance down to a depth of 400 μm. A distinct horizontal (x, y), non-random distribution pattern was also observed, with AOB generally being most abundant in the center and anammox bacteria closer to the walls of the carrier compartments. We conclude that a true three-dimensional biofilm stratification can be expected in

MBBR carrier systems, which should be taken into account when modeling and optimizing reactor performance and in the design of new MBBR carriers.

Supplementary Information

1. Reactor Medium

The synthetic reactor media consisted of the following per 100 L: 240 g $NaHCO_3$, 120 g NH_4Cl, 0.3 g pepton, 0.56 g KH_2PO_4, 70 mL 2M NaOH, 40 mL micro-nutrient solution consisting of (in g/L): 4.8 $MgSO_4 \cdot 7H_2O$; 1.6 $MnCl_2 \cdot 2H_2O$; 5.8 $CaCl_2 \cdot 2H_2O$; 0.48 $CoCl_2 \cdot 6H_2O$; 0.24 $NiCl_2 \cdot 6H_2O$; 0.26 $ZnCl_2$; 0.10 $CuSO_4 \cdot 5H_2O$; 1.44 $FeCl_2 \cdot 4H_2O$; 0.0005 BH_3O_3; 0.0022 $Na_2MoO_4 \cdot 2H_2O$; 0.00114 $Na_2SeO_3 \cdot 5H_2O$; 0.0014 $Na_3WO_3 \cdot 2H_2O$.

2. Measurement of Biofilm Carrier Compartment Size

The biofilm carrier compartment dimensions were measured using brightfield microscopy and an eyepiece grid. The area was measured at the surface, halfway to the center and at the center, *i.e.*, the point farthest from the carrier surface. Note that due to the production method the carrier area decreases with depth down to the mid-point of the carrier. That is, the total carrier compartment, from one carrier surface to the other, has a slight "waist" at the center.

3. Digital Image Analysis of Biofilms

3.1. Estimation of Biofilm Shrinkage

For estimation of biofilm shrinkage due to the cryosectioning and FISH protocols, the sides and diagonals of each assembled biofilm square were measured on the Eub338mix images, using the ruler tool in Photoshop (CS4 extended; Adobe Systems, San Jose, CA, USA). The dimensions were plotted and compared to the dimensions of the carrier compartments as determined by microscopy (see section 2 above).

3.2. Estimation of Mature Biofilm Thickness

For estimation of the mature biofilm thickness, 8 measurements, one from each side and corner of each Eub338mix image were recorded in Photoshop CS4. Obvious biofilm channels were ignored in the measurements.

3.3. Estimation of Biovolume Fraction of Populations

For estimation of the biovolume fraction of each population, an intensity threshold was set manually for each image, using the threshold function in Photoshop CS4. The resulting binary (black and white) images of probe-labeled biomass were exported as 16 megapixel TIFs to the digital image analysis software daime [39], where a noise reduction step was performed ("Noise reduction" function, number of non-zero neighbor pixels: 1). Due to the large size of the images, image resolution was reduced to 1 megapixel before measuring biovolume fractions. Following 2D segmentation of the binary images autofluorescent objects present in both the Nse1472 and Bfu613 images were removed. The biovolume fraction was calculated for each image after an additional step of artefact elimination [with the daime artifact rejection tool using a 50% congruency threshold and final congruency requirement between the specific probes and the biofilm reference (EUB338mix) of at least 90%]. Images of homogenized biofilm were analyzed through initial intensity thresholding (cutoff: 30), followed by RATS-L segmentation, followed by artifact elimination and calculation of the biovolume fraction of each specific probe-target population as part of the Eub388 probe mix with a >90% requirement for total congruency. The horizontal "wall-to-wall" population distribution was analyzed using the multidirectional Slicer tool in daime 2.0 [20], creating 100 μm thick virtual slices and applying a baseline smoothing of 20%. Relative population abundances were estimated for each virtual slice as described above.

Table S1. Characteristics of biofilm carriers. *

Carrier specifics	Type 1 [1]	Type 2 [1]
Depth of carrier compartments (*i.e.*, *z* dimension in figures)	2 mm	3 mm
Diameter of the carrier	30 mm	30 mm
Protected surface area per carrier	2.73×10^{-3} m^2	4.10×10^{-3} m^2
Number of square compartments	325	325

* Minichip, AnoxKaldnes, Lund, Sweden. [1] Two different carrier variants were used.

Table S2. Oligonucleotide probes used for fluorescence *in situ* hybridization (FISH) in this study.

Probe name [1]	Target	Sequence (5'–3')	FA [2] [%]
Eub338 [3]	*Most Bacteria*	GCT GCC TCC CGT AGG AGT	0–50
Amx820	Anaerobic ammonium-oxidizing bacteria, *Candidatus* "*Brocadia anammoxidans*" and *Candidatus* "*Kuenenia stuttgartiensis*"	AAA ACC CCT CTA CTT AGT GCC C	40
Apr820	*Candidatus Anammoxoglobus propionicus*	AAA CCC CTC TAC CGA GTG CCC	40
Ban162	*Candidatus Brocadia anammoxidans*	CGG TAG CCC CAA TTG CTT	40
BS820	*Candidatus Scalindua wagneri*, *Candidatus Scalindua sorokinii*	TAA TTC CCT CTA CTT AGT GCC C	20
Bfu613	*Candidatus Brocadia fulgida*	GGA TGC CGT TCT TCC GTT AAG CGG	30
Sca1309	Genus *Candidatus Scalindua*	TGG AGG CGA ATT TCA GCC TCC	5
Scabr1114	*Candidatus Scalindua brodae*	CCC GCT GGT AAC TAA AAA CAA G	20
Pla46	*Planctomycetales*	GAC TTG CAT GCC TAA TCC	30
Nso1225	Most beta-proteobacterial AOB	CGC CAT TGT ATT ACG TGT GA	35
Nse1472	*Nitrosomonas europea*, *N. halophila*, *N. eutropha*, *Kraftisried-Isolat Nm103*	ACC CCA GTC ATG ACC CCC	50
Ntspa662 [4]	Genus *Nitrospira*	GGA ATT CCG CGC TCC TCT	35
Ntspa1151	Sublineage II of the genus *Nitrospira*	TTC TCC TGG GCA GTC TCT CC	35–40
Ntspa1431	Sublineage I of the genus *Nitrospira*	TTG GCT TGG GCG ACT TCA	35
Ntg840	*Nitrotoga arctica*	CTA AGG AAG TCT CCT CCC	10–20
Nit3 [4]	Genus *Nitrobacter*	CCT GTG CTC CAT GCT CCG	40
NmII	*Nitrosomonas communis* lineage	TTA AGA CAC GTT CCG ATG TA	35

Table S2. *Cont.*

Probe name [1]	**Target**	**Sequence (5'–3')**	**FA** [2] **[%]**
Alf968	*Alphaproteobacteria*, except of *Rickettsiales*	GGT AAG GTT CTG CGC GTT	20
CF319a	Most *Flavobacteria*, some *Bacteroidetes*, some *Sphingobacteria*	TGG TCC GTG TCT CAG TAC	35
Gam42a [4]	*Gammaproteobacteria*	GCC TTC CCA CAT CGT TT	35

[1] For probe specifications, see ProbeBase (http://www.microbial-ecology.net/probebase/) [40]; [2] FA = Formamide; [3] Used in a mix together with Eub338 II, III and IV; and [4] Used together with an unlabeled oligonucleotide competitor as indicated in the reference.

Figure S1. Scatter plot of area *versus* depth (*z*) for the assembled confocal micrographs (open circles) and Microchip compartment area as determined by light microscopy (closed circles). N.B. due to the production method the carrier area decreases with depth down to the mid-point of the carrier. That is, the total carrier compartment, from one carrier surface to the other, has a slight "waist" at the center. The area of the confocal micrographs was measured in the biofilm reference (Eub338) channel. Error bars = 95% confidence interval.

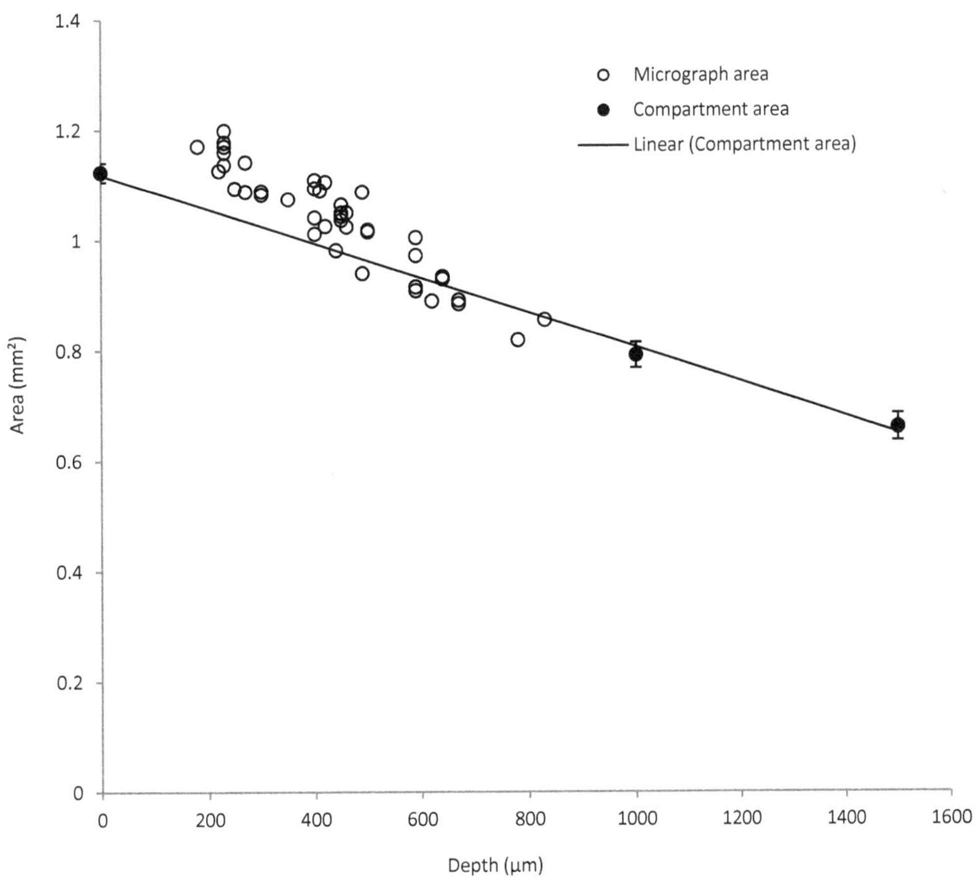

Figure S2. Scatter plot of total biomass areal density fraction *vs.* depth into the carrier. Each data point corresponds to the percentage coverage of the Eub338 probe mix, as determined by FISH, of the community in an assembled "wall-to-wall" micrograph from the indicated depth in the mature biofilm.

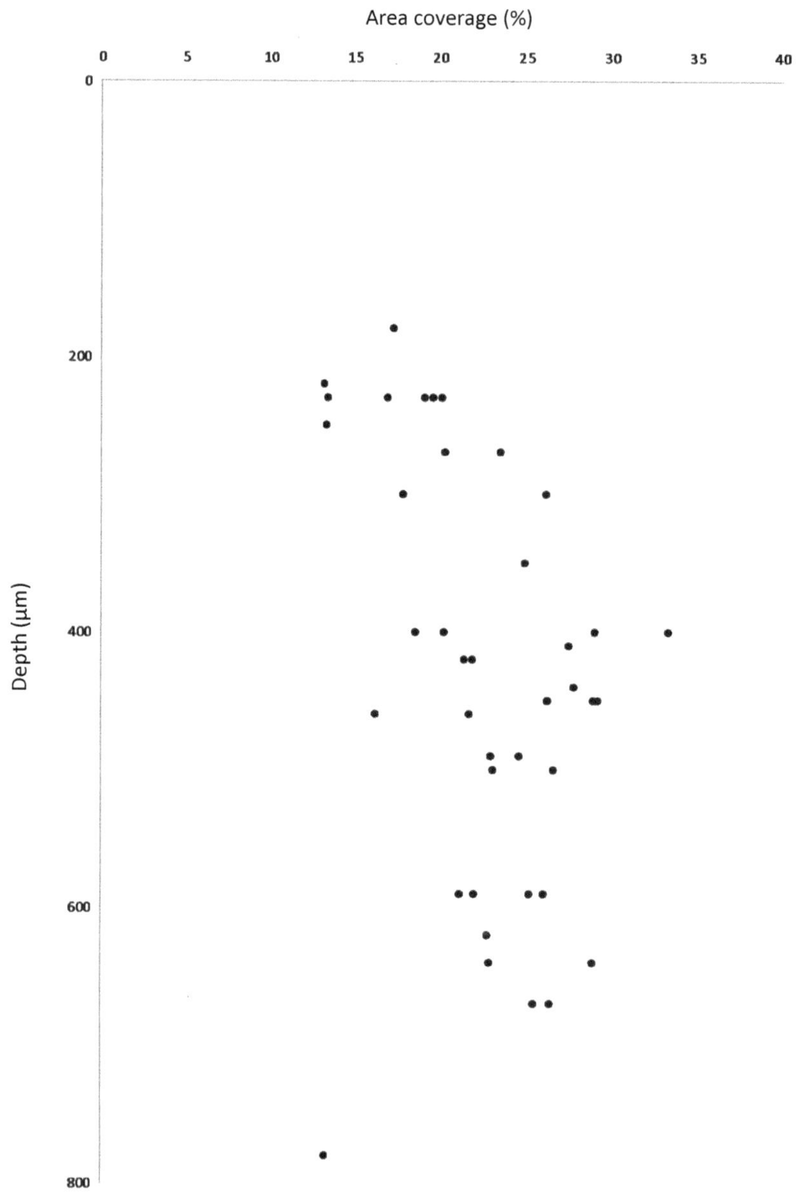

Acknowledgments

We acknowledge financial support from The Swedish Research Council for Environment, Agricultural Sciences and Spatial Planning (Formas, contract no. 214-2004-1678, 243-2008-131 and 2010-2259, 211-2010-140); The Royal Society of Sciences and Letters in Gothenburg (Kungliga Vetenskaps- och Vitterhetssamhället i Göteborg); Göteborgs universitets donationsfonder (contract B432 28/06); and Adlerbertska forskningsfonden (contract B432 835/07). We would like to thank the technical staff at AnoxKaldnes for sampling assistance and pilot plant maintenance. We also thank Alma Masić for valuable suggestions and discussions.

Author Contributions

M.H., M.C., R.A., F.S. initiated the project. M.H., R.A., F.S. designed the experiment and R.A., M.E., M.C. carried out the laboratory work. RA did biofilm retrieval and preparation, microscopy and image analyses, R.A., H.D., M.H., F.P., F.S. analyzed data and R.A., F.P., M.H., H.D,. B.M.W. wrote the manuscript.

References

1. Vlaeminck, S.E.; de Clippeleir, H.; Verstraete, W. Microbial resource management of one-stage partial nitritation/anammox. *Microb. Biotechnol.* **2012**, *5*, 433–448.
2. Sliekers, A.O.; Derwort, N.; Gomez, J.L.C.; Strous, M.; Kuenen, J.G.; Jetten, M.S.M. Completely autotrophic nitrogen removal over nitrite in one single reactor. *Water Res.* **2002**, *36*, 2475–2482.
3. Pynaert, K.; Smets, B.F.; Wyffels, S.; Beheydt, D.; Siciliano, S.D.; Verstraete, W. Characterization of an autotrophic nitrogen-removing biofilm from a highly loaded lab-scale rotating biological contactor. *Appl. Environ. Microbiol.* **2003**, *69*, 3626–3635.
4. Vlaeminck, S.E.; Terada, A.; Smets, B.F.; de Clippeleir, H.; Schaubroeck, T.; Bolca, S.; Demeestere, L.; Mast, J.; Boon, N.; Carballa, M.; Verstraete, W. Aggregate size and architecture determine microbial activity balance for one-stage partial nitritation and anammox. *Appl. Environ. Microbiol.* **2010**, *76*, 900–909.
5. Vazquez-Padin, J.; Mosquera-Corral, A.; Campos, J.L.; Mendez, R.; Revsbech, N.P. Microbial community distribution and activity dynamics of granular biomass in a canon reactor. *Water. Res.* **2010**, *44*, 4359–4370.
6. Winkler, M.K.H.; Kleerebezem, R.; Kuenen, J.G.; Yang, J.J.; van Loosdrecht, M.C.M. Segregation of biomass in cyclic anaerobic/aerobic granular sludge allows the enrichment of anaerobic ammonium oxidizing bacteria at low temperatures. *Environ. Sci. Technol.* **2011**, *45*, 7330–7337.
7. Ni, B.J.; Chen, Y.P.; Liu, S.Y.; Fang, F.; Xie, W.M.; Yu, H.Q. Modeling a granule-based anaerobic ammonium oxidizing (anammox) process. *Biotechnol. Bioeng.* **2009**, *103*, 490–499.
8. Volcke, E.I.P.; Picioreanu, C.; de Baets, B.; van Loosdrecht, M.C.M. The granule size distribution in an anammox-based granular sludge reactor affects the conversion-implications for modeling. *Biotechnol. Bioeng.* **2012**, *109*, 1629–1636.
9. Vlaeminck, S.E.; Terada, A.; Smets, B.F.; van der Linden, D.; Boon, N.; Verstraete, W.; Carballa, M. Nitrogen removal from digested black water by one-stage partial nitritation and anammox. *Environ. Sci. Technol.* **2009**, *43*, 5035–5041.
10. Qiao, S.; Nishiyama, T.; Fujii, T.; Bhatti, Z.; Furukawa, K. Rapid startup and high rate nitrogen removal from anaerobic sludge digester liquor using a snap process. *Biodegradation* **2012**, *23*, 157–164.
11. Liu, T.; Li, D.; Zeng, H.; Li, X.; Liang, Y.; Chang, X.; Zhang, J. Distribution and genetic diversity of functional microorganisms in different canon reactors. *Bioresour. Technol.* **2012**, *123*, 574–580.

12. Chen, H.; Liu, S.; Yang, F.; Xue, Y.; Wang, T. The development of simultaneous partial nitrification, anammox and denitrification (snad) process in a single reactor for nitrogen removal. *Bioresour. Technol.* **2009**, *100*, 1548–1554.
13. Cho, S.; Fujii, N.; Lee, T.; Okabe, S. Development of a simultaneous partial nitrification and anaerobic ammonia oxidation process in a single reactor. *Bioresour. Technol.* **2011**, *102*, 652–659.
14. Tsushima, I.; Ogasawara, Y.; Kindaichi, T.; Satoh, H.; Okabe, S. Development of high-rate anaerobic ammonium-oxidizing (anammox) biofilm reactors. *Water Res.* **2007**, *41*, 1623–1634.
15. Plaza, E.; Stridh, S.; Örnmark, J.; Kanders, L.; Trela, J. Swedish experiences of the deammonification process in a biofilm system. In Proceedings of WEF-IWA Conference on Nutrient Recovery and Management, Miami, FL, USA, 9–12 January 2011.
16. Christensson, M.; Ekström, S.; Andersson Chan, A.; Le Vaillant, E.; Lemaire, R. Experience from start-ups of the first anita mox plants. *Water Sci. Technol.* **2013**, *67*, 2677–2684.
17. Persson, F.; Sultana, R.; Suarez, M.; Hermansson, M.; Plaza, E.; Wilén, B.-M. Structure and composition of biofilm communities in a moving bed biofilm reactor for nitritation-anammox at low temperatures. *Bioresour. Technol.* **2014**, *154*, 267–273.
18. Alpkvist, E.; Bengtsson, J.; Overgaard, N.C.; Christensson, M.; Heyden, A. Simulation of nitrification of municipal wastewater in a Moving Bed™ biofilm process: A bottom-up approach based on a 2D-continuum model for growth and detachment. *Water Sci. Technol.* **2007**, *55*, 247–255.
19. Masic, A.; Bengtsson, J.; Christensson, M. Measuring and modeling the oxygen profile in a nitrifying moving bed biofilm reactor. *Math. Biosci.* **2010**, *227*, 1–11.
20. Almstrand, R.; Daims, H.; Persson, F.; Sorensson, F.; Hermansson, M. New methods for analysis of spatial distribution and coaggregation of microbial populations in complex biofilms. *Appl. Environ. Microbiol.* **2013**, *79*, 5978–5987.
21. Kindaichi, T.; Ito, T.; Okabe, S. Ecophysiological interaction between nitrifying bacteria and heterotrophic bacteria in autotrophic nitrifying biofilms as determined by microautoradiography-fluorescence *in situ* hybridization. *Appl. Environ. Microbiol.* **2004**, *70*, 1641–1650.
22. Okabe, H.; Kindaichi, T.; Nakamura, K.; Ito, T. Eco-physiology of autotrophic nitrifying biofilms. *Water Sci. Technol.* **2005**, *52*, 225–232.
23. Strous, M.; Heijnen, J.J.; Kuenen, J.G.; Jetten, M.S.M. The sequencing batch reactor as a powerful tool for the study of slowly growing anaerobic ammonium-oxidizing microorganisms. *Appl. Microbiol. Biotechnol.* **1998**, *50*, 589–596.
24. De Beer, D.; Stoodley, P.; Roe, F.; Lewandowski, Z. Effects of biofilm structures on oxygen distribution and mass transport. *Biotechnol. Bioeng.* **1994**, *43*, 1131–1138.
25. Daims, H.; Wagner, M. *In situ* techniques and digital image analysis methods for quantifying spatial localization patterns of nitrifiers and other microorganisms in biofilm and flocs. *Method Enzymol.* **2011**, *496*, 185–215.
26. Juretschko, S.; Timmermann, G.; Schmid, M.; Schleifer, K.H.; Pommerening-Roser, A.; Koops, H.P.; Wagner, M. Combined molecular and conventional analyses of nitrifying bacterium diversity in activated sludge: Nitrosococcus mobilis and nitrospira-like bacteria as dominant populations. *Appl. Environ. Microbiol.* **1998**, *64*, 3042–3051.
27. Koops, H.P.; Pommerening-Roser, A. Distribution and ecophysiology of the nitrifying bacteria emphasizing cultured species. *FEMS Microbiol. Ecol.* **2001**, *37*, 1–9.

28. Bollmann, A.; Bar-Gilissen, M.J.; Laanbroek, H.J. Growth at low ammonium concentrations and starvation response as potential factors involved in niche differentiation among ammonia-oxidizing bacteria. *Appl. Environ. Microbiol.* **2002**, *68*, 4751–4757.
29. Park, H.; Rosenthal, A.; Jezek, R.; Ramalingam, K.; Fillos, J.; Chandran, K. Impact of inocula and growth mode on the molecular microbial ecology of anaerobic ammonia oxidation (anammox) bioreactor communities. *Water Res.* **2010**, *44*, 5005–5013.
30. Van der Star, W.R.; Miclea, A.I.; van Dongen, U.G.; Muyzer, G.; Picioreanu, C.; van Loosdrecht, M.C. The membrane bioreactor: A novel tool to grow anammox bacteria as free cells. *Biotechnol. Bioeng.* **2008**, *101*, 286–294.
31. Egli, K.; Bosshard, F.; Werlen, C.; Lais, P.; Siegrist, H.; Zehnder, A.J.B.; van der Meer, J.R. Microbial composition and structure of a rotating biological contactor biofilm treating ammonium-rich wastewater without organic carbon. *Microb. Ecol.* **2003**, *45*, 419–432.
32. Isaka, K.; Date, Y.; Sumino, T.; Yoshie, S.; Tsuneda, S. Growth characteristic of anaerobic ammonium-oxidizing bacteria in an anaerobic biological filtrated reactor. *Appl. Microbiol. Biotechnol.* **2006**, *70*, 47–52.
33. Terada, A.; Lackner, S.; Tsuneda, S.; Smets, B.F. Redox-stratification controlled biofilm (rescobi) for completely autotrophic nitrogen removal: The effect of co- *versus* counter-diffusion on reactor performance. *Biotechnol. Bioeng.* **2007**, *97*, 40–51.
34. McQuarrie, J.P.; Boltz, J.P. Moving bed biofilm reactor technology: Process applications, design, and performance. *Water Environ. Res.* **2011**, *83*, 560–575.
35. Almstrand, R.; Lydmark, P.; Sorensson, F.; Hermansson, M. Nitrification potential and population dynamics of nitrifying bacterial biofilms in response to controlled shifts of ammonium concentrations in wastewater trickling filters. *Bioresour. Technol.* **2011**, *102*, 7685–7691.
36. Lydmark, P.; Lind, M.; Sorensson, F.; Hermansson, M. Vertical distribution of nitrifying populations in bacterial biofilms from a full-scale nitrifying trickling filter. *Environ. Microbiol.* **2006**, *8*, 2036–2049.
37. Manz, W.; Amann, R.; Ludwig, W.; Wagner, M.; Schleifer, K.H. Phylogenetic oligodeoxynucleotide probes for the major subclasses of proteobacteria: Problems and solutions. *Syst. Appl. Microbiol.* **1992**, *15*, 593–600.
38. Almstrand, R.; Lydmark, P.; Lindgren, P.E.; Sorensson, F.; Hermansson, M. Dynamics of specific ammonia-oxidizing bacterial populations and nitrification in response to controlled shifts of ammonium concentrations in wastewater. *Appl. Microbiol. Biotechnol.* **2013**, *97*, 2183–2191.
39. Daims, H.; Lucker, S.; Wagner, M. Daime, a novel image analysis program for microbial ecology and biofilm research. *Environ. Microbiol.* **2006**, *8*, 200–213.
40. Loy, A.; Horn, M.; Wagner, M. Probebase: An online resource for rRNA-targeted oligonucleotide probes. *Nucl. Acids Res.* **2003**, *31*, 514–516.

13

Biofilm Matrix and Its Regulation in *Pseudomonas aeruginosa*

Qing Wei and Luyan Z. Ma *

State Key Laboratory of Microbial Resources, Institute of Microbiology, Chinese Academy of Sciences, No.3, 1st Beichen West Road, Chaoyang District, Beijing 100101, China;
E-Mail: vubwqing@hotmail.com

* Author to whom correspondence should be addressed. Email: luyanma27@im.ac.cn;

Abstract: Biofilms are communities of microorganisms embedded in extracellular polymeric substances (EPS) matrix. Bacteria in biofilms demonstrate distinct features from their free-living planktonic counterparts, such as different physiology and high resistance to immune system and antibiotics that render biofilm a source of chronic and persistent infections. A deeper understanding of biofilms will ultimately provide insights into the development of alternative treatment for biofilm infections. The opportunistic pathogen *Pseudomonas aeruginosa*, a model bacterium for biofilm research, is notorious for its ability to cause chronic infections by its high level of drug resistance involving the formation of biofilms. In this review, we summarize recent advances in biofilm formation, focusing on the biofilm matrix and its regulation in *P. aeruginosa*, aiming to provide resources for the understanding and control of bacterial biofilms.

Keywords: *Pseudomonas aeruginosa*; biofilm matrix; exopolysaccharides; gene regulation; anti-biofilm

---

1. Introduction

Biofilms are microbial communities encased in extracellular polymeric substances (EPS) [1]. Biofilm formation represents a protective mode of growth that allows microorganisms to survive in hostile environments and disperse seeding cells to colonize new niches under desirable conditions. Biofilms can form on a variety of surfaces and are prevalent in natural, industrial, and hospital niches. These sessile microbial communities are physiologically distinct from free-living planktonic counterparts [2,3]. Clinically, biofilms are responsible for many persistent and chronic infections due to their inherent resistance to antimicrobial agents and the selection for phenotypic variants. A better understanding of the genetic and molecular mechanisms of biofilm formation may provide strategies for the control of chronic infections and problems related to biofilm formation.

The EPS of biofilm is a mixture of polysaccharides, extracellular DNA (eDNA), and proteins, which function as matrix, or glue, holding microbial cells together. The biofilm matrix contributes to the overall architecture and the resistance phenotype of biofilms [4,5]. Uncovering roles played by EPS matrices in biofilm formation will be beneficial for the design of targeted molecules to control biofilm formation. In this review, advances in biofilm formation and regulation are presented with a focus on the biofilm matrix in *P. aeruginosa*, a model organism for biofilm research.

2. Matrices of *P. aeruginosa* Biofilms

Colonization of the lungs of cystic fibrosis (CF) patients by *P. aeruginosa* is the major cause of morbidity and mortality in CF populations. These infections generally persist despite the use of long term aggressive antimicrobial therapy and have been associated with the formation of antibiotic-resistant biofilms, whereby bacterial communities form microcolonies embedded in a hydrated EPS matrix [6,7]. The relative importance of the EPS matrix is dependent on the genetic background of strains, nutritional conditions and developmental phases of biofilms [8]. It is generally acknowledged that the EPS of biofilms functions as both a structural scaffold and/or a protective barrier to harsh environments [9,10]. At least three polysaccharides (Psl, Pel and alginate) have been identified in *P. aeruginosa* that play important roles in structure maintenance and antibiotic resistance of biofilm [9,11–15]. The respective nature and functions in biofilm formation and development of biofilm matrix components including exopolysaccharides (Psl, Pel, and alginate), eDNA, proteins, and proteinaceous surface appendages such as fimbriae, type IV pili (T4P), and flagellum will be discussed below.

2.1. Psl Polysaccharide

Polysaccharide synthesis locus of *P. aeruginosa* PAO1 was identified in 2004 by three different groups [11,13,16]. The Psl cluster consists of 15 co-transcribed genes (*pslA* to *pslO*, PA2231-2245) encoding proteins to synthesize Psl, which is important to initiate and maintain biofilm structure by providing cell-cell and cell-surface interactions [15,17–19]. It was further demonstrated that only 11 of the 15 *psl* genes are required for the synthesis of Psl-dependent biofilm [20]. A deletion from *pslA* to *pslD* in *P. aeruginosa* strain PA14 leads to the incapability of producing Psl polysaccharide [21]. It was previously shown that Psl is a galactose-rich and mannose-rich exopolysaccharide with relatively lower amounts of glucose and xylose [17]. Recently, Psl was found to contain a repeating pentasaccharide consisting of D-mannose, D-glucose, and L-rhamnose [20]. Additionally, functions of individual *psl* genes such as *pslA*, *pslB* and *pslD* have been studied at genetic and biochemical levels, revealing that those genes are crucial for the biofilm formation and Psl synthesis [22–24].

The roles of Psl in biofilm formation have been thoroughly investigated due to its major contribution to the biofilm formation in *P. aeruginosa* [19]. Firstly, overproduction of the Psl polysaccharide led to enhanced cell-surface and intercellular adhesion of *P. aeruginosa*, suggesting its importance in adhesion, which is critical for initiation and maintenance of the biofilm structure [18,20]. Afterwards, using fluorescently labeled lectins, Ma and colleagues could directly visualize Psl exopolysaccharide formation at different stages of biofilm development (Figure 1). They found that Psl polysaccharide anchors on bacterial cell surface in a helical shape, which promotes strong bacterial cell-cell interactions. This results in the assembly of a biofilm and its matrix at the early stage of biofilm development. Later on, Psl polysaccharide accumulated on the periphery of three dimension-structured macrocolonies during biofilm maturation. This localization pattern provided the structure support and allowed for later biofilm dispersion [15]. In addition, Psl staining demonstrated that Psl can form a fiber-like matrix that enmeshes bacteria within biofilms. Recently, it was found that Psl fiber matrix was formed via a T4P-dependent migration strategy, a way similar to spider web formation [25]. Independently, Zhao and colleagues also showed that *P. aeruginosa* can deposit Psl trails during migration on a surface, which influences the surface motility of subsequent cells, leading to the biofilm initiation [26]. These recent discoveries further expand our understanding on the biology of Psl polysaccharide.

Figure 1. Scheme of biofilm development in *P. aeruginosa*. Selected images showed how the matrix of Psl polysaccharide (**red** fluorescence) enmeshes bacterial cells (**green** fluorescence) within bacterial communities during biofilm development (**I**: initial attachment; **II**: irreversible attachment; **III**: microcolony formation; **IV**: biofilm maturation; **V**: biofilm dispersion). The figure was used with the permission of the authors [15,27] and modified herein.

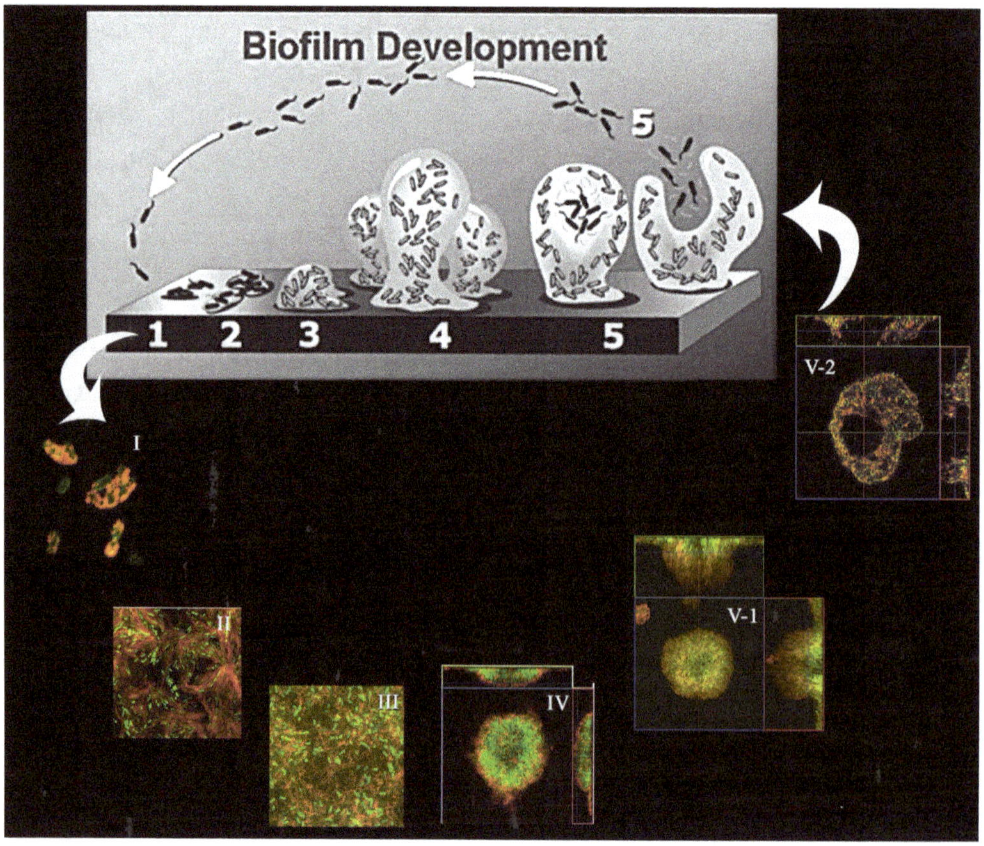

Strikingly, the function of Psl was further characterized to have a signaling role in stimulating two diguanylate cyclase, SiaD and SadC, to produce more of the intracellular second messenger molecule c-di-GMP, thus ultimately increasing the production of Psl itself and forming a unique positive feedback regulatory circuit [28].

In addition, Psl was found to have roles in pathogenesis and protection against the immune system. It was shown that by increasing contact between bacterial cells and epithelial cells, Psl polysaccharide indirectly stimulates NF-κB activity and facilitates flagellin-mediated proinflammatory signaling [29]. Furthermore, Mishra *et al.* found that *P. aeruginosa* Psl polysaccharide could reduce neutrophil phagocytosis and the oxidative response via limiting complement-mediated opsonization [30]. This study clearly presented Psl as a survival advantage *in vivo* and provided evidence for the clearance of persistent infections using Psl as a target.

Reports showed that Psl also had roles in antibiotics resistance. It was recently revealed that Psl can promote resistance to the biofilm inhibitor Polysorbate 80 [31]. Yang and colleagues reported that the formation of tight microcolony structure, mainly by Psl, in *P. aeruginosa* contributed to the resistance to antibiotic treatment [19]. By using fluorescent labeling of antibiotics, the extracellular matrix was also found to protect *P. aeruginosa* biofilms from killing by limiting the penetration of tobramycin; this sequestration happened at the periphery of biofilms [32]. However, no clear evidence was identified in which biofilm matrix is critical for this protection and interaction. Furthermore, another group showed that Psl provided a generic first line of defense against both cationic and anionic antibiotics during the initial stages of biofilm development and the Psl-mediated protection was extendable to *Escherichia coli* and *Staphylococcus aureus* in co-culture biofilms [33].

In summary, Psl not only functions as a scaffold for biofilm development, but also serves as a signaling molecule or track for the subsequent events leading to the formation of biofilms. This positive feedback circuit presents an evolutionary survival advantage for *P. aeruginosa* to colonize different niches. Besides, Psl also functions as a barrier for immune and antibiotic attacks.

2.2. Pel Polysaccharide

Pel polysaccharide, which is synthesized by the products of the *pel* gene cluster (*pelA-F*, PA3058-PA3064), is a glucose-rich and cellulase-sensitive extracellular matrix [12]. This gene cluster is conserved in other Gram-negative bacteria [34]. A *pel* mutant appeared to be deficient in the formation of pellicles at the air-liquid interface in standing cultures. Pel is also required for the formation of solid surface-associated biofilms [11]. Interestingly, in a non-piliated *P. aeruginosa* PAK strain, the *pel* mutation was discovered to have severe defects in the initial attachment process on solid surfaces, suggesting that Pel polysaccahride can compensate as an attachment factor in the absence of other adhesins such as type IV pili [34]. However, the precise role of Pel polysaccharide in attachment in other *P. aeruginosa* strains needs further investigation. In addition, Pel polysaccharide can serve as a primary structure scaffold for the community of cells by maintaining the cell-to-cell interactions in PA14 biofilms and play a protective role by enhancing resistance to aminoglycoside antibiotics in biofilms [14]. The role of Pel in biofilm formation was further investigated and it was revealed that Pel could function together with other types of EPS throughout biofilm development in *P. aeruginosa* PAO1, although in a less important role as compared to Psl. [19].

One remaining question about Pel is its biochemical composition. Although the original study identified Pel as a glucose-rich matrix material [12], its defined structure and feature have not been completely understood. Recently, the main glucose-containing carbohydrate of the extracellular matrix of *P. aeruginosa* PA14 was characterized to be glycerophosphorylated cyclic β-(1,3)-glucans, which was synthesized by *ndvB* locus but not the *pel* operon [35]. A systematic analysis of the extracellular carbohydrates produced by *P. aeruginosa* PA14 elucidated the structure of a LPS *O*-antigen polysaccharide, which suggested that the *pel* locus might be involved in the production of the LPS [36]. Recently, biochemical studies identified PelF as a cytosolic glycosyltransferase that utilizes UDP-glucose as substrate for Pel synthesis [37]. In addition, Colvin and his coworkers examined the deacetylase activity of PelA and found that this function is important for the production of the Pel polysaccharide [38]. Both studies contributed to the understanding of the synthesis of Pel in *P. aeruginosa.*

2.3. Alginate

Alginate is the exopolysaccharide that is often and mainly produced by *P. aeruginosa* clinical isolates from the lungs of CF patients [39]. The typical mucoid phenotype is due to the overproduction of this polysaccharide, which protects *P. aeruginosa* from harsh environments in CF lungs by providing an extracellular matrix in biofilms. However, it is not absolutely required during the formation of nonmucoid biofilms *in vitro* [40]. Alginate plays important roles in structural stability and protection of biofilms. It is necessary for water and nutrient retention in biofilms [41]. Interestingly, it was recently found that mucoid *P. aeruginosa* strains also relied on Psl to form biofilms [42,43]. Alginate has been identified to have functions in persistence and immune evasion [44]. Overproduction of alginate could provide resistance to antibiotics as well as opsonophagocytosis [45,46]. Alginate also has the ability to scavenge free radicals released from neutrophils and activate macrophages *in vitro* that are commonly used to kill pathogens [47]. Bragonzi and colleagues reported that overproduction of alginate did not provide increased persistence in a murine lung infection model, perhaps likely due to the reversion of mucoid phenotype to non-mucoid phenotype during infections [48].

2.4. Extracellular DNA

Extracellular DNA (eDNA) constitutes another important component of the *P. aeruginosa* biofilm matrix [16,49,50]. The eDNA appears to be generated from random chromosomal DNA that serves as a cell-to-cell interconnecting component in the biofilm. Biofilm eDNA staining suggests that the eDNA is located primarily in high concentrations within the stalks of mushroom-shaped microcolonies [49]. In addition, bacterial cells also undergo autolysis in biofilm microcolonies, but it is unclear whether autolysis contributes to eDNA or biofilm development [51]. Several hypotheses have been proposed such as direct secretion, lysis of sub-population by prophage and release of small membrane vesicles [52].

Similar to other types of biofilm matrices, eDNA also has multifaceted roles in biofilm formation, such as contribution to cation gradients, genomic DNA release and antibiotic resistance [53]. eDNA also acts as a nutrient source for bacteria during starvation [54]. Interestingly, it was recently shown that

P. aeruginosa produces an extracellular deoxyribonuclease (PA3909) that is required for utilization of eDNA as a nutrient source, further amplifying the role of eDNA during *P. aeruginosa* biofilm formation [55]. In addition, eDNA was shown to form bundles with F-actin liberated from necrotic neutrophils and the presence of those bundles could stimulate the initial development of *P. aeruginosa* biofilms [56]. eDNA can also activate neutrophils through a CpG- and TLR9-independent mechanism and serve as a major proinflammatory component of *P. aeruginosa* biofilms [57]. eDNA was further found to facilitate the twitching motility-mediated biofilm expansion by maintaining coherent cell alignments to coordinate the movement of cells in the leading edge vanguard rafts [58].

2.5. Proteins and Proteinaceous Bacterial Surface Appendages

Aside from exopolysaccharides and eDNA, extracellular proteins and several proteinaceous components are also considered to be matrix components, including type IV pili, flagella, and fimbriae. These components were found to mainly play auxiliary functions as adhesion factors and structural support in the biofilm formation of *P. aeruginosa* [59].

Flagella mediate swimming and swarming motility of *P. aeruginosa*. It can also act as an adhesin and play critical roles in the initial cell-to-surface interactions [60]. T4P is a linear actuator critical for twitching motility that involves an extension-grip-retraction mechanism [61]. T4P plays important roles in microcolony formation of *P. aeruginosa* biofilms by forming typical mushroom caps [60]. As mentioned above, T4P-driven bacterial motilities can help to form Psl fiber matrix [25]. Another group, however, drew the conclusion that flagella and pili are not required for the initial attachment or biofilm formation [62], indicating that flagella- and pili-mediated biofilm formation could be conditional and nutritional. Recently, another newly identified adhesin named CdrA was demonstrated to be a key protein functioning as a structural component of *P. aeruginosa* EPS matrix [63]. This adhesin could directly interact with Psl polysaccharide to mediate bacterial auto-aggregation and increase biofilm stability.

The *P. aeruginosa* Cup fimbriae constitute one class of appendages that facilitate the biofilm formation and assemble through chaperone/usher pathway [64]. In *P. aeruginosa* PAO1, at least four Cup systems (CupA, B, C, and E) have been identified and vary in organizations and functions [65]. It was demonstrated that Cup fimbriae are critical for the initial stage of biofilm development, particularly in cell-to-cell interaction and microcolony formation [66,67].

Numerous investigations on biofilm formation, especially the biofilm matrices of *P. aeruignosa* have provided insights into the importance and nature of each matrix component and laid foundations for the efficient treatment of biofilm-related *P. aeruginosa* infections. The roles of matrix components in *P. aeruginosa* biofilms are summarized in Table 1.

Table 1. Roles of different exopolysaccharides (EPS) in *Pseudomonas aeruginosa*

EPS	Locus *	Roles	References
Psl	PA2231–PA2245	Initial attachment and adhesion	[15,17–19]
		Primary biofilm scaffold	[15,19]
		Proinflammatory signaling	[20]
		Antibiotics resistance	[19]
		Avoidance of host defence mechanisms	[30]
		Signaling molecule to stimulate biofilm formation	[28]
		Resistance to biofilm inhibitor Polysorbate 80	[31]
		Guide of exploration and microcolony formation	[26]
Pel	PA3058–PA3064	Pellicle formation and solid surface-associated biofilm formation	[11]
		Aggregating of bacterial cells	[14]
		Aminoglycosides antibiotic resistance	[14]
		Initial attachment in the absence of type IV pili	[34]
Alginate	PA3540–PA3548	Persistence and immune evasion	[44]
		Resistance to antibiotics as well as opsonophagocytosis	[45,46]
		ROS scavenge	[47]
		Leading to mucoid	[39]
		Water and nutrient retention	[41]
eDNA		Nutrient	[54,55]
		Scaffold	[49]
		Antibiotics resistance	[53]
		Major proinflammatory component	[57]
		Promoting self-organization of bacterial biofilms	[58]
Proteinaceous component	Flagella	Initial cell-to-surface interactions	[60]
	Pili	The formation of mushroom-like microcolony	[60]
	CdrA	Mediate aggregation and increase biofilm stability	[63]
	Cup fimbriae	Cell-to-cell interaction and microcolony formation	[66,67]

* All loci were extracted from *Pseudomonas* Genome Database (www.pseudomonas.com) [21]. ROS, reactive oxygen species.

3. Regulation of Biofilm Matrix in *P. aeruginosa*

Gene regulation is important for our understanding of biofilm formation. Generally, organisms form a biofilm in response to several factors including nutritional cues, secondary messengers, host-derived signals or, in some cases, to sub-inhibitory concentrations of antibiotics [1,68]. When a cell switches to the biofilm mode of growth, it undergoes a phenotypic shift in behavior whereby a large array of genes is differentially regulated [69].

Biofilm formation is a multicellular process involving environmental signals and a concerted regulation combining both environmental signals and regulatory networks. Due to the major roles of EPS matrix in biofilm formation, its regulation is discussed.

3.1. c-di-GMP

Bis-(3'-5')-cyclic dimeric guanosine monophosphate (c-di-GMP), a ubiquitous intracellular second messenger widely distributed in bacteria, was discovered in 1987 as an allosteric activator of the cellulose synthase complex in *Gluconacetobacter xylinus* [70]. In general, c-di-GMP stimulates the biosynthesis of adhesins and exopolysaccharide mediated biofilm formation and inhibits bacterial motilities, which controls the switch between the motile planktonic and sessile biofilm-associated lifestyle of bacteria (Figure 2). Moreover, c-di-GMP controls the virulence of animal and plant pathogens, progression through the cell cycle, antibiotic production and other cellular functions [71–73].

Figure 2. Schematic presentation of physiological functions of c-di-GMP. In bacterial cells, c-di-GMP is generated by diguanylate cyclases (DGC) and broken down by specific phosphodiesterases (PDE). As a second messenger, low levels of c-di-GMP can promote motility by upregulating flagellar expression, assembly or interfering with flagellar motor function and are required for the expression of acute virulence genes. High levels of c-di-GMP however favor sessility and stimulate the synthesis of various matrix exopolysaccharides, such as Pel (mediated by PelD) and alginate (mediated by Alg44) [71,74,75].

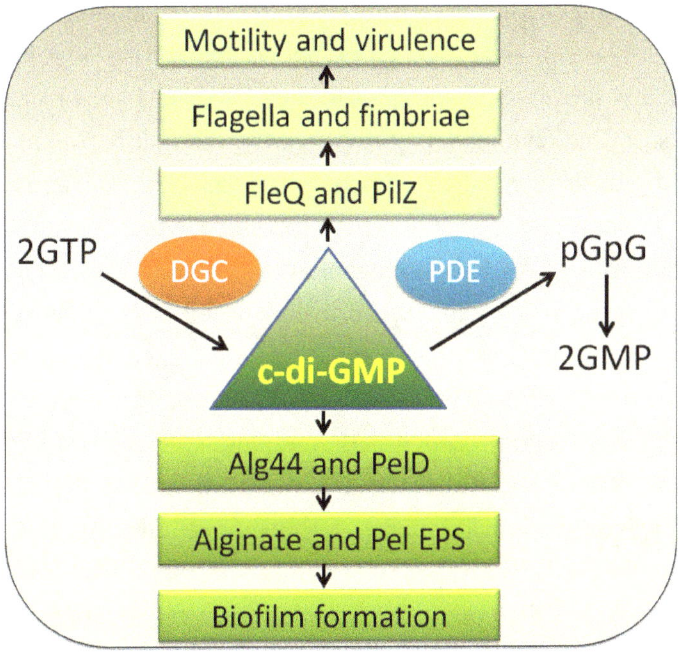

C-di-GMP is synthesized from two molecules of GTP by diguanylate cyclases (DGC) containing GGDEF domains and is broken down into 5'-phosphoguanylyl-(3'-5')-guanosine (pGpG) by specific phosphodiesterases (PDE) containing EAL or HD-GYP domains; pGpG is subsequently split into two GMP molecules (Figure 2). These conserved domains are essential for their corresponding enzymatic activities [71,73]. Whole genome sequencing analysis has revealed that *E. coli* has 19 GGDEF and 17 EAL domain proteins while *Vibrio vulnificus* encodes up to 100 of those proteins [73]. In *P. aeruginosa*, there are 41 of such proteins, including 3 HD-GYP, 17 GGDEF, and 5 EAL domain proteins, as well as 16 proteins with both GGDEF and EAL domains [76,77]. Most proteins that contain these domains show a multi-modular arrangement. In addition to GGDEF, EAL, or HD-GYP domains, there are a variety of sensory domains such as signal receiver or transmission domains and response regulator domains that are likely to receive signals from the environment [73]. These signals are proposed to be translated as an alteration of the enzymatic activity that would result in local or global fluctuations in c-di-GMP levels, which in turn would lead to behavioral adjustments [1].

The mechanism of c-di-GMP signal transduction generally involves the first recognition of c-di-GMP signal and the subsequent phenotypic regulation. To exert its function, c-di-GMP binds to effectors (c-di-GMP receptors), and by allostery, alters their structure and output function [71]. Those c-di-GMP effectors are highly diverse, among which the PilZ family of proteins are the best-studied. In *P. aeruginosa*, at least four c-di-GMP effectors are present including FleQ, PelD, Alg44, and PilZ. FleQ is a c-di-GMP-binding transcription factor, which generally functions as an activator in flagella biosynthesis. Yet it can also act as a repressor for the transcription of the *pel* gene cluster in the absence of c-di-GMP and an activator upon c-di-GMP binding [78]. PelD is part of *pel* operon of *P. aeruginosa* activated by direct binding c-di-GMP through a site that resembles the I site motif in GGDEF domain proteins [79]. Alg44 is another PilZ family protein involved in alginate synthesis [80]. PilZ is a type IV fimbrial biogenesis protein involved in twitching [81]. However, it is still a mystery if there is any receptor protein of c-di-GMP accounting for Psl polysaccharide production.

Intriguingly, c-di-GMP signaling has been shown to be involved in mediating the formation of small colony variants (SCV) in *P. aeruginosa* [82–85]. A phenotypic variant regulator (PvrR), containing a conserved EAL domain, involved in the hydrolysis of c-di-GMP, has been identified to control the wild type-like, antibiotics susceptible revertants [7]. Importantly, the link between antibiotic resistance and biofilm formation was firstly evidenced by the study of such RSCV (rugose small colony variant) phenotypes. The *wspF* mutant, also displays a SCV phenotype and the underlying mechanism was attributed to the activation of a GGDEF domain containing protein WspR [82,83]. Another interesting characteristic of *P. aeruginosa* SCVs with regard to c-di-GMP is the overexpression of Pel and Psl polysaccharides, leading to the auto-aggregation and hyper adherence phenotypes [86–88]. This feature seems to be a common theme for SCV of *P. aeruginosa*.

As we mentioned above, *P. aeruginosa* PAO1 contains 41 DGCs and PDEs that cooperatively mediate the overall concentration of c-di-GMP and finally modulate the EPS production. Original studies have led to extensive functional characterization of c-di-GMP-modulating enzymes and their roles in biofilm formation [7,68,76,77,83,84,88–94]. Based on these results, an emerging model holds that distinct or localized c-di-GMP pools may exist to reciprocally regulate motility and biofilm formation. Different lines of evidences have added bonus points to this hypothesis. One example is *arr* (aminoglycoside response regulator) which is predicted to encode an inner-membrane PDE and seems to be essential for the induction of biofilm formation while contributing to biofilm-specific aminoglycoside resistance [68]. The other example is RoeA, a DGC that plays different roles in regulating motility and biofilm formation as compared to another DGC SadC [94]. Specifically, RoeA contributes to biofilm formation by mainly controlling polysaccharide production, whereas SadC strongly impacts on flagella motility. The studies of these proteins suggested that there were subcellular pools of c-di-GMP in the cell and such pools could be produced from several ways such as specific localization and/or activation of DGCs, limitation of c-di-GMP diffusion through its effectors and/or degraders and the availability of c-di-GMP effectors [71,94].

3.2. GacA/GacS Two-Component Systems

Expression of the *pel and psl* genes for exopolysaccharide production in *P. aeruginosa* can be regulated by GacA/GacS two-component system. One of the mechanisms involves two histidine kinases, RetS and LadS that act in opposing ways on the GacA/GacS two-component system (Figure 3). The GacA/GacS system subsequently controls the transcription of two small regulatory RNAs (sRNAs), *rsmY* and *rsmZ*, leading to the decrease or increase in the translation of the *pel* or *psl* operon [95,96]. Transcriptomic analysis showed that GacS directly controls the transcription of *rsmY* and *rsmZ*, thereby antagonizing the activities of RNA-binding translational regulator, RsmA, to control the expression of over 500 genes [97,98]. It was further proved that upon binding of RsmA with the promoter of the *psl* operon, the region spanning the ribosome binding site of *psl* mRNA forms a secondary stem-loop structure that prevents ribosome access and the subsequent translation, without affecting transcription [99]. This translational control of Psl expression constitutes a novel example of translational repression by RsmA.

Furthermore, analyses of the mRNA levels using microarray analysis have shown that RetS is required for the expression of genes involved in virulence such as the type III secretion system (T3SS), yet acts as a repressor for the type VI secretion system (T6SS) and genes involved in exopolysaccharide synthesis, leading to the inhibition of biofilm formation [96]. This defines RetS as a pleiotropic regulator of multiple virulence phenotypes that mediates the activation of genes involved in acute infections and the repression of genes associated with chronic persistence [96]. A recent report showed that RetS could directly interact with GacS to modulate its phosphorylation state. [100]. During the acute infection phase, RetS interacts with GacS to form heterodimers, blocking GacS autophosphorylation and leading to reduction in *rsmZ* expression. Finally, RsmA lacking RsmZ will promote the translation of genes required for acute virulence factors. While sensing unknown environmental signals, GacS and RetS each form homodimers, allowing GacS autophosphorylation and subsequent phosphorylation of GacA, finally resulting in the expression of genes involved in chronic infections (Figure 3) [100]. Very recently, a novel RetS interacting protein, PA1611 was identified and characterized as able to both promote biofilm formation and repress T3SS and swarming motility [101], adding complexity to the classical GacS/GacA regulatory cascade. On the other hand, LadS was found to antagonize the effect of RetS, contributing to the repression of T3SS and the activation of genes necessary for exopolysaccharide production promoting biofilm formation [95] (Figure 3). However, there is a paucity of information detailing whether LadS affects the GacS or RetS. Interestingly, one *P. aeruginosa* reference strain PA14 was found to have a natural *ladS* mutation, explaining why PA14 exhibits increased virulence and displays attenuated biofilm formation as compared to PAO1 [102].

In addition, the histidine phosphotransfer (Hpt) protein HptB signaling pathway was found to control biofilm formation and T3SS, and fine-tunes *P. aeruginosa* pathogenesis [103]. Typically, Hpt protein acts as a phosphorylation relay that transits the activation signal from a sensor kinase to the cognate response regulator. Bordi and co-workers found that in contrast to the double control of *rsmYZ* expression by RetS, HptB exclusively regulates *rsmY* expression. Importantly, in this study, they demonstrated a redundant effect of the two sRNAs on T3SS gene expression, while the influence on *pel* gene expression is additive, further underpinning the novel mechanism of fine-tuned regulation of gene expression [103].

Figure 3. Schematic diagram of the two-component system regulation of biofilm formation and virulence in *P. aeruginosa* [95,96,100,104]. Unknown environmental cues received by the input domains of the three membrane-associated sensor kinases (GacS, LadS and RetS) activate or repress the expression of genes necessary for acute or chronic infection. Free regulatory protein RsmA can bind to the promoter regions of multiple genes, thus repressing expression of biofilm associated genes such as *psl* locus and enhancing bacterial motility and the production of several acute virulence factors (**Red lines**). When the response regulator GacA is phosphorylated by the upstream sensor kinase GacS, the production of small regulatory RNAs RsmZ and RsmY are stimulated, followed by the titrating to RsmA protein, which ultimately de-represses the expression of biofilm-related genes and represses the production of virulence-related factors (**Green lines**). The signaling cascade going through RetS, operating in an opposite manner to that of GacS and LadS, generates more free RsmA, resulting in T3SS activation and biofilm repression (**Red lines**). T3SS, type 3 secretion system. EPS, exopolysacchrides. P means phosphorylated state of GacA. **Dark red circle** indicates RsmA protein, which binds to the genomic DNA without environmental stimulation and binds to RsmY or RsmZ when the upstream pathways are activated.

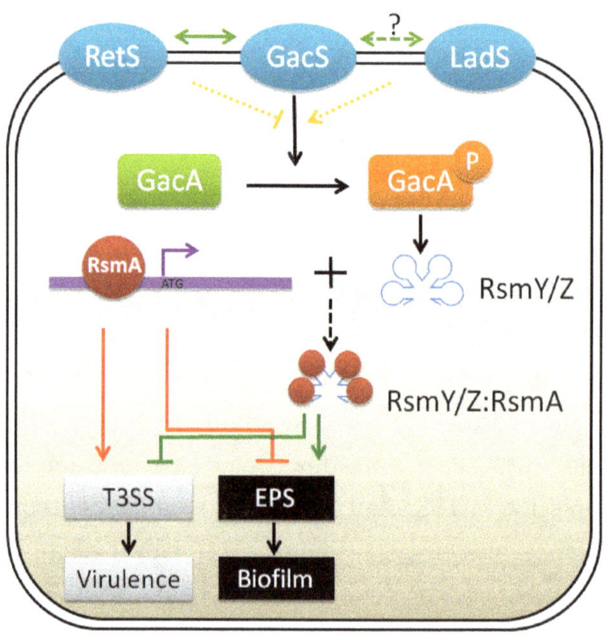

3.3. Quorum Sensing

Quorum sensing (QS), known as bacterial cell-cell communication system, represents another multicellular activity which involves the production, secretion, and detection of molecules called autoinducers (AIs) to modulate behaviors of the bacterial population [105]. QS provides a strategy for bacteria to detect each other's presence and to regulate gene expression in response to changes of bacterial population density. Up to now, many biological processes have been found to be controlled by QS, such as bioluminescence, biofilm formation, virulence factor expression, antibiotics production, sporulation, and competence for DNA uptake [106,107].

P. aeruginosa employs three quorum sensing signaling systems (LasR/LasI, RhlR/RhlI and PQS) to control cellular processes involved in the production of extracellular virulence factors and to control biofilm formation [108,109]. A large number of genes (as many as 200~300, about 6% of the genome size), including virulence factor genes and genes involved in biofilm development, are activated by two typical, interconnected and homologous acyl-homoserine lactone (AHL) quorum sensing systems, namely the LasR/LasI and RhlR/RhlI systems .

It has been shown that signal molecule 3-oxo-C12-HSL (synthesized by LasI) is necessary for the establishment of a differentiated *P. aeruginosa* biofilm since a *lasI* mutant forms flat, undifferentiated biofilms unlike wild-type biofilms [2] and *lasI* is expressed in a large number of cells during the initial stage of biofilm formation [110]. In contrast, the RhlR/RhlI system was found to be activated during the maturation stage of *P. aeruginosa* biofilm development [111], and might be important for the survival of bacterial cells growing in anaerobic conditions in biofilms [112,113].

In *P. aeruginosa*, quorum sensing regulation of exopolysaccharide was revealed to be mediated by a tyrosine phosphatase named TpbA (PA3885) that is controlled by LasR/I system and negatively regulates Pel polysaccharide production through dephosphorylation of a GGDEF-motif protein, TpbB (PA1120) [114]. This study has generated a common theme that QS seems to be a negative regulator of c-di-GMP signaling. Recently, it was shown that Psl itself could also function as a signaling molecule to stimulate its own expression via two diguanylate cyclases [28], generating a positive feedback circuit that allows efficient biofilm formation. In addition, Gilbert and colleagues found that the QS regulator LasR could bind to the promoter region of the *psl* operon, suggesting that QS can regulate *psl* expression [115]. Furthermore, the release of eDNA was demonstrated to be controlled via AHL- and *Pseudomonas* quinolone signal (PQS)-mediated quorum sensing systems [49].

3.4. Other Types of Regulation

In addition to the typical regulation of biofilm development, the biofilm matrix is also under control by other types of regulation. One example is the metabolic regulation mediated by AlgC, a checkpoint enzyme that coordinates the total amount of exopolysaccharides in *P. aeruginosa* by control of sugar precursors pool for exopolysaccharides synthesis [116] (Figure 4). It was demonstrated that overexpression of one exoplysaccharide could reduce the production of the other. For example, overproduction of Psl led to reduced level of alginate, Pel overexpression resulted in less Psl production, and overproduction of alginate and Psl caused decreased levels of B-band LPS. The enzymatic regulation of exopolysaccharide provided us a very interesting clue about the survival strategy used by *P. aeruginosa* in diverse conditions. It is easily speculated that *P. aeruginosa* produces one major type of exopolysaccharide in certain phases whereas it generates another major type of exopolysaccharide upon changing environments.

As a key polysaccharide for biofilm formation, Psl expression is regulated at multiple levels. In addition to the aforementioned mechanisms, Psl was found to be regulated by RpoS transcriptionally, and post-transcriptionally by RsmA, an RNA binding protein [99]. The transcriptional regulator AmrZ, previously shown to positively regulate twitching motility and alginate synthesis [117], was also shown to directly bind to the promoter region of the *psl* operon to repress its transcription [118]. The

AmrZ-mediated switch from Psl production to alginate production provides another example of acute-to-chronic transition strategy used by *P. aeruginosa*.

Figure 4. Diagram of AlgC-dependent enzymatic regulation of the production of exopolysaccharides in *P. aeruginosa* PAO1 [116]. The product of *algC* gene (AlgC) is a bifunctional enzyme with phosphomannomutase (PMM) and phosphoglucomutase (PMG) activities that can catalyze the conversion of mannose-6-phosphate (mannose-6-P) and glucose-6-phosphate (glucose-6-P) into mannose-1-phosphate (mannos-1-P) and glucose-1-phosphate (glucose-1-P), respectively. The PMM/PMG activity of AlgC is required for the biosynthesis pathways of four exopolysaccharides (Psl, alginate, LPS and Pel) in *P. aeruginosa*. Experimental data showed that overproduction of individual exopolysaccharides reduces synthesis of the other exopolysaccharides, indicating that AlgC is the checkpoint enzyme that limits the production of *P. aeruginosa* exopolysaccharides, influencing the biofilm formation.

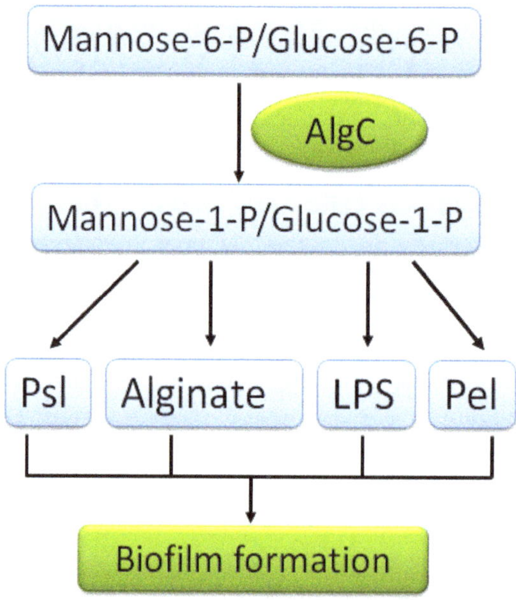

4. Matrix-Driven Strategies against Biofilms

Once biofilms develop into a mature stage, they become extremely difficult to eradicate from infections sites with traditional antimicrobial agents [6]. Agents that inhibit biofilm formation or transform bacteria from biofilm life style to free-living individuals are ideal to eradicate biofilm. The strategies used for anti-biofilm mainly stem from two basic ways: matrix synthesis and its regulatory mechanisms.

For example, disruption of the initial attachment that is dependent on a large array of adhesins would contribute to inhibition of the establishment of biofilms, while the digestion of the EPS matrix may be another method to interfere with biofilm formation. As we mentioned before, DNase I treatment has already shown efficacy in the inhibition of the early development of biofilm [50]. It was also reported that alginate lyase could enhance antibiotic killing of mucoid *P. aeruginosa* in biofilms [119]. In addition, the macrolide antibiotic azithromycin was shown to block alginate formation and quorum sensing signaling [120] and was further reported to improve lung function of CF patients, especially in the subgroup chronically colonized by *Pseudomonas* [121].

Antagonizing the intracellular signaling molecules to control biofilm formation has also been investigated. One example is the identification of furanones, which have shown their ability to inhibit the biofilm formation of *P. aeruginosa in vitro* [122,123]. Molecules of this type have been reported to function through inhibiting the AHL-dependent QS systems in *P. aeruginosa.* Iron has also been employed in distinct aspects to control the formation of biofilms. Singh and his colleagues have identified an innate immunity component, lactoferrin, which prevents *P. aeruginosa* biofilm formation by chelating iron and stimulating the type IV pili-mediated twitching motility [124]. Furthermore, iron salts such as ferric ammonium citrate were found to not only perturb biofilm formation but also disrupt existing biofilms by *P. aeruginosa* [125]. In a screen of co-therapy of antibiotics against *P. aeruginosa,* 14-alpha-lipoyl andrographolide (AL-1), a diterpenoid lactone derivative from the herb *Andrographis paniculata* appeared to inhibit biofilm formation by decreasing EPS production and to sensitize the bacterium to a variety of antibiotics [126].

Recently, it was found that Gram-positive bacterium *Bacillus subtilis* produced a factor that prevented biofilm formation and could break down existing biofilms. The factor was identified to be a mixture of D-leucine, D-methionine, D-tyrosine, and D-tryptophan that could disassemble at nanomolar concentrations. D-amino acid treatment subserved the release of amyloid fibers that linked cells together in the biofilm. In addition, D-amino acids also prevented biofilm formation by *Staphylococcus aureus* and *P. aeruginosa,* indicating it may be a widespread signal for biofilm disassembly [127]. Furthermore, the same group identified another biofilm disassembly compound, norspermidine, which targets directly and specifically with the exopolysaccharide matrix and this biofilm inhibition effect could be enhanced together with D-amino acids and is effective in other bacterial species [128].

5. Perspectives

Accumulating data presented in the recent literature provides valuable insights into the novel roles of the biofilm matrix and its regulatory mechanism in *P. aeruginosa* biofilm formation. A deep understanding of the mechanisms involved in biofilm formation will ultimately shed light on the generation of alternative treatments for *P. aeruginosa* infections. There is no doubt that future studies will reveal additional biofilm matrix components and identify more elaborate regulatory circuits for biofilm formation. Finally, the interaction of the biofilm matrix and the synergistic effects of different anti-biofilm strategies should also be regarded as major concerns.

Acknowledgments

We thank Shiwei Wang, Di Wang, Tianhu Zhao and other students from Ma Lab at Institute of Microbiology, Chinese Academy of Sciences for their help in manuscript preparation. We are also grateful to reviewers for their critical and constructive suggestions. This work was supported by Chinese Academy of Science grant KSCXZ-YW-BR-5 (L.M.), National Natural Science Foundation of China grant 31270177, 31140041 (L.M.), and 31300066 (Q.W.).

References

1. Karatan, E.; Watnick, P. Signals, regulatory networks, and materials that build and break bacterial biofilms. *Microbiol. Mol. Biol Rev.* **2009**, *73*, 310–347.
2. Davies, D.G.; Parsek, M.R.; Pearson, J.P.; Iglewski, B.H.; Costerton, J.W.; Greenberg, E.P. The involvement of cell-to-cell signals in the development of a bacterial biofilm. *Science* **1998**, *280*, 295–298.
3. Hall-Stoodley, L.; Costerton, J.W.; Stoodley, P. Bacterial biofilms: From the natural environment to infectious diseases. *Nat. Rev.* **2004**, *2*, 95–108.
4. Sutherland, I.W. The biofilm matrix—An immobilized but dynamic microbial environment. *Trends Microbiol.* **2001**, *9*, 222–227.
5. Branda, S.S.; Vik, S.; Friedman, L.; Kolter, R. Biofilms: The matrix revisited. *Trends Microbiol.* **2005**, *13*, 20–26.
6. Costerton, J.W.; Stewart, P.S.; Greenberg, E.P. Bacterial biofilms: A common cause of persistent infections. *Science* **1999**, *284*, 1318–1322.
7. Drenkard, E.; Ausubel, F.M. *Pseudomonas* biofilm formation and antibiotic resistance are linked to phenotypic variation. *Nature* **2002**, *416*, 740–743.
8. Flemming, H.C.; Neu, T.R.; Wozniak, D.J. The EPS matrix: The "house of biofilm cells". *J. Bacteriol.* **2007**, *189*, 7945–7947.
9. Ryder, C.; Byrd, M.; Wozniak, D.J. Role of polysaccharides in *Pseudomonas aeruginosa* biofilm development. *Curr. Opin. Microbiol.* **2007**, *10*, 644–648.
10. Parsek, M.R.; Singh, P.K. Bacterial biofilms: An emerging link to disease pathogenesis. *Annu. Rev. Microbiol.* **2003**, *57*, 677–701.
11. Friedman, L.; Kolter, R. Two genetic loci produce distinct carbohydrate-rich structural components of the *Pseudomonas aeruginosa* biofilm matrix. *J. Bacteriol.* **2004**, *186*, 4457–4465.
12. Friedman, L.; Kolter, R. Genes involved in matrix formation in *Pseudomonas aeruginosa* PA14 biofilms. *Mol. Microbiol.* **2004**, *51*, 675–690.
13. Jackson, K.D.; Starkey, M.; Kremer, S.; Parsek, M.R.; Wozniak, D.J. Identification of *psl*, a locus encoding a potential exopolysaccharide that is essential for *Pseudomonas aeruginosa* PAO1 biofilm formation. *J. Bacteriol.* **2004**, *186*, 4466–4475.
14. Colvin, K.M.; Gordon, V.D.; Murakami, K.; Borlee, B.R.; Wozniak, D.J.; Wong, G.C.; Parsek, M.R. The *pel* polysaccharide can serve a structural and protective role in the biofilm matrix of *Pseudomonas aeruginosa*. *PLoS Pathog.* **2010**, *7*, e1001264.
15. Ma, L.; Conover, M.; Lu, H.; Parsek, M.R.; Bayles, K.; Wozniak, D.J. Assembly and development of the *Pseudomonas aeruginosa* biofilm matrix. *PLoS Pathog.* **2009**, *5*, e1000354.
16. Matsukawa, M.; Greenberg, E.P. Putative exopolysaccharide synthesis genes influence *Pseudomonas aeruginosa* biofilm development. *J. Bacteriol.* **2004**, *186*, 4449–4456.
17. Ma, L.; Lu, H.; Sprinkle, A.; Parsek, M.R.; Wozniak, D.J. *Pseudomonas aeruginosa* Psl is a galactose- and mannose-rich exopolysaccharide. *J. Bacteriol.* **2007**, *189*, 8353–8356.
18. Ma, L.; Jackson, K.D.; Landry, R.M.; Parsek, M.R.; Wozniak, D.J. Analysis of *Pseudomonas aeruginosa* conditional *psl* variants reveals roles for the *psl* polysaccharide in adhesion and maintaining biofilm structure postattachment. *J. Bacteriol.* **2006**, *188*, 8213–8221.

19. Yang, L.; Hu, Y.F.; Liu, Y.; Zhang, J.D.; Ulstrup, J.; Molin, S. Distinct roles of extracellular polymeric substances in *Pseudomonas aeruginosa* biofilm development. *Environ. Microbiol.* **2011**, *13*, 1705–1717.
20. Byrd, M.S.; Sadovskaya, I.; Vinogradov, E.; Lu, H.P.; Sprinkle, A.B.; Richardson, S.H.; Ma, L.Y.; Ralston, B.; Parsek, M.R.; Anderson, E.M.; *et al.* Genetic and biochemical analyses of the *Pseudomonas aeruginosa* Psl exopolysaccharide reveal overlapping roles for polysaccharide synthesis enzymes in Psl and LPS production. *Mol. Microbiol.* **2009**, *73*, 622–638.
21. Winsor, G.L.; Lam, D.K.; Fleming, L.; Lo, R.; Whiteside, M.D.; Yu, N.Y.; Hancock, R.E.; Brinkman, F.S. *Pseudomonas* Genome Database: Improved comparative analysis and population genomics capability for *Pseudomonas* genomes. *Nucleic Acids Res.* **2011**, *39*, D596–D600.
22. Campisano, A.; Schroeder, C.; Schemionek, M.; Overhage, J.; Rehm, B.H. PslD is a secreted protein required for biofilm formation by *Pseudomonas aeruginosa*. *Appl. Environ. Microbiol.* **2006**, *72*, 3066–3068.
23. Overhage, J.; Schemionek, M.; Webb, J.S.; Rehm, B.H. Expression of the *psl* operon in *Pseudomonas aeruginosa* PAO1 biofilms: PslA performs an essential function in biofilm formation. *Appl. Environ. Microbiol.* **2005**, *71*, 4407–4413.
24. Lee, H.J.; Chang, H.Y.; Venkatesan, N.; Peng, H.L. Identification of amino acid residues important for the phosphomannose isomerase activity of PslB in *Pseudomonas aeruginosa* PAO1. *FEBS Lett.* **2008**, *582*, 3479–3483.
25. Wang, S.; Parsek, M.R.; Wozniak, D.J.; Ma, L.Z. A spider web strategy of type IV pili-mediated migration to build a fibre-like Psl polysaccharide matrix in *Pseudomonas aeruginosa* biofilms. *Environ. Microbiol.* **2013**, *15*, 2238–2253.
26. Zhao, K.; Tseng, B.S.; Beckerman, B.; Jin, F.; Gibiansky, M.L.; Harrison, J.J.; Luijten, E.; Parsek, M.R.; Wong, G.C. Psl trails guide exploration and microcolony formation in *Pseudomonas aeruginosa* biofilms. *Nature* **2013**, *497*, 388–391.
27. Stoodley, P.; Sauer, K.; Davies, D.G.; Costerton, J.W. Biofilms as complex differentiated communities. *Annu. Rev. Microbiol.* **2002**, *56*, 187–209.
28. Irie, Y.; Borlee, B.R.; O'Connor, J.R.; Hill, P.J.; Harwood, C.S.; Wozniak, D.J.; Parsek, M.R. Self-produced exopolysaccharide is a signal that stimulates biofilm formation in *Pseudomonas aeruginosa*. *Proc. Natl. Acad. Sci. USA* **2012**, *109*, 20632–20636.
29. Byrd, M.S.; Pang, B.; Mishra, M.; Swords, W.E.; Wozniak, D.J. The *Pseudomonas aeruginosa* exopolysaccharide Psl facilitates surface adherence and NF-kappaB activation in A549 cells. *mBiosphere* **2010**, *1*, doi:10.1128/mBio.00140-10.
30. Mishra, M.; Byrd, M.S.; Sergeant, S.; Azad, A.K.; Parsek, M.R.; McPhail, L.; Schlesinger, L.S.; Wozniak, D.J. *Pseudomonas aeruginosa* Psl polysaccharide reduces neutrophil phagocytosis and the oxidative response by limiting complement-mediated opsonization. *Cell Microbiol.* **2012**, *14*, 95–106.
31. Zegans, M.E.; Wozniak, D.; Griffin, E.; Toutain-Kidd, C.M.; Hammond, J.H.; Garfoot, A.; Lam, J.S. *Pseudomonas aeruginosa* exopolysaccharide Psl promotes resistance to the biofilm inhibitor polysorbate 80. *Antimicrob. Agents Chemother.***2012**, *56*, 4112–4122.

32. Tseng, B.S.; Zhang, W.; Harrison, J.J.; Quach, T.P.; Song, J.L.; Penterman, J.; Singh, P.K.; Chopp, D.L.; Packman, A.I.; Parsek, M.R. The extracellular matrix protects *Pseudomonas aeruginosa* biofilms by limiting the penetration of tobramycin. *Environ. Microbiol.* **2013**, *15*, 2865–2878.
33. Billings, N.; Ramirez Millan, M.; Caldara, M.; Rusconi, R.; Tarasova, Y.; Stocker, R.; Ribbeck, K. The extracellular matrix component Psl provides fast-acting antibiotic defense in *Pseudomonas aeruginosa* biofilms. *PLoS Pathog.* **2013**, *9*, e1003526.
34. Vasseur, P.; Vallet-Gely, I.; Soscia, C.; Genin, S.; Filloux, A. The *pel* genes of the *Pseudomonas aeruginosa* PAK strain are involved at early and late stages of biofilm formation. *Microbiology* **2005**, *151*, 985–997.
35. Sadovskaya, I.; Vinogradov, E.; Li, J.J.; Hachani, A.; Kowalska, K.; Filloux, A. High-level antibiotic resistance in *Pseudomonas aeruginosa* biofilm: The *ndvB* gene is involved in the production of highly glycerol-phosphorylated beta-(1->3)-glucans, which bind aminoglycosides. *Glycobiology* **2010**, *20*, 895–904.
36. Coulon, C.; Vinogradov, E.; Filloux, A.; Sadovskaya, I. Chemical analysis of cellular and extracellular carbohydrates of a biofilm-forming strain *Pseudomonas aeruginosa* PA14. *PLoS One* **2010**, *5*, e14220.
37. Ghafoor, A.; Jordens, Z.; Rehm, B.H. Role of PelF in Pel polysaccharide biosynthesis in *Pseudomonas aeruginosa*. *Appl. Environ. Microbiol.* **2013**, *79*, 2968–2978.
38. Colvin, K.M.; Alnabelseya, N.; Baker, P.; Whitney, J.C.; Howell, P.L.; Parsek, M.R. PelA deacetylase activity is required for Pel polysaccharide synthesis in *Pseudomonas aeruginosa*. *J. Bacteriol.* **2013**, *195*, 2329–2339.
39. Govan, J.R.; Deretic, V. Microbial pathogenesis in cystic fibrosis: Mucoid *Pseudomonas aeruginosa* and *Burkholderia cepacia*. *Microbiol. Rev.* **1996**, *60*, 539–574.
40. Wozniak, D.J.; Wyckoff, T.J.; Starkey, M.; Keyser, R.; Azadi, P.; O'Toole, G.A.; Parsek, M.R. Alginate is not a significant component of the extracellular polysaccharide matrix of PA14 and PAO1 *Pseudomonas aeruginosa* biofilms. *Proc. Natl. Acad. Sci. USA* **2003**, *100*, 7907–7912.
41. Sutherland, I.W. Biofilm exopolysaccharides: A strong and sticky framework. *Microbiology* **2001**, *147*, 3–9.
42. Ma, L.; Wang, S.; Wang, D.; Parsek, M.R.; Wozniak, D.J. The roles of biofilm matrix polysaccharide Psl in mucoid *Pseudomonas aeruginosa* biofilms. *FEMS Immunol. Med. Microbiol.* **2012**, *65*, 377–380.
43. Yang, L.; Hengzhuang, W.; Wu, H.; Damkiaer, S.; Jochumsen, N.; Song, Z.; Givskov, M.; Hoiby, N.; Molin, S. Polysaccharides serve as scaffold of biofilms formed by mucoid *Pseudomonas aeruginosa*. *FEMS Immunol. Med. Microbiol.* **2012**, *65*, 366–376.
44. Leid, J.G.; Willson, C.J.; Shirtliff, M.E.; Hassett, D.J.; Parsek, M.R.; Jeffers, A.K. The exopolysaccharide alginate protects *Pseudomonas aeruginosa* biofilm bacteria from IFN-gamma-mediated macrophage killing. *J. Immunol.* **2005**, *175*, 7512–7518.
45. Simpson, J.A.; Smith, S.E.; Dean, R.T. Alginate inhibition of the uptake of *Pseudomonas aeruginosa* by macrophages. *J. Gen. Microbiol.* **1988**, *134*, 29–36.
46. Simpson, J.A.; Smith, S.E.; Dean, R.T. Alginate may accumulate in cystic-fibrosis lung because the enzymatic and free-radical capacities of phagocytic-cells are inadequate for its degradation. *Biochem. Mol. Biol Int.* **1993**, *30*, 1021–1034.

47. Simpson, J.A.; Smith, S.E.; Dean, R.T. Scavenging by alginate of free-radicals released by macrophages. *Free Radic. Bio. Med.* **1989**, *6*, 347–353.

48. Bragonzi, A.; Paroni, M.; Nonis, A.; Cramer, N.; Montanari, S.; Rejman, J.; Di Serio, C.; Doring, G.; Tummler, B. *Pseudomonas aeruginosa* microevolution during cystic fibrosis lung infection establishes clones with adapted virulence. *Am. J. Resp. Crit. Care* **2009**, *180*, 138–145.

49. Allesen-Holm, M.; Barken, K.B.; Yang, L.; Klausen, M.; Webb, J.S.; Kjelleberg, S.; Molin, S.; Givskov, M.; Tolker-Nielsen, T. A characterization of DNA release in *Pseudomonas aeruginosa* cultures and biofilms. *Mol. Microbiol.* **2006**, *59*, 1114–1128.

50. Whitchurch, C.B.; Tolker-Nielsen, T.; Ragas, P.C.; Mattick, J.S. Extracellular DNA required for bacterial biofilm formation. *Science* **2002**, *295*, 1487.

51. Webb, J.S.; Thompson, L.S.; James, S.; Charlton, T.; Tolker-Nielsen, T.; Koch, B.; Givskov, M.; Kjelleberg, S. Cell death in *Pseudomonas aeruginosa* biofilm development. *J. Bacteriol.* **2003**, *185*, 4585–4592.

52. Spoering, A.L.; Gilmore, M.S. Quorum sensing and DNA release in bacterial biofilms. *Curr. Opin. Microbiol.* **2006**, *9*, 133–137.

53. Mulcahy, H.; Charron-Mazenod, L.; Lewenza, S. Extracellular DNA chelates cations and induces antibiotic resistance in *Pseudomonas aeruginosa* biofilms. *PLoS Pathog.* **2008**, *4*, e1000213.

54. Finkel, S.E.; Kolter, R. DNA as a nutrient: Novel role for bacterial competence gene homologs. *J. Bacteriol.* **2001**, *183*, 6288–6293.

55. Mulcahy, H.; Charron-Mazenod, L.; Lewenza, S. *Pseudomonas aeruginosa* produces an extracellular deoxyribonuclease that is required for utilization of DNA as a nutrient source. *Environ. Microbiol.* **2010**, *12*, 1621–1629.

56. Parks, Q.M.; Young, R.L.; Poch, K.R.; Malcolm, K.C.; Vasil, M.L.; Nick, J.A. Neutrophil enhancement of *Pseudomonas aeruginosa* biofilm development: Human F-actin and DNA as targets for therapy. *J. Med. Microbiol.* **2009**, *58*, 492–502.

57. Fuxman Bass, J.I.; Russo, D.M.; Gabelloni, M.L.; Geffner, J.R.; Giordano, M.; Catalano, M.; Zorreguieta, A.; Trevani, A.S. Extracellular DNA: A major proinflammatory component of *Pseudomonas aeruginosa* biofilms. *J. Immunol.* **2010**, *184*, 6386–6395.

58. Gloag, E.S.; Turnbull, L.; Huang, A.; Vallotton, P.; Wang, H.; Nolan, L.M.; Mililli, L.; Hunt, C.; Lu, J.; Osvath, S.R.; *et al.* Self-organization of bacterial biofilms is facilitated by extracellular DNA. *Proc. Natl. Acad. Sci. USA* **2013**, *110*, 11541–11546.

59. Mann, E.E.; Wozniak, D.J. *Pseudomonas* biofilm matrix composition and niche biology. *Fems. Microbiol. Rev.* **2012**, *36*, 893–916.

60. O'Toole, G.A.; Kolter, R. Flagellar and twitching motility are necessary for *Pseudomonas aeruginosa* biofilm development. *Mol. Microbiol.* **1998**, *30*, 295–304.

61. Skerker, J.M.; Berg, H.C. Direct observation of extension and retraction of type IV pili. *Proc. Natl. Acad. Sci. USA* **2001**, *98*, 6901–6904.

62. Klausen, M.; Heydorn, A.; Ragas, P.; Lambertsen, L.; Aaes-Jorgensen, A.; Molin, S.; Tolker-Nielsen, T. Biofilm formation by *Pseudomonas aeruginosa* wild type, flagella and type IV pili mutants. *Mol. Microbiol.* **2003**, *48*, 1511–1524.

63. Borlee, B.R.; Goldman, A.D.; Murakami, K.; Samudrala, R.; Wozniak, D.J.; Parsek, M.R. *Pseudomonas aeruginosa* uses a cyclic-di-GMP-regulated adhesin to reinforce the biofilm extracellular matrix. *Mol. Microbiol.* **2010**, *75*, 827–842.
64. Vallet, I.; Olson, J.W.; Lory, S.; Lazdunski, A.; Filloux, A. The chaperone/usher pathways of *Pseudomonas aeruginosa*: Identification of fimbrial gene clusters (*cup*) and their involvement in biofilm formation. *Proc. Natl. Acad. Sci. USA* **2001**, *98*, 6911–6916.
65. Giraud, C.; Bernard, C.S.; Calderon, V.; Yang, L.; Filloux, A.; Molin, S.; Fichant, G.; Bordi, C.; De Bentzmann, S. The PprA-PprB two-component system activates CupE, the first non-archetypal *Pseudomonas aeruginosa* chaperone-usher pathway system assembling fimbriae. *Environ. Microbiol.* **2011**, *13*, 666–683.
66. Kulasekara, H.D.; Ventre, I.; Kulasekara, B.R.; Lazdunski, A.; Filloux, A.; Lory, S. A novel two-component system controls the expression of *Pseudomonas aeruginosa* fimbrial *cup* genes. *Mol. Microbiol.* **2005**, *55*, 368–380.
67. Ruer, S.; Stender, S.; Filloux, A.; De Bentzmann, S. Assembly of fimbrial structures in *Pseudomonas aeruginosa*: Functionality and specificity of chaperone-usher machineries. *J. Bacteriol.* **2007**, *189*, 3547–3555.
68. Hoffman, L.R.; D'Argenio, D.A.; MacCoss, M.J.; Zhang, Z.; Jones, R.A.; Miller, S.I. Aminoglycoside antibiotics induce bacterial biofilm formation. *Nature* **2005**, *436*, 1171–1175.
69. Whiteley, M.; Bangera, M.G.; Bumgarner, R.E.; Parsek, M.R.; Teitzel, G.M.; Lory, S.; Greenberg, E.P. Gene expression in *Pseudomonas aeruginosa* biofilms. *Nature* **2001**, *413*, 860–864.
70. Ross, P.; Weinhouse, H.; Aloni, Y.; Michaeli, D.; Weinberger-Ohana, P.; Mayer, R.; Braun, S.; De Vroom, E.; Van der Marel, G.A.; Van Boom, J.H.; *et al.* Regulation of cellulose synthesis in *Acetobacter xylinum* by cyclic diguanylic acid. *Nature* **1987**, *325*, 279–281.
71. Hengge, R. Principles of c-di-GMP signalling in bacteria. *Nat. Rev.* **2009**, *7*, 263–273.
72. Cotter, P.A.; Stibitz, S. c-di-GMP-mediated regulation of virulence and biofilm formation. *Curr. Opin. Microbiol.* **2007**, *10*, 17–23.
73. Jenal, U.; Malone, J. Mechanisms of cyclic-di-GMP signaling in bacteria. *Annu. Rev. Genet.* **2006**, *40*, 385–407.
74. Merighi, M.; Lee, V.T.; Hyodo, M.; Hayakawa, Y.; Lory, S. The second messenger bis-(3'-5')-cyclic-GMP and its PilZ domain-containing receptor Alg44 are required for alginate biosynthesis in *Pseudomonas aeruginosa*. *Mol. Microbiol.* **2007**, *65*, 876–895.
75. Romling, U.; Galperin, M.Y.; Gomelsky, M. Cyclic di-GMP: The first 25 years of a universal bacterial second messenger. *Microbiol. Mol. Biol Rev.* **2013**, *77*, 1–52.
76. Ryan, R.P.; Lucey, J.; O'Donovan, K.; McCarthy, Y.; Yang, L.; Tolker-Nielsen, T.; Dow, J.M. HD-GYP domain proteins regulate biofilm formation and virulence in *Pseudomonas aeruginosa*. *Environ. Microbiol.* **2009**, *11*, 1126–1136.
77. Kulasakara, H.; Lee, V.; Brencic, A.; Liberati, N.; Urbach, J.; Miyata, S.; Lee, D.G.; Neely, A.N.; Hyodo, M.; Hayakawa, Y.; *et al.* Analysis of *Pseudomonas aeruginosa* diguanylate cyclases and phosphodiesterases reveals a role for bis-(3'-5')-cyclic-GMP in virulence. *Proc. Natl. Acad. Sci. USA* **2006**, *103*, 2839–2844.

78. Hickman, J.W.; Harwood, C.S. Identification of FleQ from *Pseudomonas aeruginosa* as a c-di-GMP-responsive transcription factor. *Mol. Microbiol.* **2008**, *69*, 376–389.
79. Lee, V.T.; Matewish, J.M.; Kessler, J.L.; Hyodo, M.; Hayakawa, Y.; Lory, S. A cyclic-di-GMP receptor required for bacterial exopolysaccharide production. *Mol. Microbiol.* **2007**, *65*, 1474–1484.
80. Remminghorst, U.; Rehm, B.H. Alg44, a unique protein required for alginate biosynthesis in *Pseudomonas aeruginosa. FEBS Lett.* **2006**, *580*, 3883–3888.
81. Alm, R.A.; Bodero, A.J.; Free, P.D.; Mattick, J.S. Identification of a novel gene, *pilZ*, essential for type 4 fimbrial biogenesis in *Pseudomonas aeruginosa. J. Bacteriol.* **1996**, *178*, 46–53.
82. D'Argenio, D.A.; Calfee, M.W.; Rainey, P.B.; Pesci, E.C. Autolysis and autoaggregation in *Pseudomonas aeruginosa* colony morphology mutants. *J. Bacteriol.* **2002**, *184*, 6481–6489.
83. Hickman, J.W.; Tifrea, D.F.; Harwood, C.S. A chemosensory system that regulates biofilm formation through modulation of cyclic diguanylate levels. *Proc. Natl. Acad. Sci. USA* **2005**, *102*, 14422–14427.
84. Kuchma, S.L.; Brothers, K.M.; Merritt, J.H.; Liberati, N.T.; Ausubel, F.M.; O'Toole, G.A. BifA, a cyclic-Di-GMP phosphodiesterase, inversely regulates biofilm formation and swarming motility by *Pseudomonas aeruginosa* PA14. *J. Bacteriol.* **2007**, *189*, 8165–8178.
85. Meissner, A.; Wild, V.; Simm, R.; Rohde, M.; Erck, C.; Bredenbruch, F.; Morr, M.; Romling, U.; Haussler, S. *Pseudomonas aeruginosa cupA*-encoded fimbriae expression is regulated by a GGDEF and EAL domain-dependent modulation of the intracellular level of cyclic diguanylate. *Environ. Microbiol.* **2007**, *9*, 2475–2485.
86. Kirisits, M.J.; Prost, L.; Starkey, M.; Parsek, M.R. Characterization of colony morphology variants isolated from *Pseudomonas aeruginosa* biofilms. *Appl. Environ. Microbiol.* **2005**, *71*, 4809–4821.
87. Starkey, M.; Hickman, J.H.; Ma, L.; Zhang, N.; De Long, S.; Hinz, A.; Palacios, S.; Manoil, C.; Kirisits, M.J.; Starner, T.D.; *et al. Pseudomonas aeruginosa* rugose small-colony variants have adaptations that likely promote persistence in the cystic fibrosis lung. *J. Bacteriol.* **2009**, *191*, 3492–3503.
88. Malone, J.G.; Jaeger, T.; Spangler, C.; Ritz, D.; Spang, A.; Arrieumerlou, C.; Kaever, V.; Landmann, R.; Jenal, U. YfiBNR mediates cyclic di-GMP dependent small colony variant formation and persistence in *Pseudomonas aeruginosa. PLoS Pathog.* **2010**, *6*, e1000804.
89. Choy, W.K.; Zhou, L.; Syn, C.K.; Zhang, L.H.; Swarup, S. MorA defines a new class of regulators affecting flagellar development and biofilm formation in diverse *Pseudomonas* species. *J. Bacteriol.* **2004**, *186*, 7221–7228.
90. Kazmierczak, B.I.; Lebron, M.B.; Murray, T.S. Analysis of FimX, a phosphodiesterase that governs twitching motility in *Pseudomonas aeruginosa. Mol. Microbiol.* **2006**, *60*, 1026–1043.
91. Klebensberger, J.; Birkenmaier, A.; Geffers, R.; Kjelleberg, S.; Philipp, B. SiaA and SiaD are essential for inducing autoaggregation as a specific response to detergent stress in *Pseudomonas aeruginosa. Environ. Microbiol.* **2009**, *11*, 3073–3086.
92. Merritt, J.H.; Brothers, K.M.; Kuchma, S.L.; O'Toole, G.A. SadC reciprocally influences biofilm formation and swarming motility via modulation of exopolysaccharide production and flagellar function. *J. Bacteriol.* **2007**, *189*, 8154–8164.
93. Roy, A.B.; Petrova, O.E.; Sauer, K. The phosphodiesterase DipA (PA5017) is essential for *Pseudomonas aeruginosa* biofilm dispersion. *J. Bacteriol.* **2012**, *194*, 2904–2915.

94. Merritt, J.H.; Ha, D.G.; Cowles, K.N.; Lu, W.; Morales, D.K.; Rabinowitz, J.; Gitai, Z.; O'Toole, G.A. Specific control of *Pseudomonas aeruginosa* surface-associated behaviors by two c-di-GMP diguanylate cyclases. *mBiosphere* **2010**, *1*, doi:10.1128/mBio.00183-10.
95. Ventre, I.; Goodman, A.L.; Vallet-Gely, I.; Vasseur, P.; Soscia, C.; Molin, S.; Bleves, S.; Lazdunski, A.; Lory, S.; Filloux, A. Multiple sensors control reciprocal expression of *Pseudomonas aeruginosa* regulatory RNA and virulence genes. *Proc. Natl. Acad. Sci. USA* **2006**, *103*, 171–176.
96. Goodman, A.L.; Kulasekara, B.; Rietsch, A.; Boyd, D.; Smith, R.S.; Lory, S. A signaling network reciprocally regulates genes associated with acute infection and chronic persistence in *Pseudomonas aeruginosa*. *Dev. cell* **2004**, *7*, 745–754.
97. Brencic, A.; McFarland, K.A.; McManus, H.R.; Castang, S.; Mogno, I.; Dove, S.L.; Lory, S. The GacS/GacA signal transduction system of *Pseudomonas aeruginosa* acts exclusively through its control over the transcription of the RsmY and RsmZ regulatory small RNAs. *Mol. Microbiol.* **2009**, *73*, 434–445.
98. Brencic, A.; Lory, S. Determination of the regulon and identification of novel mRNA targets of *Pseudomonas aeruginosa* RsmA. *Mol. Microbiol.* **2009**, *72*, 612–632.
99. Irie, Y.; Starkey, M.; Edwards, A.N.; Wozniak, D.J.; Romeo, T.; Parsek, M.R. *Pseudomonas aeruginosa* biofilm matrix polysaccharide Psl is regulated transcriptionally by RpoS and post-transcriptionally by RsmA. *Mol. Microbiol.* **2010**, *78*, 158–172.
100. Goodman, A.L.; Merighi, M.; Hyodo, M.; Ventre, I.; Filloux, A.; Lory, S. Direct interaction between sensor kinase proteins mediates acute and chronic disease phenotypes in a bacterial pathogen. *Genes Dev.* **2009**, *23*, 249–259.
101. Kong, W.; Chen, L.; Zhao, J.; Shen, T.; Surette, M.G.; Shen, L.; Duan, K. Hybrid sensor kinase PA1611 in *Pseudomonas aeruginosa* regulates transitions between acute and chronic infection through direct interaction with RetS. *Mol. Microbiol.* **2013**, *88*, 784–797.
102. Mikkelsen, H.; McMullan, R.; Filloux, A. The *Pseudomonas aeruginosa* reference strain PA14 displays increased virulence due to a mutation in *ladS*. *PLoS One* **2011**, *6*, e29113.
103. Bordi, C.; Lamy, M.C.; Ventre, I.; Termine, E.; Hachani, A.; Fillet, S.; Roche, B.; Bleves, S.; Mejean, V.; Lazdunski, A.; *et al.* Regulatory RNAs and the HptB/RetS signalling pathways fine-tune *Pseudomonas aeruginosa* pathogenesis. *Mol. Microbiol.* **2010**, *76*, 1427–1443.
104. Jimenez, P.N.; Koch, G.; Thompson, J.A.; Xavier, K.B.; Cool, R.H.; Quax, W.J. The multiple signaling systems regulating virulence in *Pseudomonas aeruginosa*. *Microbiol. Mol. Biol Rev.* **2012**, *76*, 46–65.
105. Camilli, A.; Bassler, B.L. Bacterial small-molecule signaling pathways. *Science* **2006**, *311*, 1113–1116.
106. Fuqua, C.; Winans, S.C.; Greenberg, E.P. Census and consensus in bacterial ecosystems: The LuxR-LuxI family of quorum-sensing transcriptional regulators. *Annu. Rev. Microbiol.* **1996**, *50*, 727–751.
107. Waters, C.M.; Bassler, B.L. Quorum sensing: Cell-to-cell communication in bacteria. *Annu. Rev. Cell Dev. Bio.* **2005**, *21*, 319–346.
108. Schuster, M.; Greenberg, E.P. A network of networks: Quorum-sensing gene regulation in *Pseudomonas aeruginosa*. *Int. J. Med. Microbiol.* **2006**, *296*, 73–81.

109. Venturi, V. Regulation of quorum sensing in *Pseudomonas*. *FEMS Microbiol. Rev.* **2006**, *30*, 274–291.
110. De Kievit, T.R.; Gillis, R.; Marx, S.; Brown, C.; Iglewski, B.H. Quorum-sensing genes in *Pseudomonas aeruginosa* biofilms: Their role and expression patterns. *Appl. Environ.Microbiol.* **2001**, *67*, 1865–1873.
111. Sauer, K.; Camper, A.K.; Ehrlich, G.D.; Costerton, J.W.; Davies, D.G. *Pseudomonas aeruginosa* displays multiple phenotypes during development as a biofilm. *J. Bacteriol.* **2002**, *184*, 1140–1154.
112. Yoon, S.S.; Hennigan, R.F.; Hilliard, G.M.; Ochsner, U.A.; Parvatiyar, K.; Kamani, M.C.; Allen, H.L.; DeKievit, T.R.; Gardner, P.R.; Schwab, U.; *et al. Pseudomonas aeruginosa* anaerobic respiration in biofilms: Relationships to cystic fibrosis pathogenesis. *Dev. Cell* **2002**, *3*, 593–603.
113. Singh, P.K.; Schaefer, A.L.; Parsek, M.R.; Moninger, T.O.; Welsh, M.J.; Greenberg, E.P. Quorum-sensing signals indicate that cystic fibrosis lungs are infected with bacterial biofilms. *Nature* **2000**, *407*, 762–764.
114. Ueda, A.; Wood, T.K. Connecting quorum sensing, c-di-GMP, pel polysaccharide, and biofilm formation in *Pseudomonas aeruginosa* through tyrosine phosphatase TpbA (PA3885). *PLoS Pathog.* **2009**, *5*, e1000483.
115. Gilbert, K.B.; Kim, T.H.; Gupta, R.; Greenberg, E.P.; Schuster, M. Global position analysis of the *Pseudomonas aeruginosa* quorum-sensing transcription factor LasR. *Mol. Microbiol.* **2009**, *73*, 1072–1085.
116. Ma, L.; Wang, J.; Wang, S.; Anderson, E.M.; Lam, J.S.; Parsek, M.R.; Wozniak, D.J. Synthesis of multiple *Pseudomonas aeruginosa* biofilm matrix exopolysaccharides is post-transcriptionally regulated. *Environ. Microbiol.* **2012**, *14*, 1995–2005.
117. Tart, A.H.; Blanks, M.J.; Wozniak, D.J. The AlgT-dependent transcriptional regulator AmrZ (AlgZ) inhibits flagellum biosynthesis in mucoid, nonmotile *Pseudomonas aeruginosa* cystic fibrosis isolates. *J. Bacteriol.* **2006**, *188*, 6483–6489.
118. Jones, C.J.; Ryder, C.R.; Mann, E.E.; Wozniak, D.J. AmrZ modulates *Pseudomonas aeruginosa* biofilm architecture by directly repressing transcription of the *psl* operon. *J. Bacteriol.* **2013**, *195*, 1637–1644.
119. Alkawash, M.A.; Soothill, J.S.; Schiller, N.L. Alginate lyase enhances antibiotic killing of mucoid *Pseudomonas aeruginosa* in biofilms. *Acta Pathol. Microbiol. Immunol. Scand.* **2006**, *114*, 131–138.
120. Hoffmann, N.; Lee, B.; Hentzer, M.; Rasmussen, T.B.; Song, Z.; Johansen, H.K.; Givskov, M.; Hoiby, N. Azithromycin blocks quorum sensing and alginate polymer formation and increases the sensitivity to serum and stationary-growth-phase killing of *Pseudomonas aeruginosa* and attenuates chronic *P. aeruginosa* lung infection in $Cftr^{-/-}$ mice. *Antimicrobial. Agents Chemother.* **2007**, *51*, 3677–3687.
121. Florescu, D.F.; Murphy, P.J.; Kalil, A.C. Effects of prolonged use of azithromycin in patients with cystic fibrosis: A meta-analysis. *Pulm. Pharmaco. Ther.* **2009**, *22*, 467–472.
122. Hentzer, M.; Riedel, K.; Rasmussen, T.B.; Heydorn, A.; Andersen, J.B.; Parsek, M.R.; Rice, S.A.; Eberl, L.; Molin, S.; Hoiby, N.; *et al.* Inhibition of quorum sensing in *Pseudomonas aeruginosa* biofilm bacteria by a halogenated furanone compound. *Microbiology* **2002**, *148*, 87–102.

123. Givskov, M.; de Nys, R.; Manefield, M.; Gram, L.; Maximilien, R.; Eberl, L.; Molin, S.; Steinberg, P.D.; Kjelleberg, S. Eukaryotic interference with homoserine lactone-mediated prokaryotic signalling. *J. Bacteriol.* **1996**, *178*, 6618–6622.
124. Singh, P.K.; Parsek, M.R.; Greenberg, E.P.; Welsh, M.J. A component of innate immunity prevents bacterial biofilm development. *Nature* **2002**, *417*, 552–555.
125. Musk, D.J.; Banko, D.A.; Hergenrother, P.J. Iron salts perturb biofilm formation and disrupt existing biofilms of *Pseudomonas aeruginosa*. *Chem. Biol.* **2005**, *12*, 789–796.
126. Zeng, X.; Liu, X.; Bian, J.; Pei, G.; Dai, H.; Polyak, S.W.; Song, F.; Ma, L.; Wang, Y.; Zhang, L. Synergistic effect of 14-alpha-lipoyl andrographolide and various antibiotics on the formation of biofilms and production of exopolysaccharide and pyocyanin by *Pseudomonas aeruginosa*. *Antimicrob. Agents Chemother.* **2011**, *55*, 3015–3017.
127. Kolodkin-Gal, I.; Romero, D.; Cao, S.; Clardy, J.; Kolter, R.; Losick, R. D-amino acids trigger biofilm disassembly. *Science* **2010**, *328*, 627–629.
128. Kolodkin-Gal, I.; Cao, S.; Chai, L.; Bottcher, T.; Kolter, R.; Clardy, J.; Losick, R. A self-produced trigger for biofilm disassembly that targets exopolysaccharide. *Cell* **2012**, *149*, 684–692.

14

The Role of Bacterial Biofilms and Surface Components in Plant-Bacterial Associations

Pablo C. Bogino [1], **María de las Mercedes Oliva** [2], **Fernando G. Sorroche** [1,†] **and Walter Giordano** [1,*]

[1] Department of Molecular Biology, National University of Río Cuarto, Ruta 36 Km 601, Río Cuarto, Córdoba X5804BYA, Argentina; E-Mails: pbogino@exa.unrc.edu.ar (P.C.B.); fernando.sorroche@biol.lu.se (F.G.S.)

[2] Department of Microbiology and Immunology, National University of Río Cuarto, Ruta 36 Km 601, Córdoba X5804BYA, Argentina; E-Mail: moliva@exa.unrc.edu.ar

[†] Current address: Department of Cell and Organism Biology, Lund University, Sölvegatan 35, Lund 22362, Sweden

[*] Author to whom correspondence should be addressed; E-Mail: wgiordano@exa.unrc.edu.ar;

Abstract: The role of bacterial surface components in combination with bacterial functional signals in the process of biofilm formation has been increasingly studied in recent years. Plants support a diverse array of bacteria on or in their roots, transport vessels, stems, and leaves. These plant-associated bacteria have important effects on plant health and productivity. Biofilm formation on plants is associated with symbiotic and pathogenic responses, but how plants regulate such associations is unclear. Certain bacteria in biofilm matrices have been found to induce plant growth and to protect plants from phytopathogens (a process termed biocontrol), whereas others are involved in pathogenesis. In this review, we systematically describe the various components and mechanisms involved in bacterial biofilm formation and attachment to plant surfaces and the relationships of these mechanisms to bacterial activity and survival.

Keywords: biofilms; autoaggregation; plant-associated bacteria; bacterial surface compounds; bacterial exopolymeric compounds

---

1. Introduction

Biofilms are bacterial communities in which cells are embedded in a matrix of extracellular polymeric compounds attached to a surface [1]. Living in biofilms helps protect bacteria from deleterious conditions [2] and the formation of biofilms appears to be an important factor in the disease cycle of bacterial pathogens in both animals and plants.

Bacterial surface components and extracellular compounds [primarily flagella, lipopolysaccharides (LPSs), and exopolysaccharides (EPSs)], in combination with environmental and quorum-sensing signals, are crucial for autoaggregation and biofilm development in most bacterial species studied to date [3,4]. In the generally accepted model of biofilm formation, environmental signals trigger the process, and flagella are required for the biofilm community to approach and move across the surface. The initial steps of attachment are mediated by outer membrane proteins (e.g., calcium-binding proteins), pili, or LPSs. After the formation of microcolonies, the production of quorum-sensing signals is required for the formation of a mature biofilm [5]. EPSs provide the architectural form of biofilms and stabilize their 3-dimensional structure. Biofilms are often permeated by channels that act as a circulatory system, allowing the bacteria to exchange water, nutrients, enzymes, and signals, dispose of potentially toxic metabolites, and display enhanced metabolic cooperativity [4,6]. The dispersal of biofilms allows bacteria to colonize other surfaces or substrates, thus completing a sequential developmental process.

The composition of biofilms varies depending on the system. The major components are typically water and the bacterial cells, followed by the EPSs of the matrix [7], which provides (i) a physical barrier against the diffusion of antibiotics, defense substances, or other important compounds from the host; and (ii) protection against environmental stress factors, such as UV radiation, pH changes, osmotic stress, and desiccation [8,9]. In *Agrobacterium tumefaciens*, a plant pathogen that persists as surface-associated populations on plants or soil particles, cellulose overproduction resulted in increased biofilm formation on roots [10]. Minor biofilm components include macromolecules such as proteins, DNA, and various lysis products [1], which affect the overall properties of the biofilm.

Bacterial biofilms are widely distributed and play important roles in many environments. The environments occupied by soil bacteria range from rhizospheres rich in nutrients and root exudates to bulk soil deficient in nitrogen, phosphates, water, and other nutrients. The size of bacterial aggregates varies from small to large as a function of the nutrient availability at a given site [11]. A hypothetical model of various 3-dimensional shapes of root-biofilm structures determined by nutrient availability has been presented [12].

Many species of beneficial soil bacteria, including rhizobia, form microcolonies or biofilms when they colonize roots. We recently summarized data on surface attachment and/or biofilm formation by rhizobacteria [5]. Biofilm development also contributes to the virulence of phytopathogenic bacteria through various mechanisms, including blockage of xylem vessels, increased resistance to plant antimicrobial compounds, and/or enhanced colonization of specific habitats [13]. The processes of autoaggregation and biofilm development are relevant to both bacterial survival and host plant colonization (Figure 1). A variety of environmental, genetic, and structural factors affect bacterial adhesion, cell-cell interactions, and plant colonization, and ultimately plant-bacterial interactions in general. In this article, we review recent findings on the mechanisms involved in attachment, cell aggregation, and biofilm formation on plant surfaces by bacteria.

Figure 1. Bacterial autoaggregation and biofilm development, and their relationship with plant colonization. Cell aggregation and biofilm formation in plant-bacterial associations are regulated by environmental signals, nutrient limitation of growth, quorum sensing, EPSs, flagella, LPSs, and other factors.

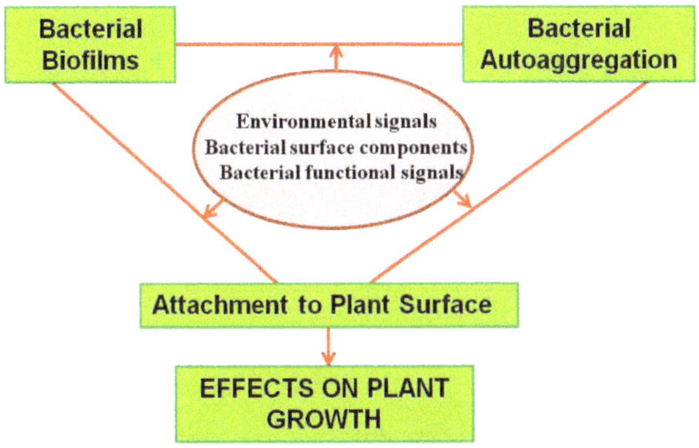

2. Cell–Cell Adhesive Interactions: Bacterial Autoaggregation

Bacteria were studied for many years as isolated cell entities. However, like many microorganisms, they have a strong tendency to congregate or aggregate. A common phenotypic manifestation of this behavior is autoaggregation, which is based on adhesive interactions among bacteria. Autoaggregation can be visualized macroscopically by the typical clumping or "fluffing" of cells in liquid cultures, followed by sedimentation of the clumps under static conditions [3,14,15].

Unfavorable growth conditions or low metabolic activity have been found to induce aggregative behavior in bacteria that normally grow in a dispersed, non-aggregated manner. In this context, autoaggregation may reflect a survival strategy that is triggered under hostile environmental conditions [16–19].

The autoaggregative characteristic of bacteria has important implications for the production of bacteria-based inoculants for agriculture. Bacterial aggregates can be produced on a large scale and then separated more easily from the culture medium as compared with dispersed bacteria. The biomass of aggregated bacterial cells in bioreactors remains more constant, and the survival of such bacteria during the inoculant storage period is enhanced, in comparison with non-aggregated cells [17,18,20,21].

2.1. Surface Factors Involved in Bacterial Autoaggregation

Because of their strategic location on the cell surface, LPSs, outer membrane proteins, and proteinaceous structures such as pili have been reported to affect adhesion among bacteria and consequently the autoaggregation phenotype.

LPS is an important surface structural component of Gram-negative bacteria and covers ~75% of the surface area of the outer membrane. It is a tripartite molecule consisting of lipid A, core oligosaccharides, and *O*-antigen, structurally formed by amphiphilic glycoconjugates whose composition varies within and between species. LPS molecules are positioned among the proteins and phospholipids of the outer bacterial membrane and contribute to the structural properties of the membrane; e.g., they act as a permeability barrier against various types of molecules. The structural heterogeneity of the *O*-antigen, the most external portion of the LPS molecule, confers versatility and adaptability to bacteria that are exposed to variable environmental conditions [22]. Changes of LPS structure usually affect adhesive forces among bacteria, possibly through alteration of cell surface hydrophobicity. For example, in rhizobacteria such as *Rhizobium leguminosarum* and *R. etli*, LPS modifications typically alter the autoaggregation phenotype [23–25]. *Ensifer* lpsB mutants, which have a truncated LPS core, display a more strongly autoaggregative phenotype as compared with wild-type parental strains (Sorroche *et al.*, unpubl. data). The rhamnose-rich *O*-antigen in the outermost part of the LPS of the xylem-limited phytopathogen *Xylella fastidiosa* is involved in cell-cell aggregation [26]. The autoaggregative ability of *X. fastidiosa* appears to be an important virulence mechanism because the bacterial clusters block the passage of water and nutrients from the roots to the leaves of the host plant [27–29].

Bioassay studies of the impact of biological factors on *Azospirillum brasilense* autoaggregation indicated that outer membrane proteins promote bacterial flocculation [30]. A 67-kDa outer membrane lectin on the bacterial surface specifically recognized an EPS synthesized by the aggregated cells. The interaction between the lectin and the EPS may be responsible in part for cell-cell interactions leading to autoaggregation of this species [31].

Pili, fimbriae, and flagella are proteinaceous polymeric appendages acting as bacterial surface organelles. Their numerous functions include mediation of motility, interbacterial interactions, bacterial-host interactions, and surface colonization [32]. Pili are associated with autoaggregation in *X. fastidiosa*. This process depends on the presence of polarly located type I and type IV pili, each of which plays a specific role in the structural dynamics of the bacterial aggregates [33].

2.2. Extracellular Factors Involved in Bacterial Autoaggregation

Some bacteria secrete molecules that promote autoaggregation. Extracellular polymeric materials have been shown to act as "molecular glue" that initiates and maintains contact between cells, causing flocculation. The main extracellular compounds are EPSs, which are linear or branched molecules formed by one repeated sugar (homopolysaccharides) or by a mixture of different sugars (heteropolysaccharides). An example is galactoglucan (EPS II) from the symbiotic rhizobacterium *Ensifer meliloti*. This extracellular EPS is secreted in two major fractions, low molecular weight (LMW) and high molecular weight (HMW), according to the degree of polymerization [34]. Mutant strains that are unable to synthesize EPS II fail to autoaggregate under static conditions. A mucR mutant secreting almost exclusively the HMW fraction of EPS II showed a weak aggregative phenotype, suggesting that the LMW fraction plays the active role in autoaggregation. Aggregation of the non-EPS-producing strains and the mucR strain was restored by resuspending the cells in culture medium containing EPS II [14].

Cellulose is an exopolymer with agglutinating activity in *R. leguminosarum*. Upon contact with the host plant, this rhizobacterium aggregates on the root surface using cellulose microfibrils [35–37].

Hostile environmental conditions or low metabolic activity can induce an autoaggregative phenotype via the synthesis of EPSs that have agglutinating activity. For example, the aggregative phenotype of the rhizobacterium *A. brasilense* depends on the production of an arabinose-rich extracellular polysaccharide that is synthesized in cultures in stationary and programmed cell death phases. Aggregative behavior in *Pseudomonas aeruginosa* is triggered by the presence of the toxic detergent SDS. This behavior depends on production of the EPS PsI. Inactivation of either the psi gene or the c-di-GMP-mediated signaling system that activates the gene results in reduced autoaggregation [19,38,39].

3. Cell–Cell and Cell–Surface Interactions: Bacterial Biofilm Formation

Surface and extracellular bacterial components have been extensively studied because they involve molecules that play crucial roles during the process of infection of the host plant, independently of the development of a beneficial or pathogenic relationship. Such bacterial components are also key molecules in the establishment, maturation, and dispersal of biofilms. We will summarize in this section the structure and function of bacterial compounds that play a role in the development of biofilms by beneficial or pathogenic bacteria associated with the surface or interior of plant tissues.

3.1. Structural and Functional Components Involved in Biofilm Formation

All bacteria live as a multicellular conglomerate encased in a protective matrix of polymeric substances produced by the bacteria themselves. The highly organized and dynamic social structure of bacteria requires intercellular communication via quorum sensing [40]. Bacterial aggregates frequently adhere at a solid-liquid interface prior to adsorption on a thin film of organic molecules that constitutes the adhesion site. The transport of cells to this interface may be mediated by passive mechanisms or by the intrinsic motility of planktonic bacteria. The accumulation of bacterial cells at the interface is biphasic, consisting of (i) a non-specific reversible stage mediated by hydrophobic and electrostatic interactions between cells and adjacent surfaces; and (ii) an irreversible stage in which the adhesion process is completed and a bacterial microcolony is established [41]. Surface bacterial components such as flagella, pili, fimbriae, and LPSs play a crucial role in physical processes during the initial stages of biofilm formation on surfaces. The growth, maturation, and disassembly phases of biofilms depend primarily on the biosynthesis of extracellular biopolymers such as EPSs, proteins, and extracellular DNA (eDNA). These polymers promote or provide immobilization of bacterial cells into the matrix, mechanical stability of the biofilm structure, cohesive interaction with the interface, and the architecture and functionality of the encased microbial community [42].

3.1.1. Surface Bacterial Factors

Because of their exposure to the external environment and their chemical properties, LPSs on the outer membrane of Gram-negative bacteria are capable of undergoing adhesive interactions with both biotic and abiotic surfaces.

Structural changes of LPSs have been shown to alter biofilm formation or structure in beneficial plant-associated bacteria, including *Pseudomonas fluorescens* [43]. Mutant strains of various beneficial rhizobacteria with altered LPS structure display changes in the biofilm formation process. For example, a *Bradyrhizobium japonicum O*-antigen mutant showed enhanced adhesion to plastic supports [44]. A *R. leguminosarum* lipid a mutant showed increased lateral interactions with an abiotic surface;

this effect had no effect on the ability of the bacteria to form a biofilm on the surface [45]. A mutant of the alfalfa symbiont *E. meliloti* that synthesized a structurally modified LPS because of mutations of the *lpsB* and *bacA* genes showed reduced biofilm formation ability [46,47]. *E. meliloti lpsB* mutants showed reduced nodulation abilities because of delays in the invasion steps; however, their nitrogen-fixing capacity was similar to that of wild-type [48,49], and the modified LPS molecule increased bacterial adsorption to alfalfa roots [50].

In phytopathogenic bacteria, as in beneficial bacteria, LPSs play crucial roles during the early stages of interaction with the host and development of virulence. These phenomena coincide with the normal stages of biofilm formation. Various mutations related to LPS synthesis in phytopathogenic bacteria such as *P. aeruginosa* [51], *Pseudomonas syringae* [52], *Xanthomonas axonopodis* [53], and *Xanthomonas citri* [54] caused reductions in both biofilm formation ability and virulence.

Because of their surface cell location and physico-chemical properties, LPSs play a key role during the initial steps of biofilm formation (e.g., adherence to surfaces) and the development of mature biofilm through interactions of cells with other cells and with matrix components.

Proteinaceous appendages (pili and flagella) are bacterial virulence factors that lead to pathogenesis in plant, animal, and human hosts and play a key role during colonization steps [55]. The initial stages of biofilm formation are dependent on bacterial motility mediated by the polar flagellum and multiple type IV pili (TfP), which enable the free-swimming phenotype to reach a suitable surface and the surface-motile phenotype to adhere to and move on the surface [56]. These appendages thus have a dual role as motile machines and adhesins that move and fix bacteria to surfaces and among surfaces [57]. Various types of movement (e.g., crawling, pulling, walking) have been associated with TfP [56]. The coordinated TfP pulling associated with release and with flagellar rotation-translation has been presented as a model whereby bacteria are able to travel through the biofilm matrix [58]. Motility-defective mutants of *P. aeruginosa* and *E. coli* were unable to attach to surfaces or to develop a normal biofilm [59,60].

Pili structure is crucial for adhesion and biofilm formation in certain phytopathogenic bacteria, including *Acidovorax citrulli*, the causal agent of bacterial fruit blotch in cucurbits [61], and *X. fastidiosa*, the causal agent of Pierce's disease in grapes [62]. *A. citrulli* uses TfP and *X. fastidiosa* uses type I pili to colonize and move upstream against sap flow in xylem vessels while oriented parallel to the surface, prior to the complete development of biofilm and plant disease. TfP is also important for the pathogenesis of *Ralstonia solanacearum* and *Xanthomonas oryzae* pv. *oryzicola* [63,64]. Pili are well-established virulence factors for phytopathogenic bacteria because of their important roles in adherence and plant colonization.

Motility mediated by flagella is manifested as either "swimming" of free cells in aqueous environments and coordinated "swarming" of bacterial populations on solid moist surfaces [65]. Both swimming and swarming are essential for various stages of biofilm development, e.g., the search for a favorable habitat, attachment to a surface, architectural assembly, structural disassembly, and release from the biofilm matrix [66]. The complex association between motility and biofilm formation involves the use of a particular structure for different functions at different stages and requires the precise integration of environmental and cellular signals [67].

Flagella-mediated motility in rhizobacteria is essential for biofilm establishment and therefore for plant colonization. *E. meliloti*, a paradigm of beneficial symbiotic interaction between rhizobacteria and legume plants, showed reduced biofilm formation ability on abiotic surfaces as a result of mutations on genes related to flagella synthesis [68]. An association between flagella-mediated motility and biofilm formation and an effect of quorum-sensing signals on both of these processes were demonstrated in peanut-nodulating *Bradyrhizobium* sp. strains [69]. Both aflagellate and flagellated but nonmotile mutants of the well-studied pathogen *A. tumefaciens*, the causal agent of crown gall, showed reduced biofilm formation ability under static conditions because of defects in surface attachment. However, the aflagellate mutants were able to quickly develop an unusually dense and tall biofilm under flow conditions [70]. In contrast to results of other studies that suggest a role of flagella as an adhesin in *Aeromonas* spp. [71], these results for *A. tumefaciens* are consistent with those for other bacterial species [72], which suggest that flagella do not function as an adhesin and that other surface structures can be involved in attachment and subsequent biofilm formation [70].

X. axonopodis pv. *citri*, a phytopathogenic bacterium that establishes itself on the leaves (phyllosphere) of citrus species and produces citrus canker disease [73], has the ability to form biofilms on abiotic and biotic surfaces, including the cankers of diseased plants [74]. Flagella-mediated motility plays a key role in several stages of biofilm formation, including surface adherence, maturation, and dispersal [75]. Sliding motility (not mediated by flagella) and the regulation of both swimming and sliding motility through diffusible signal factor (DSF) were reported for *X. axonopodis* [75]. Similar findings were reported for *Xanthomonas campestris* pv. *campestris*, although flagellin mutants did not show altered virulence [76]. Swimming motility was found to be essential for biofilm formation and colonization of plant tissues in vascular pathogens such as *R. solanacearum*, *Pantoea stewartii*, and *Dickeya dadantii* [77–79].

In contrast to findings in Gram-negative bacteria, it appears that motility is not essential for biofilm development in various non-motile Gram-positive bacteria. Surface proteins (e.g., Bap, Esp) have been reported to be involved in initial adherence to surfaces in these non-motile species [80,81].

Several surface structures that function as adhesins have been found to play important roles during surface attachment. Rhicadhesins and Raps (*Rhizobium*-adhering proteins) are important for root attachment in various Rhizobiaceae species [82,83]. Glucomannan, a surface polysaccharide of *R. leguminosarum* bv. *trifolli*, binds to pea and vetch lectins [84]. Two surface-associated proteins were found to be involved in biofilm formation in *Pseudomonas putida*. LapA plays a key role in the early stages of biofilm formation by mediating bacterial adherence to various surfaces (including seeds and roots), while LapF is crucial in later stages by mediating cell-cell interactions during sessile growth [85,86].

3.1.2. Extracellular Factors

The highly developed level of bacterial organization reflected by biofilm formation ability and multicellularity requires dynamic and functional microorganisms that are embedded into a complex mixture of extracellular polymeric components collectively termed the "biofilm matrix" [1]. These components form bridges, channels, avenues, and pores and support an impressively elaborate 3-dimensional architectural structure within which cellular arrangements are transiently constructed. In addition to its structural protective function, the biofilm matrix plays a key role in bacterial physiology and ecology, including cellular interactions, nutrient utilization, horizontal gene transfer, and environmental fitness of the bacterial population [42]. The composition of the biofilm matrix is highly variable, depending on the type of bacteria and biofilm interface. Water is the main component of a mature biofilm. In terms of dry mass, bacteria account for <10% of a biofilm, and the matrix accounts for >90%. The biofilm matrix consists primarily of EPSs and contains smaller proportions of other biopolymers such as proteins, nucleic acids, and lipids [7]. We will summarize here the roles of the major components of the biofilm matrix; *i.e.*, extracellular factors.

A biofilm has been defined as a multicellular bacterial conglomerate adhered to a surface and immersed into a polymeric matrix formed primarily by EPSs [4]. It is therefore not surprising that various bacterial strains, even of the same species, have the ability to synthesize, export, and modify their own characteristic EPS. The function and chemical composition of each EPS is different, depending on the bacterial species or strain. A particular strain may even have the ability to produce different EPSs depending on the environmental conditions, as demonstrated for *P. aeruginosa* [87] and *Streptococcus thermophilus* [88]. Most EPSs are polyanionic molecules because of the presence of uronic acids and sugar having substituents such as pyruvate, sulfate, or phosphate. Polycationic EPSs have also been described [89]. The presence of β-1,4 (or β-1,3) and α-1,2 (or α-1,6) linkages confers greater rigidity or flexibility, respectively, to the matrix structure. The stability of the biofilm structure is thus dependent on the physico-chemical and biological properties of EPSs and on their interactions with ions, low molecular weight solutes, and other macromolecules such as proteins and eDNA [7]. EPSs are generally required not for initial adhesion but for later architectural development of the biofilm matrix [90]. The EPS network confers mechanical stability, allows for temporary immobilization of cells, and plays a crucial role in most matrix functions, including water retention, protection from environmental stresses, adsorption of compounds, and nutrient availability [42].

The production of EPSs on plant surfaces or tissues allows bacterial colonization and biofilm formation. The biological roles (beneficial or pathogenic) of bacteria on plants are related to these abilities. The compositions and biological roles of selected EPSs produced by well-known plant-associated bacteria are summarized in Table 1.

Table 1. Production, composition, and functional roles of exopolysaccharides (EPSs) in beneficial and pathogenic plant-associated bacteria.

Bacteria/plant association	Exopolysaccharide	Chemical composition	Function
E. meliloti symbiosis with *Medicago sativa*	Succinoglycan (EPS I), LMW-HMW [91]	Octasaccharide units (glucose:galactose 7:1, bearing succinyl, acetyl, and pyruvyl substituents)	Required for biofilm formation [68] EPS I LMW symbiotically active [92]
	Galactoglucan (EPS II), LMW-HMW [93]	Disaccharide units (acetylated glucose-pyruvylated galactose)	EPS II LMW symbiotically active [34] EPS II LMW controls biofilm formation [94]
R. leguminosarum symbiosis with *Trifolium*, *Pisum*, *Vicia* and *Phaseolus* spp.	Acidic EPS [95,96]	Octasaccharide units (glucose:glucuronic acid:galactose 5:2:1, modified by acetyl, pyruvyl and 3-hydroxybutanoyl groups)	Development of a structured biofilm [83,97] Required for infection and nodulation [98,99]

Table 1. *Cont.*

Bacteria/plant association	Exopolysaccharide	Chemical composition	Function
B. japonicum symbiosis with *Glycine max*	EPS [100,101]	Pentasaccharide units (mannose:galacturonic acid:glucose:galactose 1:1:2:1)	Biofilm formation on both inert and biotic surfaces. Roles during the early stages of interaction with the host plant (initial attachment of rhizobia to root epidermal cells) [102]
M. tianshanense symbiosis with *Glycyrrhiza uralensis*	EPS	ND	Involved in biofilm formation and successful establishment of symbiosis [103]
A. tumefaciens ubiquitous plant pathogen	Succinoglycan [104]	See above	Increased production of succinoglycan results in reduced attachment and biofilm formation [105]
X. fastidiosa plant pathogen	Putative Fastidian gum [106]	Putative tetrasaccharide units (glucose-1-phosphate, glucose, mannose, and glucuronic acid)	Possibly involved in bacterial pathogenicity [106] Cell attachment and overall biofilm formation [107]
X. campestris *X. axonopodis* plant pathogens	Xanthan gum [108]	Pentasaccharide units (glucose:mannose:glucuronic acid 2:2:1 derivatized with acetyl and pyruvyl moieties)	Essential for microcolony formation [74] Formation of structured biofilms on abiotic surfaces and in infected plants [109,110]
P. stewartii plant pathogen	Stewartan [111]	Heptasaccharide units (glucose:galactose:glucuronic acid 3:3:1)	Essential for appropriate adhesion and for maturation of biofilm structure. Also a virulence factor required for effective host colonization and efficient dissemination through xylem vessels [112]
E. amylovora plant pathogen	Amylovoran [113]	Pentasaccharide units (galactose:glucose 4:1, and pyruvate residues)	Pathogenicity factor required for biofilm formation [114]
	Levan [115]	Homopolymer of fructose	Virulence factor. Also contributes to biofilm formation [114]
R. solanacearum plant pathogen	Acidic EPS I [116]	Putative structure composed by *N*-acetylgalactosamine and amino sugars (bacillosamine, galactosaminuronic acid)	Major virulence factor [117]

LMW: low molecular weight; HMW: high molecular weight; ND: not determined.

The importance of cellulose (a neutral homopolysaccharide) as a key component of the polysaccharidic matrix has been demonstrated in several Enterobacteriaceae species [118], the plant-associated bacterium

A. tumefaciens [119], and *Rhizobium* species [36,120]. Cellulose plays a key role in adherence to plant tissues, biofilm formation, and the support of matrix architecture.

Polysaccharides from *Arabidopsis* roots were recently found to serve as both signals for biofilm formation and a source of sugars for the synthesis of matrix EPSs in the beneficial Gram-positive bacterium *Bacillus subtilis* [121]. Future studies on the triggering of bacterial biofilm formation by plant root exudates will be useful.

Extracellular proteins are major constituents of the biofilm matrix, but have received relatively little study in comparison with other components such as EPSs. Proteins in the biofilm matrix have both structural and physiological functions. Some matrix proteins function as extracellular enzymes and are associated with activities such as the degradation and recycling of biopolymers for nutrient availability and the modification of other exopolymers for shaping or releasing of cells from the biofilm structure. Enzymes that play such roles in the biofilm matrix include lipases, hydrolases, lyases, and glycanases [83,122,123]. Certain enzymes released by pathogenic bacteria may act as virulence factors [124,125], but such a function has not been evaluated in the context of biofilms on plant surfaces.

Some proteins in the biofilm matrix have structural functions, e.g., as lectins that bind bacterial cells to the polymeric matrix. Examples of such extracellular carbohydrate-binding proteins include a glucan-binding protein in *Streptococcus mutans* [126], LecA and LecB in *P. aeruginosa* [127,128], TasA in *Bacillus subtilis* [129], and lectins in *A. brasilense* [31]. In *P. aeruginosa*, a large quantity of matrix proteins was found in outer membrane vesicles, a typical matrix biofilm component in this species [130]. Amyloids are another common type of matrix protein with extracellular adhesin function [131].

Biofilms provide an ideal location for the exchange of genetic material. Higher levels of conjugation have been demonstrated for bacterial populations in biofilms as compared with planktonic bacteria [132]. eDNA is an important constituent of the biofilm matrix [133] and plays a role in biofilm formation in various bacterial species, including *P. aeruginosa* [134] and *Bacillus cereus* [135]. The quantity, localization, and origin of eDNA vary depending on the bacterial species. Some studies found that eDNA is arranged in certain patterns [136] and that its release is based on the lysis of certain types of bacteria [137], suggesting the occurrence of programmed cell death in biofilms [138]. In Gram-positive bacteria, eDNA is involved in adhesion to hydrophobic surfaces and in bacterial autoaggregation [139].

Lipids are also components of the biofilm matrix [140], although they have received little study in the context of plant-bacterial associations. Lipids in biofilms generally act as biosurfactants with functions such as surface activity, dispersal and bioavailability of hydrophobic substances, antibacterial or antifungal properties, and bacterial attachment and detachment [141]. Such properties have been well characterized for the rhamnolipids of *P. aeruginosa*, which play important roles at several stages of biofilm development, including surface interaction, microcolony formation, structural maintenance, and biofilm dispersal [142].

Our knowledge of the identification and functions of extracellular proteins, eDNA, and lipids in the biofilm matrix of plant-associated bacteria remains limited and fragmentary. Further studies along this line will greatly enhance our understanding of the process of biofilm formation.

4. Relationship between Biofilm Formation and Bacterial Autoaggregation

Bacteria have the unique ability to form complex cellular assemblies on biotic surfaces (animal tissues, plant tissues, detritus) and abiotic surfaces (sediments, soil particles, medical or laboratory instruments). Depending on the quantities of cells and extracellular components, such assemblies range from random cell aggregations observed on surfaces or in liquid suspension to complex, highly developed assemblies of cells encased in a exopolymeric matrix and attached to a surface (*i.e.*, biofilms) [143]. There are several bacterial models that are useful for studying the relationships between cell aggregations and biofilms. The adhesin AIDA, a surface glycoprotein of *E. coli*, has been shown to promote strong bacterial autoaggregation, biofilm formation, and *in vitro* adhesion to human and mammalian cells [144]. Strains of the opportunistic bacterium *Myroides odoratus* that showed strong adhesion to inert supports also showed enhanced autoaggregative ability [145]. Changes in the cell surface of *Porphyromonas gingivalis* (a pathogenic bacterium involved in periodontal disease) led to increased cell-cell interactions and consequent increases in autoaggregation and biofilm formation [146].

Variants of the phytopathogen *X. fastidiosa* with mutation of a gene involved in signal transduction showed a modified transcriptional profile of genes related to biofilm formation and bacterial aggregation traits such as surface attachment, EPS synthesis, and virulence [147].

Autoaggregative behavior has also been correlated with biofilm formation ability in beneficial plant-associated bacteria. Strongly autoaggregative mutants of the rhizobacterium *A. brasilense* showed a high tendency to form biofilms on inert supports [148]. In the rhizobacterium *E. meliloti*, autoaggregation [14] and biofilm formation [5] depend on a combination of bacterial signals, surface components, and EPSs. The development of cell interactions in both sessile populations and planktonic aggregations of various native strains of *E. meliloti* showed a positive correlation with the above processes and a requirement for EPS II, indicating the involvement of the same physical adhesive forces in autoaggregation and biofilm formation [15].

Surface and extracellular factors are involved in both cell aggregation and biofilm formation, processes that depend in part on physical interactions among bacteria. It is therefore reasonable to presume that alterations in one of these processes lead to changes in the other. Cell aggregation on surfaces most likely represents a transitional state that precedes the development of a structured biofilm. We postulate that the linkage of or transition from bacterial aggregation to biofilm formation is crucial for the establishment of beneficial or pathogenic relationships between bacteria and plants. Such relationships are developed through (i) interactions among bacteria in or near the plant microenvironment (e.g., rhizosphere or phyllosphere); (ii) interactions between bacteria and plant surfaces (e.g., leaf or root epidermis, root hairs, transport vessels); and (iii) biofilm formation and their biological effects.

5. Intergeneric Adhesive Interactions: Coaggregation

The coaggregation process has been defined as adhesion among genetically different microorganisms [149]. The phenomenon of coaggregation was initially described for bacteria that inhabit the human oral cavity and subsequently extended to bacteria found in other habitats, including aquatic environments and sludge [150]. Coaggregation processes and factors that affect them have been described for plant growth-promoting rhizobacteria (PGPR) [151]. These findings are relevant to our understanding of plant-bacterial interactions and the development of commercial inoculants.

Coaggregation is an integral process in the formation of mixed biofilms and is therefore ecologically important. Two models have been proposed to explain how a mixed biofilm can support different types of planktonic phase-derived bacteria. According to the first model, planktonic free cells in suspension specifically recognize and adhere to genetically different bacteria in a biofilm. The second model involves initial coaggregation of planktonic bacteria and subsequent adhesion and integration of the coaggregate during biofilm formation [149]. Regardless of these models, the coaggregation phenomenon was found to be dependent on cell surface hydrophobicity and the partner strains that participated in the interaction [152].

On a molecular level, coaggregation of human intestinal or oral bacteria and of aquatic bacteria depends on the interaction of a lectin of one participant with a complementary glycosidic receptor of the other participant [153]. Similar mechanisms are expected in bacteria that establish associations with plants, particularly in certain habitats in which bacterial populations are highly diverse, e.g., rhizospheric soil. Cell-free culture supernatants of *E. meliloti* containing EPS II induced the autoaggregation of several rhizospheric bacteria, including strains of *Pseudomonas*, *Azospirillum*, and *Burkholderia*. These findings suggest that interactions between EPS II and rhizobacteria play an important ecological role. EPS II itself may be capable of connecting different bacterial cells. According to this hypothetical model, EPS II-producing *E. meliloti* cells may act as "bridges" during the process of coaggregation with other rhizospheric bacteria [50].

6. Concluding Remarks

The bacterial cell surface plays a key role in bacterial aggregation, which in turn promotes bacterial dispersal, survival, and the ability to adhere to plant surfaces. It has been well documented that bacterial autoaggregation and biofilm development, and the relationship of these processes with plant colonization, are dependent on both surface bacterial factors and extracellular factors.

Bacteria gain several advantages from living in biofilms, including protection from predation, desiccation, and exposure to antibacterial substances, and improved acquisition of nutrients released in the plant environment. Biofilms provide survival sites for both beneficial and opportunistic pathogenic bacteria, by providing protection as above and increasing the potential of the bacteria to survive and evolve in the plant environment. Biofilms have been shown to enhance (i) the fitness of individual bacteria and (ii) more generalized plant health and productivity as a result of the cumulative selective advantage of the individual bacteria.

Detailed elucidation of the mechanisms involved in the various stages of biofilm formation will improve our understanding of microbial adaptations to this mode of life in a wide range of environments and of the interactions of bacteria with their eukaryotic hosts. Multidisciplinary studies using new approaches will clarify the ways in which bacteria move and interact in a variety of surface microenvironments during biofilm development. Such knowledge will enhance our understanding of biofilm formation on plant surfaces and of the sophisticated processes of interaction between prokaryotes and eukaryotes.

Acknowledgments

This review and our studies described herein were supported by grants from the Secretaría de Ciencia y Técnica (UNRC), Agencia Nacional de Promoción Científica y Tecnológica (ANPCyT), and Consejo Nacional de Investigaciones Científicas y Técnicas of the República Argentina (CONICET). FGS was supported by a fellowship from the CONICET. PCB, MMO and WG are Career Members of CONICET. The authors thank S. Anderson for English editing of the manuscript.

References

1. Branda, S.S.; Vik, S.; Friedman, L.; Kolter, R. Biofilms: The matrix revisited. *Trends Microbiol.* **2005**, *13*, 20–26.
2. Davey, M.E.; O'Toole, G.A. Microbial biofilms: From ecology to molecular genetics. *Microbiol. Mol. Biol. Rev.* **2000**, *64*, 847–867.
3. Schembri, M.A.; Christianesen, G.; Klemm, P. FimH-mediated autoaggregation of *E. coli*. *Mol. Microbiol.* **2001**, *41*, 1419–1430.
4. Costerton, J.W.; Lewandowski, Z.; Caldwell, D.E.; Korber, D.R.; Lappin-Scott, H.M. Microbial biofilms. *Annu. Rev. Microbiol.* **1995**, *49*, 711–745.
5. Rinaudi, L.V.; Giordano, W. An integrated view of biofilm formation in rhizobia. *FEMS Microbiol. Lett.* **2010**, *304*, 1–11.
6. Stanley, N.R.; Lazazzera, B.A. Environmental signals and regulatory pathways that influence biofilm formation. *Mol. Microbiol.* **2004**, *52*, 917–924.
7. Sutherland, I.W. Biofilm exopolysaccharides: A strong and sticky framework. *Microbiology* **2001**, *147*, 3–9.
8. Flemming, H.C. Biofilms and environmental protection. *Water Sci. Technol.* **1993**, *27*, 1–10.
9. Gilbert, P.; Das, J.; Foley, I. Biofilms susceptibility to antimicrobials. *Adv. Dent. Res.* **1997**, *11*, 160–167.
10. Matthysse, A.G.; Marry, M.; Krall, L.; Kaye, M.; Ramey, B.E.; Fuqua, C.; White, A.R. The effect of cellulose overproduction on binding and biofilm formation on roots by *Agrobacterium tumefaciens*. *Mol. Plant Microbe Interact.* **2005**, *18*, 1002–1010.
11. Van de Mortel, M.; Halverson, L.J. Cell envelope components contributing to biofilm growth and survival of *Pseudomonas putida* in low-water-content habitats. *Mol. Microbiol.* **2004**, *52*, 735–750.
12. Rudrappa, T.; Biedrzycki, M.L.; Bais, H.P. Causes and consequences of plant-associated biofilms. *FEMS Microbiol. Ecol.* **2008**, *64*, 153–166.
13. Mansfield, J.; Genin, S.; Magori, S.; Citovsky, V.; Sriariyanum, M.; Ronald, P.; Dow, M.; Verdier, V.; Beer, S.V.; Machado, M.A.; *et al.* Top 10 plant pathogenic bacteria in molecular plant pathology. *Mol. Plant Pathol.* **2012**, *13*, 614–629.

14. Sorroche, F.G.; Rinaudi, L.V.; Zorreguieta, A.; Giordano, W. EPS II-dependent autoaggregation of *Sinorhizobium meliloti* planktonic cells. *Curr. Microbiol.* **2010**, *61*, 465–470.
15. Sorroche, F.G.; Spesia, M.B.; Zorreguieta, A.; Giordano, W. A positive correlation between bacterial autoaggregation and biofilm formation in native *Sinorhizobium meliloti* isolates from Argentina. *Appl. Environ. Microbiol.* **2012**, *78*, 4092–4101.
16. Neyra, C.; Sadasivan, L. Flocculation in *Azospirillum brasilense* and *Azospirillum lipoferum*: Exopolysacharides and cyst formation. *J. Bacteriol.* **1985**, *163*, 716–723.
17. Nikitina, V.E.; Ponomareva, E.G.; Alen'kina, S.A.; Konnova, S.A. The role of cell-surface lectins in the aggregation of *Azospirilla. Microbiology* **2001**, *70*, 471–476.
18. Bahat-Samet, E.; Castro-Sowinski, S.; Okon, Y. Arabinose content of extracellular polysaccharide plays a role in cell aggregation *of Azospirillum brasilense*. *FEMS Microbiol. Lett.* **2004**, *237*, 195–203.
19. Klebensberger, J.; Birkenmaier, A.; Geffers, R.; Kjelleberg, S.; Philipp, B. SiaA and SiaD are essential for inducing autoaggregation as a specific response to detergent stress in *Pseudomonas aeruginosa*. *Environ. Microbiol.* **2009**, *11*, 3073–3086.
20. Malik, A.; Sakamoto, M.; Hanazaki, S.; Osawa, M.; Susuki, T.; Tochigi, M.; Kakii, K. Coaggregation among nonfloculating bacteria isolated from activated sludge. *Appl. Environ. Microbiol.* **2003**, *69*, 6056–6064.
21. Burdman, S.; Jurkevitch, E.; Soria-Díaz, M.E.; Serrano, A.M.; Okon, Y. Extracellular polysaccharide composition of *Azospirillum brasilense* and its relation with cell aggregation. *FEMS. Microbiol. Lett.* **2000**, *189*, 259–264.
22. Lerouge, I.; Vanderlayden, J. *O*-antigen structural variation: Mechanisms and possible roles in animal/plant-microbe interactions. *FEMS Microbiol. Rev.* **2001**, *26*, 17–47.
23. Priefer, U.B. Genes involved in lipopolysaccharide production and symbiosis are clustered on the chromosome of *Rhizobium leguminosarum* biovar *viciae* VF39. *J. Bacteriol.* **1989**, *171*, 6161–6168.
24. Cava, J.R.; Elias, P.M.; Turowski, D.A.; Noel, K.D. *Rhizobium leguminosarum* CFN42 genetic region encoding *lipopolysaccharide* structures essential for complete nodule development on bean plants. *J. Bacteriol.* **1989**, *171*, 8–15.
25. García de los Santos, A.; Brom, S. Characterization of two plasmid-borne lps beta loci of *Rhizobium etli* required for lipopolysaccharide synthesis and for optimal interaction with plants. *Mol. Plant Microbe Interact.* **1997**, *10*, 891–902.
26. Clifford, J.C.; Rapicavoli, J.N.; Roper, M.C. A rhamnose-rich *O*-antigen mediates adhesion, virulence and host colonization for the xylem-limited phytopathogen, *Xylella fastidiosa*. *Mol. Plant. Microbe Interact.* **2013**, *26*, *676*–685.
27. Hopkins, D.L. Seasonal concentration of the Pierce's disease bacterium in grapevine stems, petioles, and leaf veins. *Phytopathology* **1981**, *71*, 415–418.
28. Tyson, G.E.; Stojanovic, B.J.; Kuklinski, R.F.; Divittorio, T.J.; *S*ullivan, M.L. Scanning electron microscopy of Pierce's disease bacterium in petiolar xylem of grape leaves. *Phytopathology* **1985**, *75*, 264–269.
29. Newman, K.L.; Almeida, R.P.P.; Purcell, A.H.; Lindow, S.E. Use of a green fluorescent strain for analysis of *Xylella fastidiosa* colonization of *Vitis vinifera*. *Appl. Environ. Microbiol.* **2003**, *69*, 7319–7327.

30. Burdman, S.; Jurkevitch, E.; Schwartsburd, B.; Okon, Y. Involvement of outer membrane proteins in the aggregation of *Azospirillum brasilense*. *Microbiology* **1999**, 145, 1145–1152.
31. Mora, P.; Rosconi, F.; Franco Fraguas, L.; Castro-Sowinski, S. *Azospirillum brasilense* Sp7 produces an outer-membrane lectin that specifically binds to surface-exposed extracellular polysaccharide produced by the bacterium. *Arch. Microbiol.* **2008**, *189*, 519–524.
32. Craig, L.; Pique, M.E.; Tainer, J.A. Type IV pilus structure and bacterial pathogenicity. *Nature* **2004**, *2*, 363–378.
33. De La Fuente, L.; Burr, T.J.; Hoch, H.C. Autoaggregation of *Xilella fastidiosa* cells is influenced by type I and type IV pili. *Appl. Environ. Microbiol.* **2008**, *74*, 5579–5581.
34. González, J.E.; Reuhs, B.L.; Walker, G.C. Low molecular weight EPS II of *Rhizobium meliloti* allows nodule invasion in *Medicago sativa*. *Proc. Natl. Acad. Sci. USA* **1996**, *93*, 8636–8641.
35. Ausmees, N.; Jonsson, H.; Höglund, S.; Ljunggren, H.; Lindberg, M. Structural and putative regulatory genes involved in cellulose synthesis in *Rhizobium leguminosarum* bv. *trifolii*. *Microbiology* **1999**, *145*, 1253–1262.
36. Laus, M.C.; van Brussel, A.A.; Kijne, J.W. Role of cellulose fibrils and exopolysaccharides of *Rhizobium leguminosarum* in attachment to and infection of Vicia sativa root hairs. *Mol. Plant Microbe Interact.* **2005**, *18*, 533–538.
37. Williams, A.; Wilkinson, A.; Krehenbrink, M.; Russo, D.M.; Zorreguieta, A.; Downie, J.A. Glucomannan-mediated attachment of *Rhizobium leguminosarum* to pea root hairs is required for competitive nodule infection. *J. Bacteriol.* **2008**, *190*, 4706–4715.
38. Klebensberger, J.; Rui, O.; Fritz, E.; Schink, B.; Philipp, B. Cell aggregation of *Pseudomonas aeruginosa* strain PAO1 as an energy-dependent stress response during growth with sodium dodecyl sulfate. *Arch. Microbiol.* **2006**, *185*, 417–427.
39. Klebensberger, J.; Lautenschlager, K.; Bressler, D.; Wingender, J.; Philipp, B. Detergent-induced cell aggregation in subpopulations of *Pseudomonas aeruginosa* as a preadaptive survival strategy. *Environ. Microbiol.* **2007**, *9*, 2247–2259.
40. Hooshangi, S.; Bentley, W.E. From unicellular properties to multicellular behavior: Bacteria quorum sensing circuitry and applications. *Curr. Opin. Biotech.* **2008**, *19*, 550–555.
41. Klausen, M.; Gjermansen, M.; Kreft, J.U.; Tolker-Nielsen, T. Dynamics of development and dispersal in sessile microbial communities: Examples from *Pseudomonas aeruginosa* and *Pseudomonas putida* model biofilms. *FEMS Microbiol. Lett.* **2006**, *261*, 1–11.
42. Flemming, H.C.; Wingender, J. The biofilm matrix. *Nature Rev. Microbiol.* **2010**, *8*, 623–633.
43. Spiers, A.J.; Rainey, P.B. The *Pseudomonas fluorescens* SBW25 wrinkly spreader biofilm requires attachment factor, cellulose fibre and LPS interactions to maintain strength and integrity. *Microbiology* **2005**, *151*, 2829–2839.
44. Lee, Y.W.; Jeong, S.Y.; In, Y.H.; Kim, K.Y.; So, J.S.; Chang, W.S. Lack of *O*-polysaccharide enhances biofilm formation by *Bradyrhizobium japonicum*. *Lett. Appl. Microbiol.* **2010**, *50*, 452–456.
45. Vanderlinde, E.M.; Muszynski, A.; Harrison, J.J.; Koval, S.F.; Foreman, D.L.; Ceri, H.; Kannenberg, E.L.; Carlson, R.W.; Yost, C.K. *Rhizobium leguminosarum* biovar *viciae* 3841, deficient in 27-hydroxyoctacosanoate-modified lipopolysaccharide, is impaired in desiccation tolerance, biofilm formation and motility. *Microbiology* **2009**, *155*, 3055–3069.

46. Ferguson, G.P.; Roop R.M., II; Walker, G.C. Deficiency of a *Sinorhizobium meliloti bacA* mutant in alfalfa symbiosis correlates with alteration of the cell envelope. *J. Bacteriol.* **2002**, *184*, 5625–5632.
47. Hirsch, A.M.; Lum, M.R.; Fujishige, N.A. Microbial Encounters of a Symbiotic Kind-Attaching to Roots and Other Surfaces. In *Root Hairs: Plant Cell Monographs*; Emons, A.M.C., Ketelaar, T., Eds.; Springer-Verlag: Berlin, Germany, 2009; Volume 12, pp. 295–314.
48. Lagares, A.; Caetano-Anolles, G.; Niehaus, K.; Lorenzen, J.; Ljunggren, H.D.; Pühler, A.; Favelukes, G. A *Rhizobium meliloti* lipopolysaccharide mutant altered in competitiveness for nodulation of alfalfa. *J. Bacteriol.* **1992**, *174*, 5941–5952.
49. Campbell, G.R.O.; Reuhs, B.L.; Walker, G.C. Chronic intracelular infection of alfalfa nodules by *Sinorhizobium meliloti* requires correct lipopolysaccharide core. *Proc. Natl. Acad. Sci. USA* **2002**, *99*, 3938–3943.
50. Sorroche, F.G. National University of Río Cuarto, Córdoba, Argentina. Unpublished work, 2011.
51. Lau, P.C.Y.; Lindhout, T.; Beveridge, T.J.; Dutcher, J.R.; Lam, J.S. Differential lipopolysaccharide core capping leads to quantitative and correlated modifications of mechanical and structural properties in *Pseudomonas aeruginosa* biofilms. *J. Bacteriol.* **2009**, *191*, 6618–6631.
52. Deng, W.L.; Lin, Y.C.; Lin, R.H.; Wei, C.F.; Huang, Y.C.; Peng, H.L.; Huang, H.C. Effects of *galU* mutation on *Pseudomonas syringae*—Plant interactions. *Mol. Plant Microbe Interact.* **2010**, *23*, 1184–1196.
53. Li, J.; Wang, N. The wxacO gene of *Xanthomonas citri* ssp. *citri* encodes a protein with a role in lipopolysaccharide biosynthesis, biofilm formation, stress tolerance and virulence. *Mol. Plant Pathol.* **2011**, *12*, 381–396.
54. Li, J.; Wang, N. Genome-wide mutagenesis of *Xanthomonas axonopodis* pv. *citri* reveals novel genetic determinants and regulation mechanisms of biofilm formation. *PLoS One* **2011**, *6*, e21804.
55. Dhakal, B.K.; Bower, J.M.; Mulvey, M.A. Pili, Fimbriae. In *Encyclopedia of Microbiology*; Moselio, S., Ed.; Academy Press: Oxford, UK, 2009; p. 470.
56. Conrad, J.C. Physics of bacterial near-surface motility using flagella and type IV pili: Implications for biofilm formation. *Res. Microbiol.* **2012**, *163*, 619–629.
57. Petrova, O.E.; Sauer, K. Sticky situations: Key components that control bacterial surface attachment. *J. Bacteriol.* **2012**, *194*, 2413–2425.
58. Jin, F.; Conrad, J.C.; Gibiansky, M.L.; Wong, G.C.L. Bacteria use type-IV pili to slingshot on surfaces. *Proc. Natl. Acad. Sci. USA* **2011**, *108*, 12617–12622.
59. Wood, T.K.; González Barrios, A.F.; Herzberg, M.; Lee, J. Motility influences biofilm architecture in *Escherichia coli*. *Appl. Microbiol. Biotechnol.* **2006**, *72*, 361–367.
60. Barken, K.B.; Pamp, S.J.; Yang, L.; Gjermansen, M.; Bertrand, J.J.; Klausen, M.; Givskov, M.; Whitchurch, C.B.; Engel, J.B.; Tolker-Nielsen, T. Roles of type IV pili, flagellum-mediated motility and extracellular DNA in the formation of mature multicellular structures in *Pseudomonas aeruginosa* biofilms. *Environ. Microbiol.* **2008**, *10*, 2331–2343.
61. Bahar, O.; de la Fuente, L.; Burdman, S. Assessing adhesion, biofilm formation and motility of *Acidovorax citrulli* using microfluidic flow chambers. *FEMS Microbiol. Lett.* **2010**, *312*, 33–39.

62. De la Fuente, L.; Montanes, E.; Meng, Y.; Li, Y.; Burr, T.J.; Hoch, H.C.; Wu, M. Assessing adhesion forces of type I and type IV pili of *Xylella fastidiosa* bacteria by use of a microfluidic flow chamber. *Appl. Environ. Microbiol.* **2007**, *73*, 2690–2696.

63. Kang, Y.; Liu, H.; Genin, S.; Schell, M.A.; Denny, T.P. *Ralstonia solanacearum* requires type 4 pili to adhere to multiple surfaces and for natural transformation and virulence. *Mol. Microbiol.* **2002**, *2*, 427–437.

64. Wang, L.; Makino, S.; Subedee, A.; Bogdanove, A.J. Novel candidate virulence factors in rice pathogen *Xanthomonas oryzae* pv. *oryzicola* as revealed by mutational analysis. *Appl. Environ. Microbiol.* **2007**, *73*, 8023–8027.

65. Jarrell, K.F.; McBride, M.J. The surprisingly diverse ways that prokaryotes move. *Nat. Rev. Microbiol.* **2008**, *6*, 466–476.

66. Klausen, M.; Heydorn, A.; Ragas, P.; Lambertsen, L.; Aaes-Jørgensen, A.; Molin, S.; Tolker-Nielsen, T. Biofilm formation by *Pseudomonas aeruginosa* wild type, flagella and type IV pili mutants. *Mol. Microbiol.* **2003**, *48*, 1511–1524.

67. Verstraeten, N.; Braeken, K.; Debkumari, B.; Fauvart, M.; Fransaer, J.; Vermant, J.; Michiels, J. Living on a surface: Swarming and biofilm formation. *Trends Microbiol.* **2008**, *16*, 496–506.

68. Fujishige, N.A.; Kapadia, N.N.; de Hoff, P.L.; Hirsch, A.M. Investigations of *Rhizobium* biofilm formation. *FEMS Microbiol. Ecol.* **2006**, *56*, 195–206.

69. Nievas, F.; Bogino, P.; Sorroche, F.; Giordano, W. Detection, characterization, and biological effect of quorum-sensing signaling molecules in peanut-nodulating bradyrhizobia. *Sensors* **2012**, *12*, 2851–2873.

70. Merritt, P.M.; Danhorn, T.; Fuqua, C. Motility and chemotaxis in *Agrobacterium tumefaciens* surface attachment and biofilm formation. *J. Bacteriol.* **2007**, *189*, 8005–8014.

71. Kirov, S.M.; Castrisios, M.; Shaw, J.G. *Aeromonas* flagella (polar and lateral) are enterocyte adhesins that contribute to biofilm formation on surfaces. *Infect. Immun.* **2004**, *72*, 1939–1945.

72. Lemon, K.P.; Higgins, D.E.; Kolter, R. Flagellar motility is critical for *Listeria monocytogenes* biofilm formation. *J. Bacteriol.* **2007**, *189*, 4418–4424.

73. Brunings, A.M.; Gabriel, D.W. *Xanthomonas citri*: Breaking the surface. *Mol. Plant Pathol.* **2003**, *4*, 141–157.

74. Rigano, L.A.; Siciliano, F.; Enrique, R.; Sendín, L.; Filippone, P.; Torres, P.S.; Qüesta, J.; Dow, J.M.; Castagnaro, A.P.; Vojnov, A.A.; *et al.* Biofilm formation, epiphytic fitness, and canker development in *Xanthomonas axonopodis* pv. *citri*. *Mol. Plant Microbe Interact.* **2007**, *20*, 1222–1230.

75. Malamud, F.; Torres, P.S.; Roeschlin, R.; Rigano, L.A.; Enrique, R.; Bonomi, H.R.; Castagnaro A.P.; Marano, M.R.; Vojnov, A.A. The *Xanthomonas axonopodis* pv. *citri* flagellum is required for mature biofilm and canker development. *Microbiology* **2011**, *157*, 819–829.

76. Lee, M.C.; Weng, S.F.; Tseng, Y.H. Flagellin gene fliC of *Xanthomonas campestris* is upregulated by transcription factor Clp. *Biochem. Biophys. Res. Commun.* **2003**, *307*, 647–652.

77. Tans-Kersten, J.; Huang, H.; Allen, C. *Ralstonia solanacearum* needs motility for invasive virulence on tomato. *J. Bacteriol.* **2001**, *183*, 3597–3605.

78. Herrera, C.M.; Koutsoudis, M.D.; Wang, X.; von Bodman, S.B. *Pantoea stewartii* subsp. *stewartii* exhibits surface motility, which is a critical aspect of Stewart's wilt disease development on maize. *Mol. Plant Microbe Interact.* **2008**, *21*, 1359–1370.

79. Jahn, C.E.; Willis, D.K.; Charkowski, A.O. The flagellar sigma factor fliA is required for *Dickeya dadantii* virulence. *Mol. Plant Microbe Interact.* **2008**, *21*, 1431–1442.

80. Cucarella, J.; Solano, C.; Valle, J.; Amorena, B.; Lasa, I.; Penadés, J.R. Bap, a *Staphylococcus aureus* surface protein involved in biofilm formation. *J. Bacteriol.* **2001**, *183*, 2888–2896.

81. Toledo-Arana, A.; Valle, J.; Solano, C.; Arrizubieta, M.J.; Cucarella, J.; Lamata, M. The enterococcal surface protein, Esp, is involved in *Enterococcus faecalis* biofilm formation. *Appl. Environ. Microbiol.* **2001**, *67*, 4538–4545.

82. Ausmees, N.; Jacobsson, K.; Lindberg, M. A unipolarly located, cell-surface-associated agglutinin, RapA, belongs to a family of *Rhizobium*-adhering proteins (Rap) in *Rhizobium leguminosarum* bv. *trifolii*. *Microbiology* **2001**, *147*, 549–559.

83. Russo, D.M.; Williams, A.; Edwards, A.; Posadas, D.M.; Finnie, C.; Dankert, M.; Downie, J.A.; Zorreguieta, A. Proteins exported via the PrsD-PrsE type I secretion system and the acidic exopolysaccharide are involved in biofilm formation by *Rhizobium leguminosarum*. *J. Bacteriol.* **2006**, *188*, 4474–4486.

84. Laus, M.C.; Logman, T.J.; Lamers, G.E.; van Brussel, A.A.; Carlson, R.W.; Kijne, J.W. A novel polar surface polysaccharide from *Rhizobium leguminosarum* binds host plant lectin. *Mol. Microbiol.* **2006**, *59*, 1704–1713.

85. Hinsa, S.M.; Espinosa-Urgel, M.; Ramos, J.L.; O'Toole, G.A. Transition from reversible to irreversible attachment during biofilm formation by *Pseudomonas fluorescens* WCS365 requires an ABC transporter and a large secreted protein. *Mol. Microbiol.* **2003**, *49*, 905–918.

86. Martínez-Gil, M.; Yousef-Coronado, F.; Espinosa-Urgel, M. LapF, the second largest *Pseudomonas putida* protein, contributes to plant root colonization and determines biofilm architecture. *Mol. Microbiol.* **2010**, *77*, 549–561.

87. Ryder, C.; Byrd, M.; Wozniak, D.J. Role of exopolysaccharides in *Pseudomonas aeruginosa* biofilm development. *Curr. Opin. Microbiol.* **2007**, *10*, 644–648.

88. Vaningelgem, F.; Zamfir, M.; Mozzi, F.; Adriany, T.; Vancanneyt, M.; Swings, J.; de Vuyst, L. Biodiversity of exopolysaccharides produced by *Streptococcus thermophilus* strains is reflected in their production and their molecular and functional characteristics. *Appl. Environ. Microbiol.* **2004**, *70*, 900–912.

89. Jefferson, K.K. *Bacterial Polysaccharides. Current Innovations and Future Trends*; Ullrich, M., Ed.; Caister Academic: Norfolk, UK, 2009; pp. 175–186.

90. Sutherland, I.W. Biofilm Exopolysaccharides. In *Microbial Extracellular Polymeric Substances*; Wingender, J., Neu, T.R., Flemming, H.C., Eds.; Springer: Berlin, Germany, 1999; pp. 73–92.

91. Reuber, T.L.; Walker, G.C. Biosynthesis of succinoglycan, a symbiotically important exopolysaccharide of *Rhizobium meliloti*. *Cell* **1993**, *74*, 269–280.

92. Cheng, H.P.; Walker G.C. Succinoglycan is required for initiation and elongation of infection threads during nodulation of alfalfa by *Rhizobium meliloti*. *J. Bacteriol.* **1998**, *180*, 5183–5191.

93. Her, G.R.; Glazebroock, J.; Walker, G.C.; Reinhold, V.N. Structural studies of a novel exopolysaccharide produced by a mutant of *Rhizobium meliloti* strain Rm1021. *Carbohyd. Res.* **1990**, *198*, 305–312.
94. Rinaudi, L.V.; González, J.E. The low-molecular-weight fraction of exopolysaccharide II from *Sinorhizobium meliloti* is a crucial determinant of biofilm formation. *J. Bacteriol.* **2009**, *191*, 7216–7224.
95. Robertson, B.K.; Aman, P.; Darvill, A.G.; McNeil, M.; Albersheim, P. Host-symbiont interactions. The structure of acidic extracellular polysaccharides secreted by *Rhizobium leguminosarum* and *Rhizobium trifolii*. *Plant Physiol.* **1981**, *67*, 389–400.
96. Philip-Hollingsworth, S.; Hollingsworth, R.I.; Dazzo, F.B. Host-range related structural features of the acidic extracellular polysaccharides of *Rhizobium trifolii* and *Rhizobium leguminosarum*. *J. Biol. Chem.* **1989**, *264*, 1461–1466.
97. Janczarek, M.; Kutkowska, J.; Piersiak, T.; Skorupska, A. *Rhizobium leguminosarum* bv. *trifolii* rosR is required for interaction with clover, biofilm formation and adaptation to the environment. *BMC Microbiol.* **2010**, *10*, doi:10.1186/1471-2180-10-284.
98. Rolfe, B.G.; Carlson, R.W.; Ridge, R.W.; Dazzo, R.W.; Mateos, F.B.; Pankhurst, C.E. Defective infection and nodulation of clovers by exopolysaccharide mutants of *Rhizobium leguminosarum* bv. *trifolii*. *Aust. J. Plant Physiol.* **1996**, *23*, 285–303.
99. Van Workum, W.A.T.; van Slageren, S.; van Brussel, A.A.N.; Kijne, J.W. Role of exopolysaccharides of *Rhizobium leguminosarum* bv. *viciae* as host plant specific molecules required for infection thread formation during nodulation of *Vicia sativa*. *Mol. Plant Microbe Interact.* **1998**, *11*, 1233–1241.
100. Minamisawa, K. Comparison of extracellular polysaccharide composition, rhizobiotoxine production, and hydrogenase phenotype among various strains of *Bradyrhizobium japonicum*. *Plant Cell Physiol.* **1989**, *30*, 877–884.
101. Poveda, A.; Santamaria, M.; Bernabe, M.; Prieto, A.; Briux, M.; Corzo, J.; Jimenez-Barbero, J. Studies on the structure and the solution conformation of an acidic extracellular polysaccharide isolated from *Bradyrhizobium*. *Carbohyd. Res.* **1997**, *304*, 209–217.
102. Pérez-Giménez, J.; Mongiardini, E.J.; Althabegoiti, M.J.; Covelli, J.; Quelas, J.I.; López-García, S.L.; Lodeiro, A.R. Soybean lectin enhances biofilm formation by *Bradyrhizobium japonicum* in the absence of plants. *Int. J. Microbiol.* **2009**, *2009*, doi:10.1155/2009/719367.
103. Wang, P.; Zhong, Z.; Zhou, J.; Cai, T.; Zhu, J. Exopolysaccharide biosynthesis is important for *Mesorhizobium tianshanense*: Plant host interaction. *Arch. Microbiol.* **2008**, *189*, 525–530.
104. Cangelosi, G.A.; Hung, L.; Puvanesarajah, V.; Stacey, G.; Ozga, D.A.; Leigh, J.A.; Nester, E.W. Common loci for *Agrobacterium tumefaciens* and *Rhizobium meliloti* exopolysaccharide synthesis and their roles in plant interactions. *J. Bacteriol.* **1987**, *169*, 2086–2091.
105. Tomlinson, A.D.; Ramey-Hartung, B.; Day, T.W.; Merritt, P.M.; Fuqua, C. *Agrobacterium tumefaciens* ExoR represses succinoglycan biosynthesis and is required for biofilm formation and motility. *Microbiology* **2010**, *156*, 2670–2681.
106. Da Silva, F.R.; Vettore, A.L.; Kemper, E.L.; Leite, A.; Arruda, P. Fastidian gum: The *Xylella fastidiosa* exopolysaccharide possibly involved in bacterial pathogenicity. *FEMS Microbiol. Lett.* **2001**, *203*, 165–171.

107. Roper, M.C.; Greve, L.C.; Labavitch, J.M.; Kirkpatrick, B.C. Detection and visualization of an exopolysaccharide produced by *Xylella fastidiosa in vitro* and in planta. *Appl. Environ. Microbiol.* **2007**, *73*, 7252–7258.
108. Jansson, P.E.; Kenne, L.; Lindberg, B. Structure of the extracellular polysaccharide from *Xanthomonas campestris*. *Carbohyd. Res.* **1975**, *45*, 274–282.
109. Rigano, L.A.; Payette, C.; Brouillard, G.; Marano, M.R.; Abramowicz, L.; Torres, P.S.; Yun, M.; Castagnaro, A.P.; Oirdi, M.E.; Dufour, V.; *et al.* Bacterial cyclic beta-(1,2)-glucan acts in systemic suppression of plant immune responses. *Plant Cell* **2007**, *19*, 2077–2089.
110. Torres, P.S.; Malamud, F.; Rigano, L.A.; Russo, D.M.; Marano, M.R.; Castagnaro, A.P.; Zorreguieta, A.; Bouarab, K.; Dow, J.M.; Vojnov, A.A. Controlled synthesis of the DSF cell-cell signal is required for biofilm formation and virulence in *Xanthomonas campestris*. *Environ. Microbiol.* **2007**, *9*, 2101–2109.
111. Nimtz, M.; Mort, A.; Wray, V.; Domke, T.; Zhang, Y.; Coplin, D.L.; Geider, K. Structure of stewartan, the capsular exopolysaccharide from the corn pathogen *Erwinia stewartii*. *Carbohyd. Res.* **1996**, *288*, 189–201.
112. Koutsoudis, M.D.; Tsaltas, D.; Minogue, T.D.; von Bodman, S.B. Quorum-sensing regulation governs bacterial adhesion, biofilm development, and host colonization in *Pantoea stewartii* subspecies *stewartii*. *Proc. Natl. Acad. Sci. USA* **2006**, *103*, 5983–5988.
113. Nimtz, M.; Mort, A.; Domke, T.; Wray, V.; Zhang, Y.; Qiu, F.; Coplin, D.; Geider, K. Structure of amylovoran, the capsular exopolysaccharide from the fire blight pathogen *Erwinia amylovora*. *Carbohyd. Res.* **1996**, *287*, 59–76.
114. Koczan, J.M.; McGrath, M.J.; Zhao, Y.; Sundin, G.W. Contribution of *Erwinia amylovora* exopolysaccharides amylovoran and levan to biofilm formation: Implications in pathogenicity. *Phytopathology* **2009**, *99*, 1237–1244.
115. Geier, G.; Geider, K. Characterization and influence on virulence of the levansucrase gene from the fireblight pathogen *Erwinia amylovora*. *Physiol. Mol. Plant Pathol.* **1993**, *42*, 387–404.
116. Cook, D.; Sequeira, L. Genetic and biochemical characterization of a gene cluster from *Pseudomonas solanacearum* required for extracellular polysaccharide production and for virulence. *J. Bacteriol.* **1991**, *173*, 1654–1662.
117. Chapman, M.R.; Kao, C.C. EpsR modulates production of extracellular polysaccharides in the bacterial wilt pathogen *Ralstonia* (*Pseudomonas*) *solanacearum*. *J. Bacteriol.* **1998**, *180*, 27–34.
118. Zogaj, X.; Nimtz, M.; Rohde, M.; Bokranz, W.; Römling, U. The multicellular morphotypes of *Salmonella typhimurium* and *Escherichia coli* produce cellulose as the second component of the extracellular matrix. *Mol. Microbiol.* **2001**, *39*, 1452–1463.
119. Matthysse, A.G.; White, S.; Lightfoot, R. Genes required for cellulose synthesis in *Agrobacterium tumefaciens*. *J. Bacteriol.* **1995**, *177*, 1069–1075.
120. Robledo, M.; Rivera, L.; Jiménez-Zurdo, J.I.; Rivas, R.; Dazzo, F.; Velázquez, E.; Martínez-Molina, E.; Hirsch, A.M.; Mateos, P.F. Role of *Rhizobium* endoglucanase CelC2 in cellulose biosynthesis and biofilm formation on plant roots and abiotic surfaces. *Microb. Cell Fact.* **2012**, *11*, 125.
121. Beauregard, P.B.; Chai, Y.; Vlamakis, H.; Losick, R.; Kolter, R. *Bacillus subtilis* biofilm induction by plant polysaccharides. *Proc. Natl. Acad. Sci. USA* **2013**, *110*, 1621–1630.

122. Zhang, X.; Bishop, P. Biodegradability of biofilm extracellular polymeric substances. *Chemosphere* **2003**, *50*, 63–69.
123. Laue, H.; Schenk, A.; Li, H.; Lambertsen, L.; Neu, T.R.; Molin, S.; Ullrich, M.S. Contribution of alginate and levan production to biofilm formation by *Pseudomonas syringae*. *Microbiology* **2006**, *152*, 2909–2918.
124. Dafna, T.A.; Rosenberg, T.; Navon, N.; Burdman, S. A secreted lipolytic enzyme from *Xanthomonas campestris* pv. *vesicatoria* is expressed *in planta* and contributes to its virulence. *Mol. Plant Pathol.* **2012**, *13*, 556–567.
125. Hu, J.; Qian, W.; He, C. The *Xanthomonas oryzae* pv. *oryzae eglXoB* endoglucanase gene is required for virulence to rice. *FEMS Microbiol. Lett.* **2007**, *269*, 273–279.
126. Lynch, D.J.; Fountain, T.L.; Mazurkiewicz, J.E.; Banas, J.A. Glucan-binding proteins are essential for shaping *Streptococcus mutans* biofilm architecture. *FEMS Microbiol. Lett.* **2007**, *268*, 158–165.
127. Tielker, D.; Hacker, S.; Loris, R.; Strathmann, M.; Wingender, J.; Wilhelm, S.; Rosenau, F.; Jaeger, K.E. *Pseudomonas aeruginosa* lectin LecB is located in the outer membrane and is involved in biofilm formation. *Microbiology* **2005**, *151*, 1313–1323.
128. Diggle, S.P.; Stacey, R.E.; Dodd, C.; Cámara, M.; Williams, P.; Winzer, K. The galactophilic lectin, LecA, contributes to biofilm development in *Pseudomonas aeruginosa*. *Environ. Microbiol.* **2010**, *8*, 1095–1104.
129. Branda, S.S.; Chu, F.; Kearns, D.B.; Losick, R.; Kolter, R. A major protein component of the *Bacillus subtilis* biofilm matrix. *Mol. Microbiol.* **2006**, *59*, 1229–1238.
130. Toyofuku, M.; Roschitzki, B.; Riedel, K.; Eberl, L. Identification of proteins associated with the *Pseudomonas aeruginosa* biofilm extracellular matrix. *J. Proteome Res.* **2012**, *11*, 4906–4915.
131. Larsen, P.; Nielsen, J.L.; Dueholm, M.S.; Wetzel, R.; Otzen, D.; Nielsen, P.H. Amyloid adhesins are abundant in natural biofilms. *Environ. Microbiol.* **2007**, *9*, 3077–3090.
132. Molin, S.; Tolker-Nielsen, T. Gene transfer occurs with enhanced efficiency in biofilms and induces enhanced stabilisation of the biofilm structure. *Curr. Opin. Biotechnol.* **2003**, *14*, 255–261.
133. Dominiak, D.M.; Nielsen, J.L.; Nielsen, P.H. Extracellular DNA is abundant and important for microcolony strength in mixed microbial biofilms. *Environ. Microbiol.* **2011**, *13*, 710–721.
134. Yang, L.; Barken, K.B.; Skindersoe, M.E.; Christensen, A.B.; Givskov, M.; Tolker-Nielsen, T. Effects of iron on DNA release and biofilm development by *Pseudomonas aeruginosa*. *Microbiology* **2007**, *153*, 1318–1328.
135. Vilain, S.; Pretorius, J.M.; Theron, J.; Broezel, V.S. DNA as an adhesion: *Bacillus cereus* requires extracellular DNA to form biofilms. *Appl. Environ. Microbiol.* **2009**, *75*, 2861–2868.
136. Böckelmann, U.; Janke, A.; Kuhn, R.; Neu, T.R.; Wecke, J.; Lawrence, J.R.; Szewzyk, U. Bacterial extracellular DNA forming a defined network-like structure. *FEMS Microbiol. Lett.* **2006**, *262*, 31–38.
137. Allesen-Holm, M.; Barken, K.B.; Yang, L.; Klausen, M.; Webb, J.S.; Kjelleberg, S.; Molin, S.; Givskov, M.; Tolker-Nielsen, T. A characterization of DNA release in *Pseudomonas aeruginosa* cultures and biofilms. *Mol. Microbiol.* **2006**, *59*, 1114–1128.
138. Wang, J.; Bayles, K.W. Programmed cell death in plants: Lessons from bacteria? *Trends Plant Sci.* **2013**, *18*, 133–139.

139. Das, T.; Prashant, S.K.; Busscher, H.J.; van der Mei, H.C.; Krom, B.P. Role of extracellular DNA in initial bacterial adhesion and surface aggregation. *Appl. Environ. Microbiol.* **2010**, *76*, 3405–3408.
140. Conrad, A.; Suutari, M.K.; Keinänen, M.M.; Cadoret, A.; Faure, P.; Mansuy-Huault, L.; Block, J.C. Fatty acid lipid fractions in extracellular polymeric substances of activated sludge flocs. *Lipids* **2003**, *38*, 1093–1105.
141. Ron, E.Z.; Rosenberg, E. Natural role of biosurfactants. *Environ. Microbiol.* **2001**, *3*, 229–236.
142. Pamp, S.J.; Tolker-Nielsen, T. Multiple roles of biosurfactants in structural biofilm development by *Pseudomonas aeruginosa*. *J. Bacteriol.* **2007**, *189*, 2531–2539.
143. Stoodley, P.; Sauer, K.; Davies, D.G.; Costerton, J.W. Biofilms as complex differentiated communities. *Annu. Rev. Microbiol.* **2002**, *56*, 187–209.
144. Sherlock, O.; Schembri, M.A.; Reisner, A.; Klemm, P. Novel roles for the AIDA adhesin from diarrheagenic *Escherichia coli*: Cell aggregation and biofilm formation. *J. Bacteriol.* **2004**, *186*, 8058–8065.
145. Jacobs, A.; Chenia, H.Y. Biofilm-forming capacity, surface hydrophobicity and aggregation characteristics of *Myroides odoratus* isolated from South African *Oreochromis mossambicus* fish. *J. Appl. Microbiol.* **2009**, *107*, 1957–1966.
146. Yamaguchi, M.; Sato, K.; Yukitake, H.; Noiri, Y.; Ebisu, S.; Nakayama, K. *Porphyromonas gingivalis* mutant defective in a putative glycosyltransferase exhibits defect in biosynthesis of polysaccharide portions of lipopolysaccharide, decreased gingipain activities, strong autoaggregation and increased biofilm formation. *Infec. Immun.* **2010**, *78*, 3801–3812.
147. Voegel, T.M.; Doddapaneni, H.; Cheng, D.W.; Lin, H.; Stenger, D.C.; Kirkpatrick, B.C.; Roper, M.C. Identification of a response regulator involved in surface attachment, cell–cell aggregation, exopolysaccharide production and virulence in the plant pathogen *Xylella fastidiosa*. *Mol. Plant Pathol.* **2013**, *14*, 256–264.
148. Nievas, F. National University of Río Cuarto, Córdoba, Argentina. Unpublished work, 2012.
149. Rickard, A.H.; Gilbert, P.; High, N.J.; Kolenbrander, P.E.; Handley, P.S. Bacterial Coaggregation: An integral process in the development of multi-species biofilms. *Trends Microbiol.* **2003**, *11*, 94–100.
150. Phuong, K.; Kakii, K.; Nikata, T. Intergeneric coaggregation of non-flocculating *Acinetobacter* spp. isolates with other sludge-constituting bacteria. *J. Biosci. Bioeng.* **2009**, *107*, 394–400.
151. Joe, M.M.; Jaleel, C.A.; Sivakumar, P.K.; Zhao, C.X.; Karthikeyan, B. Co-aggregation in *Azospirillum brasilensense* MTCC-125 with other PGPR strains: Effect of physical and chemical factors and stress endurance ability. *J. Taiwan Inst. Chem. E* **2009**, *40*, 491–499.
152. Malik, A.; Kakii, K. Pair dependent co-aggregation behavior of nonflocculating sludge bacteria, *Biotechnol. Lett.* **2003**, *25*, 981–986.
153. Ledder, R.G.; Timperley, A.S.; Friswell, M.K.; MacFarlane, S.; McBain, A.J. Coaggregation between and among human intestinal and oral bacteria. *FEMS Microbiol. Ecol.* **2008**, *66*, 630–636.

Permissions

All chapters in this book were first published by MDPI; hereby published with permission under the Creative Commons Attribution License or equivalent. Every chapter published in this book has been scrutinized by our experts. Their significance has been extensively debated. The topics covered herein carry significant findings which will fuel the growth of the discipline. They may even be implemented as practical applications or may be referred to as a beginning point for another development.

The contributors of this book come from diverse backgrounds, making this book a truly international effort. This book will bring forth new frontiers with its revolutionizing research information and detailed analysis of the nascent developments around the world.

We would like to thank all the contributing authors for lending their expertise to make the book truly unique. They have played a crucial role in the development of this book. Without their invaluable contributions this book wouldn't have been possible. They have made vital efforts to compile up to date information on the varied aspects of this subject to make this book a valuable addition to the collection of many professionals and students.

This book was conceptualized with the vision of imparting up-to-date information and advanced data in this field. To ensure the same, a matchless editorial board was set up. Every individual on the board went through rigorous rounds of assessment to prove their worth. After which they invested a large part of their time researching and compiling the most relevant data for our readers.

The editorial board has been involved in producing this book since its inception. They have spent rigorous hours researching and exploring the diverse topics which have resulted in the successful publishing of this book. They have passed on their knowledge of decades through this book. To expedite this challenging task, the publisher supported the team at every step. A small team of assistant editors was also appointed to further simplify the editing procedure and attain best results for the readers.

Apart from the editorial board, the designing team has also invested a significant amount of their time in understanding the subject and creating the most relevant covers. They scrutinized every image to scout for the most suitable representation of the subject and create an appropriate cover for the book.

The publishing team has been an ardent support to the editorial, designing and production team. Their endless efforts to recruit the best for this project, has resulted in the accomplishment of this book. They are a veteran in the field of academics and their pool of knowledge is as vast as their experience in printing. Their expertise and guidance has proved useful at every step. Their uncompromising quality standards have made this book an exceptional effort. Their encouragement from time to time has been an inspiration for everyone.

The publisher and the editorial board hope that this book will prove to be a valuable piece of knowledge for researchers, students, practitioners and scholars across the globe.

List of Contributors

Ali Adem Bahar and Meagan Garafalo
Department of Biomedical and Chemical Engineering, Syracuse University, Syracuse, NY 13244, USA
Syracuse Biomaterials Institute, Syracuse University, Syracuse, NY 13244, USA

Zhigang Liu and Neville Kallenbach
Department of Chemistry, New York University, New York, NY 10003, USA

Dacheng Ren
Department of Biomedical and Chemical Engineering, Syracuse University, Syracuse, NY 13244, USA
Syracuse Biomaterials Institute, Syracuse University, Syracuse, NY 13244, USA
Department of Civil and Environmental Engineering, Syracuse University, Syracuse, NY 13244, USA
Department of Biology, Syracuse University, Syracuse, NY 13244, USA

Cláudia N. H. Marques, David G. Davies and Karin Sauer
Department of Biological Sciences, Binghamton University, Binghamton, NY 13902, USA
Binghamton Biofilm Research Center (BBRC), Binghamton University, Binghamton, NY 13902, USA

Meng Chen
Nanova, Inc. Columbia, MO 65211, USA

Qingsong Yu
Department of Mechanical and Aerospace Engineering, University of Missouri, Columbia, MO 65211, USA

Hongmin Sun
Division of Cardiovascular Medicine, Department of Internal Medicine, University of Missouri, Columbia, MO 65212, USA

Vesna Svetličić, Vera Žutić, Galja Pletikapić and Tea Mišić Radić
Division for Marine and Environmental Research, Ruđer Bošković Institute, Bijenička 54, 10000 Zagreb, Croatia

Carla Duncan
Ontario Agency for Health Protection and Promotion (OAHPP), Toronto, ON M9P 3T1, Canada

Mena Abdel-Nour and Cyril Guyard
Ontario Agency for Health Protection and Promotion (OAHPP), Toronto, ON M9P 3T1, Canada
Department of Laboratory Medicine and Pathobiology, University of Toronto, Toronto, ON M5S 1A8, Canada
Mount Sinai Hospital, Toronto, ON M5G 1X5, Canada

Donald E. Low
Department of Laboratory Medicine and Pathobiology, University of Toronto, Toronto, ON M5S 1A8, Canada
Mount Sinai Hospital, Toronto, ON M5G 1X5, Canada

Marja T. Pöllänen
Institute of Dentistry, University of Turku, FI-20014 Turku, Finland

Annamari Paino and Riikka Ihalin
Department of Biochemistry and Food Chemistry, University of Turku, FI-20014 Turku, Finland

Huirong Lin
Department of Environmental Engineering, Zhejiang University, Hangzhou 310058, China
Institute of Urban Environment, Chinese Academy of Science, Xiamen 361021, China

Guangcun Chen, Shenhai Zhu, Yingxu Chen and Jiyan Shi
Department of Environmental Engineering, Zhejiang University, Hangzhou 310058, China

Dongliang Chen and Wei Xu
Institute of High Energy Physics, Chinese Academy of Science, Beijing Synchrotron Radiation Facility, Beijing 100049, China

Xiaohan Yu
Institute of Applied Physics, Chinese Academy of Science, Shanghai Synchrotron Radiation Facility, Shanghai 201800, China

Alexandru I. Petrisor
Department of Urban and Landscape Planning, School of Urban Planning, "Ion Mincu" University of Architecture and Urban Planning, str. Academiei nr. 18-20, sector 1, Bucharest 010014, Romania
National Institute for Research and Development in Constructions, Urbanism and Sustainable Spatial Development URBAN-INCERC, sos. Pantelimon, nr. 266, sector 2, Bucharest 021652, Romania

Sandra Szyjka, Tomohiro Kawaguchi, Robert Sean Norman and Alan W. Decho
Department of Environmental Health Sciences, Arnold School of Public Health, University of South Carolina, Columbia, SC 29208, USA

Pieter T. Visscher
Center for Integrative GeoSciences, University of Connecticut, 345 Mansfield Rd., U-2045 Storrs, CT 06269, USA

Suvi Manner, Malena Skogman and Adyary Fallarero
Pharmaceutical Sciences, Department of Biosciences, Abo Akademi University, Artillerigatan 6A, 3rd Floor, Biocity, Turku FI-20520, Finland

Darla Goeres
Center for Biofilm Engineering, Montana State University, Bozeman, MT 59717, USA

Pia Vuorela
Pharmaceutical Sciences, Department of Biosciences, Abo Akademi University, Artillerigatan 6A, 3rd Floor, Biocity, Turku FI-20520, Finland
Division of Pharmaceutical Biology, Faculty of Pharmacy, University of Helsinki, Viikinkaari 5E, Helsinki FI-00014, Finland

Juan Carlos Camacho-Chab, Manuel Jesús Chan-Bacab, Susana del C. De la Rosa-García, Manuela Reyes-Estebanez and Benjamín Otto Ortega-Morales
Center of Environmental Microbiology and Biotechnology, Autonomous University of Campeche, Colonia Buenavista, San Francisco de Campeche 24039, Campeche, Mexico

Jean Guézennec and Corinne Sinquin
Ifremer Laboratoire Biotechnologies et Molécules Marines, Nantes B. P. 21105, France

Elvira Ríos-Leal
Department of Biotechnology and Bioengineering, Center for Research and Advanced Studies of the National Polytechnic Institute (CINVESTAV), Mexico City 07360, Federal District, Mexico

Raquel Muñiz-Salazar
Laboratory of Molecular Ecology and Epidemiology, School of Health Sciences, Autonomous University of Baja California, Tijuana 22390, Baja California, Mexico

Robert J. C. McLean and Kavita S. Kakirde
Department of Biology, Texas State University, 601 University Drive, San Marcos, TX 78666, USA

Robert Almstrand, Fred Sörensson and Malte Hermansson
Department of Chemistry & Molecular Biology, Microbiology, University of Gothenburg, Göteborg SE-405 30, Sweden

Frank Persson and Britt-Marie Wilén
Water Environment Technology, Department of Civil and Environmental Engineering, Chalmers University of Technology, Göteborg SE-412 96, Sweden

Holger Daims
Division of Microbial Ecology, Department of Microbiology and Ecosystem Science, University of Vienna, Vienna 1090, Austria

Maria Ekenberg and Magnus Christensson
AnoxKaldnes AB, Klosterängsvägen 11A, Lund SE-226 47, Sweden

Qing Wei and Luyan Z. Ma
State Key Laboratory of Microbial Resources, Institute of Microbiology, Chinese Academy of Sciences, No.3, 1st Beichen West Road, Chaoyang District, Beijing 100101, China

Pablo C. Bogino, Fernando G. Sorroche and Walter Giordano
Department of Molecular Biology, National University of Río Cuarto, Ruta 36 Km 601, Río Cuarto, Córdoba X5804BYA, Argentina

María de las Mercedes Oliva
Department of Microbiology and Immunology, National University of Río Cuarto, Ruta 36 Km 601, Córdoba X5804BYA, Argentina

Index

www.ingramcontent.com/pod-product-compliance
Lightning Source LLC
LaVergne TN
LVHW080948210926
848485LV00004B/799
* 9 7 8 1 6 4 6 4 6 5 6 0 6 *